Lecture Notes in Artificial Intelligence 11291

Subseries of Lecture Notes in Computer Science

More information about this series at http://www.springer.com/series/1244

Robert Fullér · Silvio Giove ·
Francesco Masulli (Eds.)

Fuzzy Logic
and Applications

12th International Workshop, WILF 2018
Genoa, Italy, September 6–7, 2018
Revised Selected Papers

 Springer

Editors
Robert Fullér
Széchenyi István University
Győr, Hungary

Silvio Giove
Ca' Foscari University of Venice
Venice, Italy

Francesco Masulli
University of Genoa
Genoa, Italy

ISSN 0302-9743   ISSN 1611-3349 (electronic)
Lecture Notes in Artificial Intelligence
ISBN 978-3-030-12543-1   ISBN 978-3-030-12544-8 (eBook)
https://doi.org/10.1007/978-3-030-12544-8

Library of Congress Control Number: 2019932919

LNCS Sublibrary: SL7 – Artificial Intelligence

This Springer imprint is published by the registered company Springer Nature Switzerland AG
The registered company address is: Gewerbestrasse 11, 6330 Cham, Switzerland

Preface

The 12th International Workshop on Fuzzy Logic and Applications, WILF 2018, held in Genoa, Italy during September 6–7, 2018, covered topics related to theoretical, experimental, and applied fuzzy techniques and systems with emphasis on fuzzy decision-making.

This event represents the pursuance of an established tradition of biannual inter-disciplinary meetings. The previous editions of WILF have been held in Naples (1995), Bari (1997), Genoa (1999), Milan (2001), Naples (2003), Crema (2005), Camogli (2007), Palermo (2009), Trani (2011), Genoa (2013), and Naples (2016). Each event has focused on distinct main thematic areas of fuzzy logic and related applications. From this perspective, one of the main goals of the WILF workshops series is to bring together researchers and developers from both academia and high-tech companies and foster multidisciplinary research.

September 6, 2018, was the first anniversary of Lotfi A. Zadeh's death. On that day, WILF 2018 included a round table entitled "Zadeh and the Future of Fuzzy Logic" aimed at emphasizing what tools based on fuzzy sets and fuzzy logic are available today to address the technological challenges of today's society of big data and which others deserve to be further developed for this purpose.

After a rigorous peer-review process, we selected 16 high-quality manuscripts from the submissions received from all Europe. These were accepted for presentation at the conference and are published in this volume. In addition to regular papers, this volume comprises also a tutorial and the short abstracts of the contributions to the round table.

The success of this workshop is to be credited to the contribution of many people, in particular to the Program Committee members for their commitment to providing high-quality, constructive reviews, to the keynote speakers Antonio Di Nola (University of Salerno, Italy) and Jon Garibaldi (The University of Nottingham, UK), to the tutorial presenters Davide Ciucci (University of Milano-Bicocca, Italy) and Corrado Mencar (University of Bari, Italy), and to the local Organizing Committee for the support in the organization of the workshop events.

September 2018

Robert Fullér
Silvio Giove
Francesco Masulli

Organization

WILF 2018 was jointly organized by the DIBRIS, University of Genoa, Italy, the INNS, International Neural Network Society, SIG Italy, and the SIREN, Italian Neural Networks Society.

Conference Chairs

Robert Fullér	Óbuda University, Budapest, Hungary
Silvio Giove	Ca' Foscari University of Venice, Italy
Francesco Masulli	University of Genoa, Italy

Tutorial Chair

Marco Viviani	University of Milan Bicocca, Italy

Program Committee

Jose M. Alonso	University of Santiago de Compostela, Spain
Plamen Angelov	Lancaster University, UK
Sansanee Auephanwiriyakul	Chiang Mai University, Thailand
Valentina Emilia Bălaş	Aurel Vlaicu University of Arad, Romania
Humberto Bustince	Public University of Navarra, Pamplona, Spain
Christer Carlsson	Åbo Akademi University, Finland
Giovanna Castellano	University of Bari, Italy
Oscar Castillo	Tijuana Institute of Technology, Mexico
Shyi-Ming Chen	National Taiwan University of Science and Technology, Taiwan
Célia Da Costa Pereira	University of Nice Sophia Antipolis, France
Scott Dick	University of Alberta, Canada
Jonathan M. Garibaldi	University of Nottingham, UK
Janusz Kacprzyk	Polish Academy of Sciences, Poland
Donald Kraft	Louisiana State University, USA
Vladik Kreinovich	University of Texas at El Paso, USA
Pedro Melo-Pinto	CITAB, UTAD, Portugal
Javier Montero	Universidad Complutense de Madrid, Spain
Silvia Muzzioli	University of Modena and Reggio Emilia, Italy
Henri Prade	IRIT, CNRS, France
Marek Reformat	University of Alberta, Canada
Alexander Shostak	University of Latvia, Latvia
Eulalia Szmidt	Polish Academy of Sciences, Poland
Andrea Tettamanzi	University of Nice Sophia Antipolis, France

| Jun Yoneyama | Aoyama Gakuin University, Japan |
| Hans-Jürgen Zimmermann | RWTH Aachen University, Germany |

Scientific Secretariats

Alberto Cabri	University of Genoa, Italy
Zied Mnasri	University Tunis El Manar UTM, Tunisia
Stefano Rovetta	University of Genoa, Italy

WILF Steering Committee

Antonio Di Nola	University of Salerno, Italy
Francesco Masulli	University of Genoa, Italy
Gabriella Pasi	University of Milano Bicocca, Italy
Alfredo Petrosino	University of Naples Parthenope, Italy

Financing Institution

DIBRIS, University of Genoa, Italy

Contents

Advances in Fuzzy Logic Theory

Machine Learning for an Adaptive Rule Base

Michal Jalůvka[(⊠)] and Eva Volna

Department of Informatics and Computers,
University of Ostrava, 30. Dubna 22, 70103 Ostrava, Czech Republic
{michal.jaluvka,eva.volna}@osu.cz

Abstract. This paper deals with a design of an original approach for machine learning, which allows the rule base adaptation. This approach uses a fuzzy inference mechanism for decision making, finite-state machine for the rule base switching, and the teacher Supervisor for creating the most suitable rules for the activity (skill) which is supposed to be learned. The used fuzzy inference mechanism is the integration of LFLCore, which was developed at the Institute for Research and Applications of Fuzzy Modeling. The proposed approach of machine learning was tested in individual experiments, in which the system learns to move with its joints. How the system moves with its joints is given by patterns which are submitted before the beginning of learning. The evaluated results with possible modifications are mentioned at the end of this paper together with a formulated conclusion.

Keywords: Machine learning · Fuzzy inference system · Finite-state machine · Supervisor · Pattern · Rule base

1 Machine Learning

Machine learning is a field of computer science that uses statistical techniques to give computer systems the ability to "learn" (e.g. progressively improve performance on a specific task) with data, without being explicitly programmed. Machine learning objectives vary depending on the approach we use. According to [10], there are 4 approaches.

- The first approach is to model the mechanisms that form the basis of human learning. An example may be a recognition of perceptions from the real world and their integration into different groups (classes).
- Another way to approach machine learning is empirical. This approach aims at discovering general principles that relate to learning algorithms characteristics and general domain principles within which these algorithms operate.
- We can also approach machine learning in general. Here, an emphasis is placed on formulating and proving theorems on the workability of whole classes of learning problems and algorithms proposal to solve these problems.
- The last option to approach machine learning is application. This approach is generally related to algorithm proposal (where we solve problem formulation,

© Springer Nature Switzerland AG 2019
R. Fullér et al. (Eds.): WILF 2018, LNAI 11291, pp. 3–16, 2019.
https://doi.org/10.1007/978-3-030-12544-8_1

solution proposal, implementation). From the point of view of machine learning, we focus on formulating a problem, proposal of representation of training examples (or training knowledge), creating a training set, and generating a knowledge base using machine learning.

2   Machine Learning Approaches to an Adaptative Rule Base

An adaptative rule base is based on finding and assessing the suitability rules. We try to have a rule which provides better decision results (as much accuracy as possible). This is similar to human learning of new skills. They try to find the steps or procedures to best control their skill. Best practices will be retained in memory for use in the future.

None of the above-mentioned machine learning approaches (Table 1) addresses the issue of machine learning to adapt the rule base. This issue is dealt with in the following works: *An Adaptive Fuzzy Controller for Trajectory Tracking of Robot Manipulator* [7], *Adaptive Fuzzy Rule-based Classification Systems* [11], and *Adaptive Fuzzy Controller: Application to the control of the temperature of a dynamic room in real time* [14], whose contribution to the problem is summarised in Table 2 according to the set criteria:

Table 1. An overview of basic machine learning methods

	Algorithms	Principle
Decision trees [8]	TDIDT, ID3, ASSISTANT, C4.5	Tree nodes are evaluated according to the attributes of the instance. The decision-making process starts from the root to the nodes to the leaf. The leaves are valued binary values (YES/NO)
Neural networks [5]	Perceptron, Backpropagation	Choosing the right topology and using the training set to configure the neural network The network consists of layers (input, hidden, output) containing neurons
Bayesian learning [3]	Gibbs algorithm, Bayes classifier, EM algorithm	Classification of hypotheses based on conditional probabilities
Feedback learning [15]	Q-learning	From the set of actions is chosen such an action, thanks to it agent finds himself in a new situation and gain the highest reward represented fair value
Learning with a set of rules [15]	Learn-one-rule FOIL	A tree structure whose nodes contain a description of IF-THEN rules. The goal is to select a node/subtree with the best candidates describing the training examples
Evolutionary algorithms [4]	Genetic algorithms	Search for hypotheses (possible solutions - population) that are expressed numerically (sequence of bits). Iteratively to generate new hypotheses from existing hypotheses using crossover and mutation and maintained in the population according to the fitness function

Table 2. Summary of machine learning approaches to address adaptation of the rule base

	Classification	Presence of the fuzzy inference system	Language description in the form of natural language	Way of adaptation
Work [7]	No	Yes	No	Changing the parameter for the given component (P, D) of a controller
Work [11]	Yes	Yes	No	Changing the height of the fuzzy set, classifying into 2 classes
Work [14]	No	Yes	No	Changing the support of a fuzzy set

- The first criterion was the implementation of decision making (not classifying objects) of the fuzzy inference system based on defined IF-THEN rules.
- The second criterion was a language description that is in the form of a natural language text.
- The third criterion was how to adapt the rules. This criterion is not defined because it is not known how the rules should be adapted. Either there would be a given pattern according to which the rules would be set or the rules would be prioritised based on their frequency of use. If a suitable solution is found, this strategy could be taken into account.

The paper aims at proposing a machine learning approach for an adaptive rule base. Adaptation methods for rule bases that are described in the above-mentioned publications do not meet our defined criteria.

3 Proposal of a Machine Learning Method for Adaptive Rule Base

The proposed machine learning approach allows the adaptation of the rule base according to the training set. The rule base adaptation is represented by changing certain rules according to the required criteria. This change can be seen as deductive learning. If we are to achieve a reinforcement of the right rules, supervised learning or reinforcement learning can be used. When using the supervised machine learning approach, it is important to have a training set (patterns). As a way to get a pattern, an online approach was chosen, i.e. to get a pattern during adaptation from the motion of the monitored joints.

These machine learning approaches (deductive learning, supervised learning, online learning) form the basis of the proposed machine learning method. The system architecture (Fig. 1), which implements the proposed machine learning approach, consists of three parts [6]:

- Fuzzy inference system,
- Supervisor,
- Finite-state machine.

The basis of the proposed approach is LFLCore, which is part of the LFLC application [2], which was developed at the Institute for Research and Applications of Fuzzy Modeling.

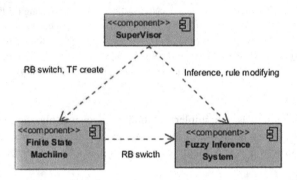

Fig. 1. System architecture

3.1 Linguistic Context

The linguistic context is defined in [13] as follows (1):

$$w = \langle v_L, v_S, v_R \rangle \quad v_L, v_S, v_R \in R \quad v_L < v_S < v_R \tag{1}$$

where v_L denotes the smallest value, v_R is the largest value, and v_S is the usual mean value to be considered in the given situation.

The construction and distribution of fuzzy sets depends on the linguistic context (specifically, on evaluating linguistic expressions). According to the default parameter settings in LFLC [2], the language "one-sided" context is defined as follows (2):

$$w = \langle 0; 0.4; 1 \rangle \tag{2}$$

The proposed machine learning approach uses two language "one-sided" contexts that are symmetric by parameter v_L. These linguistic contexts are uniformly named as language "two-sided" contexts, which are defined as follows (3):

$$
\begin{aligned}
w &= \langle v_{-R}, v_{-S}, v_L, v_S, v_R \rangle \\
v_{-R}, v_{-S}, v_L, v_S, v_R &\in \mathrm{R} \\
v_{-R} &< v_{-S} < v_L < v_S < v_R \\
|v_{-R}| &= v_R \\
|v_{-S}| &= v_S
\end{aligned}
\tag{3}
$$

The linguistic context for each input and output variable is set to (4):

$$w = \langle -1; -0.4; 0; 0.4; 1 \rangle \tag{4}$$

3.2 Expressions of Variables

As expressions of variables are used evaluating linguistic expressions [13], which are language expressions representing either a value on an ordered scale (usually a certain number), or a position on it (left/right).

They include atomic expressions (Fig. 2) and fuzzy numbers, which can be complemented by language operators, signatures (+, −), and joined by conjunctions (and, or).

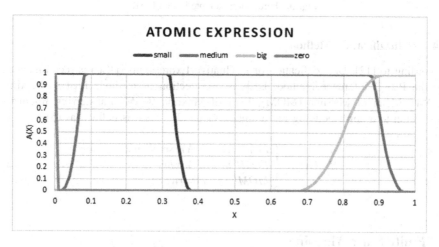

Fig. 2. Atomic expressions

3.3 Inference Method

Since the expressions of input/output variables are evaluating linguistic expressions, Perception Based Logical Deduction (PBLD) is appropriate for working with these expressions as described in [12]. This method handles the language description which is linguistically-logically interpreted. Perception (Fig. 3) means such an evaluating linguistic expression to which a value is assigned in the defined context. According to the perceived perception, an appropriate rule from the linguistic description is subsequently activated, and the result obtained in the given rule is evaluated in the form of evaluating linguistic expression.

The number of activated rules of the inference method corresponds to the number of input variables. This appears to be an advantage over traditional inference methods that process a relational interpreted language description (e.g. Mamdani fuzzy inference [9]) which activate all the defined rules.

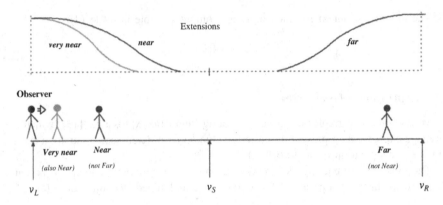

Fig. 3. Perception (adapted from [13])

3.4 Defuzzification Method

According to [12], Defuzzification of Evaluative Expressions (DEE) is recommended for the PBLD inference method. DEE is a collection of methods Last of Maxima (LOM), Mean of Maxima (MOM), First of Maxima (FOM) transferring linguistic expression to a corresponding real number. Generally, DEE is defined as (5):

$$DEE(A) = \begin{cases} LOM(A) & \text{if } A \text{ is small or zero} \\ FOM(A) & \text{if } A \text{ is big} \\ MOM(A) & \text{otherwise} \end{cases} \tag{5}$$

4 Finite-State Machine

The finite-state machine is based on a mathematical model of the language grammar, the so-called Chomsky hierarchy [1]. Finite-state machine can recognise a regular language, which is at the lowest level in the hierarchy (Type-3 grammars). The deterministic finite-state machine FA can be defined as follows (6):

$$FA = \left(Q, \sum, \delta, q_0, F \right) \tag{6}$$

where:

Q is a finite, non-empty set of states
$\sum$ is the input alphabet (a finite, non-empty set of symbols)
$\delta: Q \times \sum \rightarrow Q$ is the state-transition function
$q_0 \in Q$ is an initial state
$F \subseteq Q$ is the set of final states.

In the proposed machine learning model, states are reflected as the rule bases that are ready to perform operations of a fuzzy inference system.

- Switching the finite-state machine to the next state is decided based on the current state of the counter that acquires the values of the natural numbers.
- The state-transition function δ is reflected as a rule base (states) switching according to the respective value of the counter (symbol).
- The initial state q_0 (initial rule base) is set by the user or supervisor.
- Stop of a run of the final machine occurs when there is no state-transition function for a particular symbol.

5 Supervisor

The last block of the proposed system architecture with machine learning is the supervisor. The supervisor has access to all rules from each rule base. Each rule contained within the database has a parameter fitness. The fitness determines whether the rule is applied when using inferential methods. Rule fitness is reinforced or suppressed during adaptation depending on whether the desired system state is achieved after the performance of the operation (inference). The higher the rule fitness, the better candidate for further decision making the rule is. The fitness of all rules is initialised to 0.

Required states are submitted to the supervisor as patterns. Patterns (or sequence of patterns) are loaded onto a pattern-tape from which the supervisor reads. In addition to information on the required states, the pattern also contains information about:

- the number of a particular step,
- base rules over which the operation (inference) will be performed.

While browsing the patterns, inter-state switching occurs (such a rule base is switched, which is included in the pattern), i.e. a finite-state machine is produced. These inter-state switches are recorded by the supervisor as the transfer functions of the finite-state machine for which is valid (7):

$$\delta_i : Q_i \times \sum_i \to Q_{i+1} \tag{7}$$

where

Q_i is the rule base contained in the i-th pattern, Q_i is a subset of Q,
$\sum_i$ is the number of a particular step (counter value) contained in the i-th pattern,
Q_{i+1} is the base rule contained in the $(i + 1)$-th pattern.

The course of adaptation is divided into these basic phases (Fig. 4).

1. *Pattern loading*
   Patterns are loaded onto a pattern-tape from which the Supervisor gradually retrieves the pattern that is at the front of the queue. The pattern contains the step number (counter state), the base rule name (on which the operation to be performed), and the required state (the state to be performed).

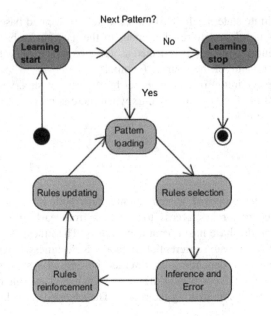

Fig. 4. The course of adaptation

2. *Rules selection*

   In the second phase, the supervisor selects such rules from the rule base that have the given antecedent, which is formed by the required and current state. The desired state value is obtained from the current pattern. The current state value is obtained from the internal state of the given device (from the joint). The values of the desired and current state are assigned to evaluating linguistic expressions to the given context (perception).

3. *Inference and Error*

   In the third phase, the Supervisor performs an inference over the selected rules, which are gradually activated, when the inference is called. After each inference, an error is calculated, e.g. how the current state after the inference differs from the reference value describing the pattern. The error is calculated by (8):

$$E(x, y) = |y - x| \tag{8}$$

   where y is the required value, x is the current value.

   As a result of this phase errors of all rules are calculated and transferred to the next phase.

4. *Rules reinforcement*

   Based on the calculated error, the supervisor will evaluate the rules either as appropriate (+1), inappropriate (−1), or almost acceptable (0). the number of almost appropriate rules MAX_k is fixed. If the supervisor identifies an almost appropriate rule, parameter k increases. Rules reinforcement is given by (9):

$$f(E,k) = \begin{cases} 1 & \text{if } E = 0 \\ 0 & \text{if } E > 0, k \leq MAX\_k \\ -1 & \text{otherwise} \end{cases} \tag{9}$$

5. *Rules updating*

In the last phase, the resulting value of the function f calculated from the previous phase is added to the fitness of each rule. This fitness determines whether the rule is correct when performing the given activity.

6 Experimental Outcomes

The proposed machine learning approach was tested on several examples in which we teach the system to perform activities consisting of certain steps [6]. Each step represents the movement of an individual joint. In this experiment, the rules for the movement of four joints are adapted. Here, the system should control the joints so that it can take one step.

Joint Specifications:
The input value "Current state" is reflected as the internal state of one joint. The angle of rotation of a given joint can take values from the interval $[-180°; 180°]$. The angle of the joints J1 and J3 is expressed by vectors $\vec{n}$ and $\vec{v_n}$ and the angle of the joints J2 and J4 is expressed by vectors $\vec{m}$ and $\vec{v_m}$ (Fig. 5). The output variable "Action hit" is reflected as a change in the state of one joint having values from the interval $[-360°, 360°]$. As required by the fuzzy inference system, these intervals are converted according to a defined context. The input variable "Requested state" reflects the state obtained from a given pattern whose value belongs to the given language context.

If the top joint (J1 and J3) is in action, the bottom joint (J2 and J4) retains the angle between vectors $\vec{m}$ and $\vec{v_m}$ (Fig. 5).

The structure of the presented pattern is shown in Table 3, where the "desired state" is the measured value corresponding to the angle between the vectors $\vec{n}$ (or $\vec{m}$) and $\vec{v}$. In this case, the Finite-state machine switches the rules of the individual joints according to the defined pattern stored in Table 3.

The structure of rule base is as follows:

- Input/Output Variables:
  - Context: $w = \langle -1; -0, 4; 0; 0, 4; 1 \rangle$
  - Expressions: ($\pm$ro ze, $\pm$vv sm, $\pm$vv me, $\pm$vv bi), where 'ro ze' is roughly zero, 'vv sm' is very very roughly small, 'vv me' is very very roughly medium, 'vv bi' is very very roughly big.
- Linguistic description
  - Current status & Desired state - > Action hit.
- Rules:
  - Number of rules: 512.
  - Rule specification: Inconsistent deactivated rules.

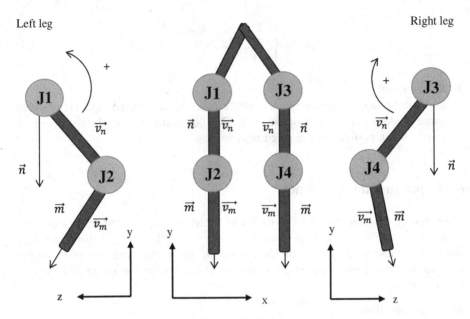

Fig. 5. Joints of left and right legs

Table 3. Sequence for movement of both legs

Step	Rule base regarding the joint	Required state	Step	Bule-base regarding the joint	Required state
1	J1	0	15	J2	−0.02
2	J3	0	16	J4	−0.02
3	J2	0	17	J1	0.138
4	J4	0	18	J3	−0.1
5	J1	0	19	J2	−0.072
6	J3	0.094	20	J4	−0.094
7	J2	0	21	J1	0.1
8	J4	−0.36	22	J3	−0.02
9	J1	−0.072	23	J2	−0.072
10	J3	0.21	24	J4	−0.27
11	J2	0.027	25	J1	0
12	J4	−0.205	26	J3	0
13	J1	−0.1	27	J2	0
14	J3	0.127	28	J4	0

- Inference method:
    - Perception-Based Logical Deduction.
- Defuzzification method:
    - Defuzzification of Evaluative Expressions.

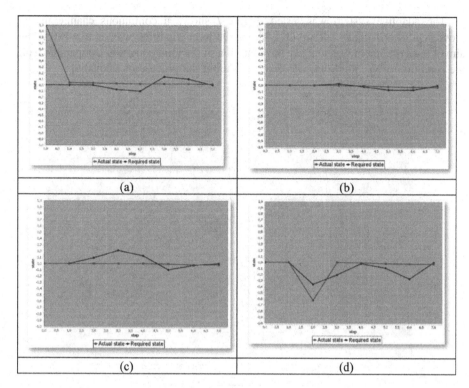

Fig. 6. (a) Continuous change in the motion of the left hip (J1). (b) Continuous change in the motion of the left knee (J2). (c) Continuous change in the motion of the right hip (J3). (d) Continuous change in the motion of the right knee (J4). (Color figure online)

Adaptation of the rule base is done in 21 iterations. This number is determined by setting the initial states for the upper joints J1 and J3 (Fig. 5). The initial value of the lower joints J2 and J4 is always set to 0°. The number of steps in one iteration of adaptive learning is set to 28. This sequence of patterns includes seven movements for each joint. A continuous change of state in one iteration is shown in Fig. 6. The initial value of the left hip is set to 180° and the initial state of the other joints is set to value 0°.

6.1   Evaluation of Experiments

During the adaptation in each experiment, we managed to adapt the individual rule base and select the most appropriate rules for the given activity. These rule bases can be uploaded to an expert system or LFLC [2] (compatibility is preserved). However, it must be noted that each iteration occurs for setting different initial values and reinforces rules that have the same antecedent and different consequence. As a result, these rules are evaluated in such a way that one of them has a positive fitness value and others have negative fitness values.

The adaptive system was tested. Figure 7 shows a continuous change in the movement of one joint - left hip (J1). The initial joint condition was set to −180°. The aim of the experiment was to make the condition of the joint condition comparable with the condition of the joint that was detected during the process of adaptation.

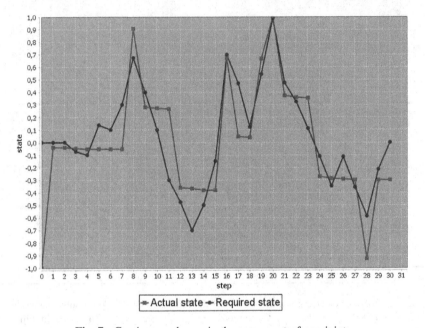

Fig. 7. Continuous change in the movement of one joint.

The red curve showing the current state of the joint at each step approaches, after adaptation, the blue curve showing the state that has to occur in a given step. In some steps, you can see it stay on a place, even if it is to move a bit. It is because when using the defined evaluating linguistic expressions, the adaptation of the system does not register small movements and evaluates the best solution to stay in a place (Fig. 6, steps 1−7).

The benefits of this approach to adaptation of the rule bases are the following:

- The user does not intervene to the rule bases during adaptation, because the supervisor solves everything in the proposed system.
- The boundaries of the language context are fixed, the transformation of the language context to the desired interval and vice versa can be done in the fuzzy inference system interface.
- Simple adaptations when changing the desired allowable value interval for a given joint.
- Simple adaptations when changing patterns of the training set.

7 Conclusions

The aim of the paper was to design a machine learning approach to adapt rule bases. Machine learning's own approach includes the initial setting of the rule bases, the inter-state switching proposal, and the proposal of the method of their adaptation. Each rule base, under which the fuzzy inference system determines, has a defined limited number of evaluating linguistic expressions, type of a fuzzification and defuzzification method, and a randomly generated set of inconsistent of rules. For the inter-state switching, a simple mechanism of the finite-state machine was proposed to allow the rule bases to be switched. Thanks to the finite-state machine, the system avoided using only one rule base that would be defined by multiple input/output variables. We have also proposed how to adapt the rule bases, e.g. what the most appropriate rules should contain according to the pattern. This proposed approach to machine learning adapting the rule bases was tested. The subject of the experimental study was the joint movement according to the presented patterns. The course of these experiments was recorded and subsequently evaluated.

The proposed system will be further developed because we have identified the following shortcomings in the adaptive learning, which will be gradually eliminated.

- The system does not react to small movements (the best option is to remain in a place). This is evident from Fig. 6.
- Redundancy of rule bases. If an antecedent has more consequences, the proposed system prefers such consequence, which fitness is the greatest.
- Evaluation of the best rule is calculated based on the difference between the desired and the current state, see Eqs. (8) and (9).
- It does not cover all combinations of rules, but the proposed rule base is sufficient for this experimental study.

This proposed approach of machine learning can be used in systems such as a humanoid robot. This robot can learn a few activities consisting of simple activities. These activities include, for example, the movement of a robot's joint. Knowledge of "joint movement" could expand the knowledge of moving one leg or both legs. If the system knows to control its legs, it could learn to walk, run, squat, jump. In doing so, these rules to adapt to learning new activities/skills would be used. After deploying the system into operation, it can perform a newly learnt activity. It can be stated that the more activities the system can do, the larger the area of the linguistic description in the rule base it will cover.

Acknowledgments. The research described here has been financially supported by University of Ostrava grant SGS04/PřF/2018. Any opinions, findings and conclusions or recommendations expressed in this material are those of the authors and do not reflect the views of the sponsors.

References

1. Chomsky, N.: Three models for the description of language. IRE Trans. Inf. Theory **2**(3), 113–124 (1956)
2. Dvorak, A., Habiballa, H., Novak, V., Pavliska, V.: The software package LFLC 2000-its specificity, recent and perspective applications. Comput. Ind. **51**, 269–280 (2003)

3. Friedman, N., Geiger, D., Goldszmidt, M.: Bayesian network classifiers. Mach. Learn. **29**(2–3), 131–163 (1997)
4. Goldberg, D.E., Holland, J.H.: Genetic algorithms and machine learning. Mach. Learn. **3**(2), 95–99 (1988)
5. Gurney, K.: An Introduction to Neural Networks. CRC Press, Boca Raton (2014)
6. Jaluvka, M.: The Machine learning for the adaptive rule base (in Czech). Master thesis, University of Ostrava, Czech Republic (2017)
7. Khalate, A.A., Leena, G., Ray, G.: An adaptive fuzzy controller for trajectory tracking of robot manipulator. Intell. Control Autom. **2**(4), 364–370 (2011)
8. Lior, R.: Data Mining with Decision Trees: Theory and Applications, vol. 81. World Scientific, Singapore (2014)
9. Mamdani, E., Assilian, S.: An experiment in linguistic synthesis with a fuzzy logic controller. an experiment in linguistic synthesis with a fuzzy logic controller. In: Readings in Fuzzy Sets for Intelligent Systems, pp. 283–289 (1993)
10. Mitchell, T.M.: Machine Learning. McGraw-Hill, Boston (1997)
11. Nozaki, K., Ishibuchi, H.: Adaptive fuzzy rule-based classification systems. IEEE Trans. Fuzzy Syst. **4**(3), 238–250 (1996)
12. Novák, V., Perfilieva, I.: On the semantics of perception-based fuzzy logic deduction. Int. J. Intell. Syst. **19**(11), 1007–1031 (2004)
13. Novák, V., Perfilieva, I., Mockor, J.: Mathematical Principles of Fuzzy Logic. Springer, Heidelberg (2012)
14. Rojas, I., et al.: Adaptive fuzzy controller: application to the control of the temperature of a dynamic room in real time. Fuzzy Sets Syst. **157**(16), 2241–2258 (2006)
15. Russell, S.J., Norvig, P.: Artificial Intelligence: A Modern Approach. Pearson Education Limited, London (2016)

Data-Driven Induction of Shadowed Sets Based on Grade of Fuzziness

Dario Malchiodi$^{(\boxtimes)}$ and Anna Maria Zanaboni

Università degli Studi di Milano, Milan, Italy
{malchiodi,zanaboni}@di.unimi.it

Abstract. We propose a procedure devoted to the induction of a shadowed set through the post-processing of a fuzzy set, which in turn is learned from labeled data. More precisely, the fuzzy set is inferred using a modified support vector clustering algorithm, enriched in order to optimize the fuzziness grade. Finally, the fuzzy set is transformed into a shadowed set through application of an optimal alpha-cut. The procedure is tested on synthetic and real-world datasets.

Keywords: Shadowed sets · Fuzzy set induction · Machine learning · Support vector clustering

1 Introduction

In the last decades fuzzy sets have been proved to be a powerful means for knowledge representation, reasoning and decision making in uncertain contexts. However, as their usage becomes widespread, the trade-off between the detailed nature of the fuzzy membership function and its symbolic interpretation is getting undisguised. A possible way to address uncertainity trying to manage this trade-off is to identify three regions of the universe of discourse, namely a belongingness region, an exclusion region and a "grey" region where genuine uncertainity holds. A lot of work has been done in this direction in different research areas (rough sets [22], fuzzy sets [4,8,23], three-valued logic [3], type-2 fuzzy logic [21], see [15,24] for a survey). We will focus on the construct of shadowed sets introduced by Pedrycz [18,19], and used in different learning contexts [5,9,10,13,25,26]. Given a shadowed set A, the domain of discourse is split into three regions, called the *core*, the *exclusion* and the *shadowed* region, where membership value to A is 1, 0 and unknown, respectively. The shadowed set is induced by a fuzzy set because its shadowed regions' position and width are determined by the constraint of preserving the amount of fuzziness of the originating fuzzy set. More precisely, the induced shadowed set is completely determined by an α-*cut*, namely a value $0 \leq \alpha \leq 1/2$ used to cut the codomain [0,1] of the fuzzy membership function into the zones $[0, \alpha]$, $(\alpha, 1 - \alpha)$, $[1 - \alpha, 1]$ where full belongingness, uncertainity and full exclusion are, respectively, assigned. This gives rise to a membership function $S_A : X \mapsto \{0, [0, 1], 1\}$,

© Springer Nature Switzerland AG 2019
R. Fullér et al. (Eds.): WILF 2018, LNAI 11291, pp. 17–28, 2019.
https://doi.org/10.1007/978-3-030-12544-8_2

where X is the universe of discourse and 1, $[0,1]$, 0 are associated to the three mentioned zones.

The search for the α-cut is an optimization problem; moreover, the particular definition of fuzziness of a fuzzy set obviously affects the resulting procedure. In [20] analytical formulas are provided to calculate the optimal α-cut using the gradual grade of fuzziness, and a comparison with other fuzziness measures is discussed.

We describe a data-driven procedure for the induction of shadowed sets based on the post-processing of a fuzzy set learned from labeled data. The procedure exploits a support vector clustering [1] algorithm in which the inference is done starting from a set of objects in X, labeled with their membership degrees to A. As a next step, a sphere S in a space H is found so that the images of objects through a function $\Phi : X \mapsto H$ are positioned w.r.t. S in function of the membership degrees. More precisely, in case of unitary membership the image of an object will belong to S, otherwise it will fall farther from the border of S as its membership to A decreases from 1 to 0 [16]. This learning algorithm is further enriched with the optimization of the fuzziness grade of A.

The paper is organized as follows: Sect. 2 is devoted to the derivation of the fuzziness degree of a piecewise linear membership function and its optimal α-cut, Sect. 3 is devoted to (a) the description of the modified support vector clustering optimization problem for learning a membership function and (b) to its enrichment with a term accounting for fuzziness degree minimization of the inferred shadowed set. In Sect. 4 we discuss experimental results on a synthetic and two real-world benchmarks. Some concluding remarks end the paper.

2   Gradual Grade of Fuzziness of a Fuzzy Set

The fuzziness grade of a fuzzy set measures the *vagueness* of the set itself. Such concept captures the amount of entropy inherently contained in a fuzzy set: indeed crisp sets have a null fuzziness grade, while on the other hand the maximal grade is attained by a fuzzy set with membership function constantly equal to $1/2$. As a rule of thumb, the sharper the boundaries of a fuzzy set, the smaller the related fuzziness. Among the proposed measures quantifying the fuzziness grade of a fuzzy set (see for instance [12,14]), we consider the one introduced in [11] quantifying the fuzziness grade of a continuous, measurable fuzzy set A whose membership function is μ_A as

$$\varphi(A) = \int_X (1 - |2\mu_A(x) - 1|)\mathrm{d}x.$$

The notion of fuzziness grade is linked to the search of an optimal α-cut transforming a fuzzy set into a shadowed set [20]. Namely, denoted by ω_1, ω_2 and ω_3 the definite exclusion, the definite belongingness and the uncertainty regions mentioned in the Introduction and restricting to them the fuzziness degree computation, the optimal α is such that

$$\varphi(\omega_1) + \varphi(\omega_2) = \varphi(\omega_3) \tag{1}$$

holds. In this way, the overall fuzziness of A is equally balanced between the shadowed (ω_3) and unshadowed ($\omega_1 \cup \omega_2$) regions.

In the rest of this paper we will focus on the family of piecewise linear functions whose general member has the following form

$$
f_{R^2,M}(x) = \begin{cases} 1 & \text{if } x \leq R^2, \\ 1 - \frac{x-R^2}{M-R^2} & \text{if } R^2 < x \leq M, \\ 0 & \text{otherwise,} \end{cases}
$$

where $R^2 > 0$ and $M > R^2$ denote the boundaries of the crisp regions of the fuzzy set (see Fig. 1(a))[1]. It is easy to show that the fuzziness degree of a fuzzy set A whose membership function has the form $\mu_A(x) = f_{R^2,M}(x)$ is

$$
\varphi(A) = \frac{M - R^2}{2} \tag{2}
$$

while, for fixed α

$$
\varphi(\omega_1) = \alpha^2 \left(M - R^2 \right),
$$
$$
\varphi(\omega_2) = \alpha^2 \left(M - R^2 \right),
$$
$$
\varphi(\omega_3) = 2(1 - \alpha^2) \left(M - R^2 \right),
$$

thus the optimal cut condition (1) reads $\alpha^2 = 1 - \alpha^2$, which corresponds to $\alpha = \sqrt{2}/2$.

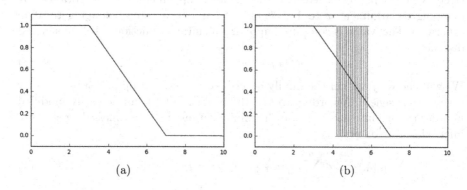

(a)   (b)

Fig. 1. Graph of (a) a membership function μ_A to a fuzzy set in the considered family, and (b) the membership function S_A to a shadowed set obtained from (a) after an optimal α-cut. Blue curve: graph of μ_A; green segments: crisp values of S_A; gray area: uncertainty zone of S_A. (Color figure online)

---

[1] The choice of R^2 and M as names for these symbols is linked to a special role they will play in Sect. 3.

3   Shadowed Set Induction

The proposed procedure learns a shadowed set in two phases: the first one infers a fuzzy set starting from a set of labeled objects, while the second phase performs on this set the α-cut described in the previous section.

Focusing on the first phase, consider a universe of discourse X, fix $n \in \mathbb{N}$ and denote by $\{x_1, \ldots, x_n\} \in X^n$ a sample of objects. Given also a set of labels $\{\mu_1, \ldots, \mu_n\} \in [0,1]^n$ whose values are the membership degrees of objects to an unknown fuzzy set A, the membership function μ_A can be learned using the approach proposed in [16] optimizing the square radius R^2 of a sphere S belonging to a space H and centered in a. Namely, the images of objects through a function $\Phi : X \mapsto H$ are such that $\Phi(x_i) \in S$ when $\mu_i = 1$, and $\Phi(x_i)$ tends to fall farther from the border of S as μ_i decreases from 1 to 0. This amounts to modifying the support vector clustering algorithm proposed in [1] as follows:

$$\min R^2 + C \sum_{i=1}^{n} (\xi_i + \tau_i) \tag{3}$$

$$\mu_i ||\Phi(x_i) - a||^2 \leq \mu_i R^2 + \xi_i \ \forall i = 1, \ldots, n, \tag{4}$$

$$(1 - \mu_i)||\Phi(x_i) - a||^2 \geq (1 - \mu_i)R^2 - \tau_i \ \forall i = 1, \ldots, n, \tag{5}$$

$$\xi_i \geq 0, \tau_i \geq 0 \quad \forall i = 1, \ldots, n. \tag{6}$$

In this formulation $C > 0$ measures the relative importance of the two components in the objective function, while ξ_i and τ_i are slack variables relaxing the constraints dealing with the positioning of points inside and outside S, respectively. Once S has been learned, the membership function μ_A is obtained by mapping its argument x to 1 if $\Phi(x) \in S$, and to a value belonging to $[0,1)$ otherwise. This value is computed applying a suitable function f to the squared distance

$$r^2(x) = ||\Phi(x) - a||^2. \tag{7}$$

We will choose f within the family described in Sect. 2, dropping subscripts for sake of conciseness. According to (2), the problem (3–6) can be easily modified in order to take into account also the minimization of the fuzziness degree of the inferred set as follows:

$$\min R^2 + C \sum_{i=1}^{n} (\xi_i + \tau_i) + D(M - R^2) \tag{8}$$

$$\mu_i ||\Phi(x_i) - a||^2 \leq \mu_i R^2 + \xi_i \ \forall i = 1, \ldots, n, \tag{9}$$

$$(1 - \mu_i)||\Phi(x_i) - a||^2 \geq (1 - \mu_i)R^2 - \tau_i \ \forall i = 1, \ldots, n, \tag{10}$$

$$||\Phi(x_i) - a||^2 \leq M\psi \ \forall i = 1, \ldots, n, \tag{11}$$

$$\xi_i \geq 0, \tau_i \geq 0 \ \forall i = 1, \ldots, n. \tag{12}$$

In this new formulation, $D > 0$ is a new hyperparameter jointly ruling with C the relative importance of the components in (8), namely devoted to the optimization of radius, slack variables, and fuzziness degree. Analogously, M is introduced as a new variable, bounded in (11) to be higher than the distance between a and any of the images $\Phi(x_i)$. Jointly considering this constraint and the objective function (8) amounts to requiring M to equal the maximum of such distances. Actually, an additional hyperparameter ψ in (11) allows to fine tune this requirement: $\psi > 1$ promotes higher values for M, and *vice versa*.

Letting $E = (1 - D(1 - 1/\psi))$ and denoting with k the kernel associated to Φ (that is, $k(x_i, x_j) = \Phi(x_i) \cdot \Phi(x_j)$), the Wolfe dual of (8–12) corresponds to the maximization of

$$\sum_{i=1}^{n}(\epsilon_i + \beta_i)k(x_i, x_i) - E^{-1} \sum_{i,j=1}^{n} (\epsilon_i + \beta_i)(\epsilon_j + \beta_j)k(x_i, x_j) \qquad (13)$$

subject to the constraints

$$\sum_{i=1}^{n} \epsilon_i = 1 - D, \qquad (14)$$

$$\sum_{i=1}^{n} \beta_i = D/\psi, \qquad (15)$$

$$-(1 - \mu_i)C \leq \epsilon_i \leq \mu_i C \; \forall i = 1, \ldots, n, \qquad (16)$$

$$\beta_i \geq 0 \; \forall i = 1, \ldots, n. \qquad (17)$$

It is easy to see that β_i is the generic Lagrangian multiplier associated to (11), while

$$\epsilon_i = \mu_i \gamma_i - (1 - \mu_i)\hat{\gamma}_i, \qquad (18)$$

being γ_i and $\hat{\gamma}_i$ the multipliers for (9) and (10). In order to be solvable, the dual problem requires $D \neq \psi/(\psi - 1)$, otherwise the objective function would not be computable.

In the experiments described later on, we will consider two kinds of kernel: the *linear* kernel defined by $k(x_i, x_j) = x_i \cdot x_j$, and the family of *Gaussian* kernels defined by

$$k(x_i, x_j) = \exp\left(-\|x_i - x_j\|^2/\sigma^2\right), \qquad (19)$$

where $\sigma > 0$ is an additional hyperparameter to be considered. This use of the so-called *kernel trick* allows to consider a universe of discourse whose members are not necessarily numbers or numerical vectors. For instance, [17] uses a similar technique in order to solve the problem of detecting a set of reliable axioms in the context of semantic Web.

Dealing with the KKT conditions [7] is a bit tricky, because these are expressed in terms of γ_i, $\hat{\gamma}_i$, β_i, and the remaining Lagrange multipliers. For sake of conciseness, we just list the salient relations linking primal and dual variables when we consider the optimal solution:

$$0 < \gamma_i < C \rightarrow R^2 = r^2(x_i), \qquad (20)$$

$$0 < \hat{\gamma}_i < C \rightarrow R^2 = r^2(x_i), \qquad (21)$$

$$\beta_i > 0 \rightarrow M = \psi^{-1}r^2(x_i), \qquad (22)$$

Table 1. Relations between the dual variables γ_i, $\hat{\gamma}_i$, and β_i.

	$\hat{\gamma}_i = 0$	$0 < \hat{\gamma}_i < C$	$\hat{\gamma}_i = C$
$\gamma_i = 0$	$\epsilon_i = 0$	$-C(1 - \mu_i) < \epsilon_i < 0$	$\epsilon_i = -C(1 - \mu_i)$
$0 < \gamma_i < C$	$0 < \epsilon_i < C\mu_i$	$-(1 - \mu_i)C < \epsilon_i < \mu_i C$	$-C(1 - \mu_i) < \epsilon_i < C(2\mu_i - 1)$
$\gamma_i = C$	$\epsilon_i = C$	$C(2\mu_i - 1) < \epsilon_i < C\mu_i$	$\epsilon_i = C(2\mu_i - 1)$

where $r^2(x) = ||\Phi(x) - a||^2$ can be obtained as

$$r^2(x) = k(x, x) - 2E^{-1} \sum_{i=1}^{n} (\epsilon_i + \beta_i) k(x, x_i)$$

$$+ E^{-2} \sum_{i,j=1}^{n} (\epsilon_i + \beta_i)(\epsilon_j + \beta_j) k(x_i, x_j).$$

The problem here is that (8–12) explicitly depend only on ϵ_i and β_i, thus only (22) is directly exploitable. By analyzing all combinations between the critical values of γ_i and $\hat{\gamma}_i$ and computing the corresponding values for ϵ_i (see Table 1), it is easy to check that

$$0 < \epsilon_i < C\mu_i \rightarrow 0 < \gamma_i < C, \tag{23}$$
$$-C(1 - \mu_i) < \epsilon_i < 0 \rightarrow 0 < \hat{\gamma}_i < C. \tag{24}$$

Jointly considering (20–24) it is therefore possible to link the optimal values of dual and primal variables:

$$0 < \epsilon_i < C\mu_i \rightarrow R^2 = r^2(x_i), \tag{25}$$
$$-C(1 - \mu_i) < \epsilon_i < 0 \rightarrow R^2 = r^2(x_i), \tag{26}$$
$$\beta_i > 0 \rightarrow M = \psi^{-1} r^2(x_i). \tag{27}$$

Once R^2 and M have been found, a shadowed set can be obtained from the corresponding fuzzy set through application of the optimal α-cut described in Sect. 2.

4   Experiments

As a first set of experiments, we tested the sensitivity of the overall learning procedure to hyperparameters[2]. Focusing on C, we considered a synthetic dataset composed by seven points whose crisp membership[3] has been fixed according

---

[2] Code and data to replicate experiments are available at https://github.com/dariomalchiodi/WILF2018.

[3] It is worth highlighting that the learning algorithm of Sect. 3 can in principle be run on objects labeled using more generic membership grades (that is, values belonging to $[0, 1]$). However, as such a rich information is normally not available in public datasets, all reported experiments rely on crisp membership labels.

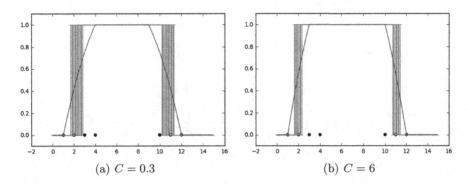

(a) $C = 0.3$ (b) $C = 6$

Fig. 2. Effect of changes of C on the membership functions learned using a linear kernel and setting $D = 0.3$ and $\psi = 1$. The fuzzy and shadowed membership functions were plotted using the same notation of Fig. 1. For each sample point, a bullet on the X-axis is drawn using black and white color when $\mu_i = 1$ and $\mu_i = 0$.

to an interval, thus suggesting a unimodal membership both to a fuzzy and a shadowed set. Using a linear kernel and fixing $D = 0.3$ and $\psi = 1$, Fig. 2 shows that rising C has the effect of sharpening the boundaries of the fuzzy set. In other words, as C grows the membership increases from 0 to 1 (and decreases from 1 to 0) in a more linear fashion.

Figures 3 and 4 describe analogous experiments focusing on the role of D and ψ which we can summarize as follows:

- D is directly related to the amount of uncertainty of the inferred shadowed set (the higher its value, the lower the fuzziness degree of the set);
- ψ primarily influences the *localization* of the inferred set, although it also affects the optimal value of M, thus it has an impact on the uncertainty described in the previous point, too.

Finally, the kernel choice obviously affects the general form of μ_A, and thus also S_A. For instance, Fig. 5 shows the effect of decreasing the parameter of a Gaussian kernel when learning S_A on the same dataset of Figs. 2 and 3. The augmented plasticity in the considered class of functions allows the procedure to find bimodal memberships concentrating around the positive points as σ decreases.

Switching to a non-synthetic dataset, Fig. 6 shows the result of the proposed technique for a sample from the veterinary domain, in which each observation is the measurement of the level of kidney function (namely, the rate of glomerular filtration, measured in mL/min/Kg), in a set of 37 dogs, each one labeled either as "ill" or "healthy". In this case the procedure relied on a Gaussian kernel with parameter $\sigma = 0.3$, fixing $C = 0.5$, $D = 0.45$ and $\psi = 1$.

As the described learning algorithm can handle objects of arbitrary dimension, we also considered the Iris dataset [6], gathering the observations of 150 iris plants, expressed as a 4-dimensional vector (sepal length, sepal width, petal length, petal width), with length and width measured in centimeters. Each observation belongs to exactly one of the classes *Setosa*, *Virginica*, and *Versicolor*

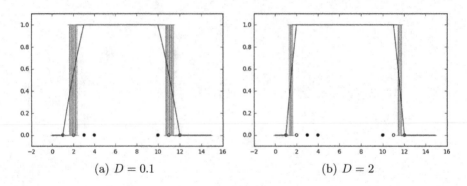

Fig. 3. Effect of changes in the parameter D on the learned unimodal membership function to a shadowed set, letting $C = 10$, $\psi = 1$ and using a linear kernel. Same notation as in Fig. 2.

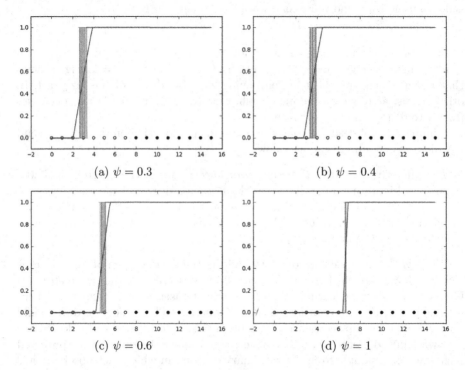

Fig. 4. Effect of changes in the parameter ψ on the membership function to a shadowed set learned with a linear kernel and setting $C = 1$ and $D = 4$. Same notation as in Fig. 2.

(where the first one is linearly separable from the remaining two, and the latter are linked by a more complex relationship). For sake of visualization, we extracted the first two principal components from the observations and performed the shadowed set inference for all the available classes, each time label-

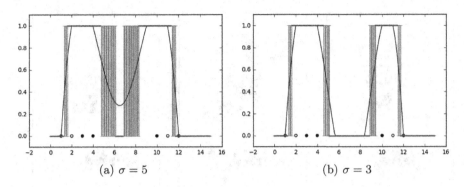

(a) $\sigma = 5$ (b) $\sigma = 3$

Fig. 5. Effect of changes in the parameter σ of the used Gaussian kernel on the bimodal membership function to a shadowed set learned when $C = 30$, $\psi = 1$ and $D = 0.8$. Same notation as in Fig. 2.

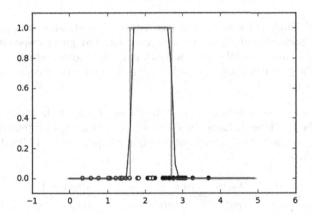

Fig. 6. Membership function for the fuzzy and shadowed sets capturing the concept of "ill dog" expressed by a real-world dataset. Same notation as in Fig. 2.

ing with $\mu_i = 1$ the observations referring to the target class and with $\mu_i = 0$ the remaining observations. Figure 7 shows the obtained results when the whole dataset is considered, using a Gaussian kernel (the related value for σ, as well as those of the remaining hyperparameters have been chosen after an exploratory procedure). In the figure, bullets are superimposed to a visualization of the membersip functions where a dark gray and white background respectively refer to positive and negative values, while a light gray background shows the uncertain areas.

In order to get quantitative results, we performed a more accurate experiment in which the following holdout scheme was iterated one hundred times.

– We randomly shuffled data and subsequently performed a stratified sampling in order to get a training and a test set gathering respectively 80% and 20% of

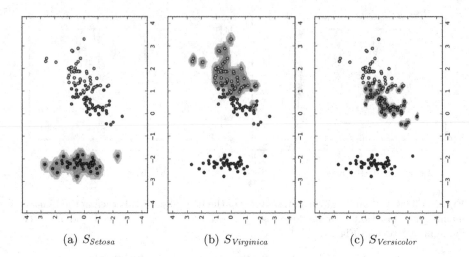

(a) S_{Setosa} (b) $S_{Virginica}$ (c) $S_{Versicolor}$

Fig. 7. Inferred shadowed sets for the Iris dataset. Bullets show the two principal components of each data item, colored in blue, red, and green respectively for the *Setosa*, *Virginica*, and *Versicolor* classes. Dark gray, light gray and white background correspond to the positive, uncertain, and negative values for the membership function. (Color figure online)

Table 2. Results of one hundred holdout iterations of a joint shadowed set learning procedure on the Iris dataset. Each row shows average, median, and standard deviation of test error, in function of the number of principal components (# PC) extracted from the original sample.

# PC	Average error	Median error	Error std
2	0.07	0.07	0.04
3	0.03	0.03	0.02
4	0	0	0

the available items. Stratification ensured training and test sets to be balanced (in the sense that the three classes are equally represented).

- We applied the inference procedure to the training set, obtaining three membership functions to shadowed sets, each linked to a specific Iris class.
- We assessed the joint performance of these three shadowed sets by assigning each object to the class maximizing the membership function value (using the obvious order $0 < [0, 1] < 1$) and comparing the result with the target label[4].

The experiments were repeated extracting two, three, and four principal components from the dataset. Table 2 summarizes the obtained results: the pro-

---

[4] Ties were resolved in favor of the correct class, when possible.

posed approach definitely learns the Iris dataset, outperforming similar techniques based on the sole induction of fuzzy sets [2,16].

5 Conclusions

Reducing the complexity of structures described in terms of fuzzy sets has the desirable effect of allowing an easier interpretation of models induced from data. With this aim, we propose a learning algorithm for shadowed sets, which are sets endowed with a three-valued membership function defining full membership, full exclusion and genuine uncertainty w.r.t. candidate points. This algoritm identifies the shadowed set according to an optimal α-cut performed on a fuzzy set, in turn inferred from data using a modified support vector clustering approach also optimizing the fuzziness degree. A preliminary set of experiments on synthetic data allowed us to gain better insights on the role of hyperparameters; we also tested the procedure on real-world datasets, getting improvements with respect to a previous approach solely based on fuzzy sets. Besides a deeper experimentation phase, we plan to extend the technique considering different families of membership functions, as well as the jointly learning of several shadowed sets.

References

1. Ben-Hur, A., Horn, D., Siegelmann, H.T., Vapnik, V.: Support vector clustering. J. Mach. Learn. Res. **2**(Dec), 125–137 (2001)
2. Cermenati, L., Malchiodi, D., Zanaboni, A.M.: Simultaneous learning of fuzzy sets. In: Proceedings of the 2018 Italian Workshop on Neural Networks. Springer, Heidelberg (2018, submitted)
3. Ciucci, D., Dubois, D., Lawry, J.: Borderline vs. unknown: comparing three-valued representations of imperfect information. Int. J. Approx. Reason. **55**(9), 1866–1889 (2014)
4. Deng, X., Yao, Y.: Decision-theoretic three-way approximations of fuzzy sets. Inf. Sci. **279**, 702–715 (2014)
5. El-Hawy, M., Wassif, K., Hefny, H., Hassan, H.: Hybrid multi-attribute decision making based on shadowed fuzzy numbers. In: 2015 IEEE 7th International Conference on Intelligent Computing and Information Systems, ICICIS 2015, pp. 514–521 (2016)
6. Fisher, R.A.: The use of multiple measurements in taxonomic problems. Ann. Eugenics **7**(2), 179–188 (1936)
7. Fletcher, R.: Practical Methods of Optimization. Wiley, Hoboken (2013)
8. Grzegorzewski, P.: Fuzzy number approximation via shadowed sets. Inf. Sci. **225**, 35–46 (2013)
9. Hryniewicz, O.: Possibilistic analysis of Bayesian estimators when imprecise prior information is described by shadowed sets. Adv. Intell. Syst. Comput. **642**, 238–247 (2018)
10. Jian, C.: Improved shadowed sets data selection method in extension neural network. J. Netw. **8**(12), 2728–2735 (2013)
11. Klir, G.J., St. Clair, U.H.S., Yuan, B.: Fuzzy Set Theory: Foundations and Applications. Prentice Hall, Upper Saddle River (1997)

12. Knopfmacher, J.: On measures of fuzziness. J. Math. Anal. Appl. **49**(3), 529–534 (1975)
13. Li, X., Geng, P., Qiu, B.: A cluster boundary detection algorithm based on shadowed set. Intell. Data Anal. **20**(1), 29–45 (2016)
14. Liu, B.: Uncertainty theory. In: Uncertainty Theory, pp. 1–79. Springer, Heidelberg (2010)
15. Lu, J., Han, J., Hu, Y., Zhang, G.: Multilevel decision-making: a survey. Inf. Sci. **346–347**, 463–487 (2016)
16. Malchiodi, D., Pedrycz, W.: Learning membership functions for fuzzy sets through modified support vector clustering. In: Masulli, F., Pasi, G., Yager, R. (eds.) WILF 2013. LNCS (LNAI), vol. 8256, pp. 52–59. Springer, Cham (2013). https://doi.org/10.1007/978-3-319-03200-9_6
17. Malchiodi, D., Tettamanzi, A.G.B.: Predicting the possibilistic score of OWL axioms through modified support vector clustering. In: Symposium on Applied Computing, SAC 2018, 9–13 April 2018, Pau, France. ACM, New York (2018). https://doi.org/10.1145/3167132.3167345
18. Pedrycz, W.: Shadowed sets: representing and processing fuzzy sets. IEEE Trans. Syst. Man Cybern. Part B: Cybern. **28**(1), 103–109 (1998)
19. Pedrycz, W.: From numeric to granular description and interpretation of information granules. Fundamenta Informaticae **127**(1–4), 399–412 (2013)
20. Tahayori, H., Sadeghian, A., Pedrycz, W.: Induction of shadowed sets based on the gradual grade of fuzziness. IEEE Trans. Fuzzy Syst. **21**(5), 937–949 (2013)
21. Wijayasekara, D., Linda, O., Manic, M.: Shadowed type-2 fuzzy logic systems. In: Proceedings of the 2013 IEEE Symposium on Advances in Type-2 Fuzzy Logic Systems, T2FUZZ 2013–2013 IEEE Symposium Series on Computational Intelligence, SSCI 2013, pp. 15–22 (2013)
22. Yao, Y.: Rough sets and three-way decisions. In: Ciucci, D., Wang, G., Mitra, S., Wu, W.-Z. (eds.) RSKT 2015. LNCS (LNAI), vol. 9436, pp. 62–73. Springer, Cham (2015). https://doi.org/10.1007/978-3-319-25754-9_6
23. Yao, Y., Wang, S., Deng, X.: Constructing shadowed sets and three-way approximations of fuzzy sets. Inf. Sci. **412–413**, 132–153 (2017)
24. Zhang, G., Lu, J., Gao, Y.: Multi-Level Decision Making, Models, Methods and Application. Springer, Berlin (2014). https://doi.org/10.1007/978-3-662-46059-7
25. Zhang, J., Shen, L.: An improved fuzzy c-means clustering algorithm based on shadowed sets and PSO. Comput. Intell. Neurosci. **2014**, 22 (2014)
26. Zhou, Y., Tian, L., Liu, L.: Improved extension neural network and its applications. Math. Probl. Eng. **2014**, 14 (2014)

Test-Cost-Sensitive Quick Reduct

Alessio Ferone[1], Tsvetozar Georgiev[2], and Antonio Maratea[1](✉)

[1] University of Naples Parthenope (IT), Naples, Italy
{alessio.ferone,antonio.maratea}@uniparthenope.it
[2] University of Ruse (BG), Ruse, Bulgaria
tgeorgiev@ecs.uni-ruse.bg

Abstract. In real-world applications, the data gathering process is necessarily bounded by costs in terms of money, time or resources that need to be spent in order to sample a sufficient amount of good quality data. From this point of view Feature Selection (FS) is essential to reduce the total sampling cost while trying to keep the information content of sampled data unaltered, and Rough Sets (RS) offer a natural representation of FS in terms of the so-called *reducts*. In this paper a modified version of the Quick Reduct (QR) algorithm is proposed, where the criterium to add features to the reduct accounts also for the costs of the features. Exploiting granular computing and the indiscernibility principle, the Test-Cost-Sensitive Quick Reduct (TCSQR) here proposed efficiently derives a close-to-optimal subset of informative and inexpensive features. Promising experimental results have been obtained on three different cost scenarios.

Keywords: Rough set theory · Cost sensitive learning · Granulation · Feature selection · Dimensionality reduction

1 Introduction

There are two assumptions that justify the crude accuracy as a measure of performance for a classifier: the balanced distribution of samples among classes and the equal misclassification cost. In most real cases, both are violated from a moderate to a severe degree. Class imbalance can be tackled at the algorithmic level, at the data level or in other hybrid ways [10]; similarly, cost-aware classification can be tackled at the algorithmic level, at the data level or in other hybrid ways [9]. Not surprisingly, both problems are often correlated (imbalance is due to high sampling costs or vice-versa), and the combination of both imbalanced and cost-aware classification techniques has been experimented [19].

When the minimization of costs becomes the main target, because gathering samples is excessively expensive, risky or time-consuming, costs for each feature must be included in the evaluation of performance of a classifier, and it may well happen that the best classifier is not the one with the highest accuracy. The typical examples are related to medical diagnosis and to security threats: in

© Springer Nature Switzerland AG 2019
R. Fullér et al. (Eds.): WILF 2018, LNAI 11291, pp. 29–42, 2019.
https://doi.org/10.1007/978-3-030-12544-8_3

the first case a medical exam (feature) can be very expensive and at the same time not particularly helpful in improving the percentage of correct diagnosis in combination with other exams; in the second case a single break in security can be very dangerous and expensive to remedy, even if the large part of normal activities is correctly classified (minority class has the highest cost). While in the first case it is of utmost importance to find the optimal subset of features, in the second case it is of utmost importance the correct classification of minority instances.

Assuming each measured sample $x \in \Re^d$ as a vector of dimension d, due to redundancy, correlation or causation it normally happens that the minimum number of features required to fully represent the data without any information loss is $d' < d$. In the scientific literature many feature selection [3] and feature extraction [4] algorithms have been proposed to reduce the actual dimensionality to its minimum—the so-called *intrinsic dimensionality* of the dataset—with the desirable effect of reducing the computational burden while at the same time increasing the generalization capabilities of the classifiers [20]. Compared to feature extraction, feature selection keeps unchanged the original semantics, values and sampling spaces of the selected features, only eliminating the redundant ones: in general it aims to find the most predictive subset of the original features for the subsequent classification.

With a rising trend in last years [22,23], granular computing (and in particular Rough Set theory [15]) has been extensively exploited for feature selection or extraction [5,17] in the Machine Learning community. *Granular computing* is based on the idea of an *information granule*—a subset of indistinguishable samples according to an equivalence relation—allowing the partition of the universe of discourse into *granules* representing a coarse approximation grid for all the embedded concepts. *Rough Set theory* on the other side can be seen as a family of methodologies that exploit granules [16] to simplify the tackled problem, using only the available data.

In this paper, granular computing has been exploited for a cost-sensitive feature selection method that is an adaptation of Quick Reduct including feature costs. The paper is organized as follows: in Sect. 2 an outline of Rough Set theory and a short overview of related work is presented; in Sect. 3 the proposed approach is described in detail; in Sect. 4 the experimental comparison of the proposed approach with competitors is presented; in Sect. 5 the main conclusions are drawn and future work is sketched.

2   Background

Dimensionality reduction techniques help to reduce the redundancy at the same time increasing the generalization capabilities of a classifier; it becomes essential in the analysis of high dimensional datasets, to cope with the well known "curse of dimensionality" (as dimension increases, the space to be explored grows exponentially and so should the amount of data points needed to keep the same sampling ratio) [1]. Most notably, many problems in Machine Learning involve high-dimensional inputs and hence a lot of effort has been put on dimensionality reduction techniques (see for example [2]). While *Feature extraction* techniques tend to destroy the underlying semantic of the features or do require *a priori* knowledge about the data, *feature selection* techniques keep the semantic of the feature (and its measured values) unchanged. In the context of feature selection, Rough Set theory can be used to discover data dependencies that help to reduce the number of attributes using only information granules contained into the dataset [15].

2.1   Reduct and Rough Sets

Let $I = (U, A)$ be an information system, where U is the finite universe of discourse and A is the finite set of attributes, with values in V_a, such that $a : U \rightarrow V_a, \forall a \in A$. Any $P \subseteq A$ can be seen as an equivalence relation $IND(P)$ defined as follows:

$$IND(P) = \{(x, y) \in U \times U | \forall a \in P, a(x) = a(y)\} \qquad (1)$$

that is the set of objects belonging to U that are not discernible by attributes in P. The equivalence relation $IND(C)$ partitions U in equivalence classes $[x]_C$, denoted by $U/IND(C)$. In the context of Rough Sets, equivalence classes, also known as information granules, are employed to approximate any subset $X \subseteq U$ by the so called *lower* and *upper* approximations of X, defined, respectively, as follows:

$$\underline{C}X = \{x | [x]_C \subseteq X\} \qquad (2)$$

$$\overline{C}X = \{x | [x]_C \bigcap X \neq \emptyset\} \qquad (3)$$

Let C and D be equivalence relations over U, then the positive region—that is the union of the lower approximations of each equivalence class defined by $X \in U/D$—is defined as follows:

$$POS_C(D) = \bigcup_{X \in U/D} \underline{C}X \qquad (4)$$

from which it is possible to derive the Rough Set *degree of dependency*, $\gamma_C(D) \in [0, 1]$, of a set of attributes D on a set of attributes C:

$$\gamma_C(D) = \frac{|POS_C(D)|}{|U|} \qquad (5)$$

Let $D \subseteq A$ be the set of decision features and $C \subseteq A$ the set of conditional attributes, a *reduct* $R \subseteq C$ is the set of minimal cardinality of the conditional attribute set C such that

$$\gamma_R(D) = \gamma_C(D) \qquad (6)$$

2.2   Feature Selection with Rough Sets

Feature selection algorithms based on Rough Set theory mostly exploit the definition of reduct to find a reduced set of attributes that preserves the degree of dependency of the full set of attributes, i.e. no attribute can be removed from the subset without lowering the dependency degree. As long as the reduced set provides identical predictive capability as the original (unreduced) set, many reducts can be derived for a given dataset, being obviously the ones with minimal cardinality the most interesting [7]: a optimal reduct is defined as the minimal set of attributes $R \subseteq C$ such that $IND(R) = IND(C)$. The reduction of attributes can be achieved through the comparison of equivalence relations generated by each subset of attributes with an iterative process that can be subtractive, additive or hybrid [24].

The Quick Reduct (QR from now on) algorithm in particular [24], shown in Algorithm 1, derives the reduct concatenating iteratively the attributes with the greatest increase in the Rough Set dependency degree, until the maximum possible dependency degree for the considered dataset $\gamma_C(D)$ has been reached. It is a greedy algorithm and it is not guaranteed to find a global minimum, because the dependency structure changes with the addition of each single attribute and, unless exploring all possible combinations, it is impossible to predict which subset will be the optimal reduct. QR however derives a subset sufficiently close to optimal in a reasonable time, resulting in general a good compromise to reduce the dataset dimensionality.

Algorithm 1. QuickReduct

1: **procedure** QUICKREDUCT(C,D)
2:      $C \leftarrow$ **the set of all conditional features**
3:      $D \leftarrow$ **the set of decision features**
4:      $R \leftarrow \emptyset$
5:      **repeat**
6:          $T \leftarrow R$
7:          **for** $f \in (C - R)$ **do**
8:              **if** $\gamma_{R \cup \{f\}}(D) > \gamma_T(D)$ **then**
9:                  $T \leftarrow R \bigcup \{f\}$
10:         $R \leftarrow T$
11:     **until** $\gamma_R(D) == \gamma_C(D)$
12: **return** R

2.3   Related Work

In general, assuming that the data gathering process does not have time or cost constraints is misleading, as most of the times the amount of available data is strictly related to the allocated data gathering budged, and it is therefore critical to select the best set of features that fulfills the budget constraint. When reasoning in terms of information granules and Rough Sets, the problem of selecting the cheapest reduct [18] has been called the *minimal test cost reduct* (MTR from now on) [11]. A number of algorithms have been proposed to this purpose (for example [6,11,12,14]), all trying to keep enough information for classification at the same time reducing costs. When the allocated data gathering budget is not sufficient to keep all necessary information, or when it is a strict requirement, the problem has been reformulated as the *feature selection with test cost constraint* (FSTC from now on) problem [13], where the upper bound of the allocated budget acts as a constraint in the feature selection process and an optimal *sub-reduct* can be derived. When the budget is not less than the budged required for the optimal reduct, FSTC reduces to MTR.

In [11] the test cost problem is formulated as selecting the set of attributes that satisfies a minimal cost criterium and a classification accuracy constraint. Inadequacy of plain reducts (a subset of attributes that are jointly sufficient and individually necessary for preserving a particular property of the given information table) from from rough sets theory [15] is highlighted, as attribute reduction to improve classification is inherently different from attribute reduction to minimize costs: total test cost is independent from the performance of the subsequent classifier. Assuming costs independent from each other, a selection criterium based on information gain and a weight function for test costs are proposed, where a non-positive exponent λ acts as parameter to tune the influence of the tests. Adjusting λ is different from setting a significance weight to each attribute in the reduction process, and with respect to [21] it is less bound to the specific application domain. Three probability distributions for simulated costs, three performance metrics and four UCI datasets test the proposed algorithm and its associated metrics. While in some case producing a non-optimal reduct, in most cases a minimal test cost reduct is obtained.

In [12] a backtracking approach to the FSTC problem has been proposed for small and medium-sized datasets. Being exponential in the number of features, backtracking is severely limited by the dataset size, so an heuristic based on information gain and the addition-deletion approach [24] has also been proposed, with polynomial time complexity. The proposed heuristic is based on an user-defined weight λ that induces preferences for low-cost features, similarly to [11]. To choose λ testing different values is suggested to generate different feature subsets, of which the best one is finally selected. Even if in this way the choice of λ is no more delegated to the user, the number of tested values for λ acts as a multiplier for computational complexity.

3   Test-Cost-Sensitive Quick Reduct

Like QR, the Test-Cost-Sensitive Quick Reduct algorithm (TCSQR from now on), shown in Algorithm 2, performs reduct construction by addition, starting with an empty set and adding iteratively the attributes representing the local optimal choice (greedy choice); like QR, the stopping criterium is reached at the maximum possible dependency degree for the dataset $\gamma_C(D)$, i.e. when no other attribute could improve the dependency degree. Differently from QR, the criterium to add features to the reduct accounts also for the cost of the features; differently from QR, a deletion step is introduced in order to remove the redundant most costly attributes.

The optimality criterium in this case is the maximization of the weighted sum of Rough Set dependency degree and $(1 - tc_R)$, where tc_R is the total cost (normalized between 0 and 1) of the features included in the reduct and the two terms are weighted through a parameter α:

$$\alpha\gamma_R(D) + (1 - \alpha)(1 - tc_R) \tag{7}$$

TCSQR considers dependency degree, but, at the same time, evaluates the cost of each feature to be added: attributes resulting in the greatest increase in the weighted sum of Rough Set dependency degree and at the same time in the greatest decrease of the total costs are added iteratively to the reduct. The choice of parameter α will be discussed in in Sect. 4.

Furthermore TCSQR sorts attributes in order of decreasing costs and tries to remove the most expensive ones keeping the dependency degree unchanged, in order to obtain an equivalent reduct with minor total cost.

Also TCSQR is a greedy algorithm and hence it is not guaranteed to find a global minimum, because the dependency structure changes with the addition or deletion of single attributes and, unless exploring all possible combinations, it is impossible to predict which subset will be the optimal reduct according to cost and dependency degrees. TCSQR however derives a close-to-optimal subset in a reasonable time. How critical this aspect is will be discussed in Sect. 4.

4   Experimental Results

4.1   Data

Experiments have been performed on the *Zoo* dataset from the UCI Repository for Machine Learning [8]. The dataset has been chosen to ease comparisons with the related literature: it has 16 attributes and 101 instances divided in 7 classes, with 33 reducts [11].

4.2   Cost Generation

Most datasets from the UCI library [8] have no intrinsic test costs. Following [11,12] three different schemata have been adopted to simulate test costs, following a discrete uniform, discrete bounded normal, and discrete bounded Pareto distribution respectively. Test costs are integers ranging from M to N, and are generated independently. These distributions and their respective generation process are briefly discussed hereafter.

For each distribution ten test costs have been generated. Tables 1 and 2 show the test costs generated and corresponding reducts, respectively.

Algorithm 2. TCSQuickReduct

1: **procedure** TCSQUICKREDUCT(C,D,Q(),α)
2:     $C \leftarrow$ **the set of all conditional features**
3:     $D \leftarrow$ **the set of decision features**
4:     $q() \leftarrow$ **cost function of decision features**
5:     $R \leftarrow \emptyset$
6: **repeat**
7:      $T \leftarrow R$
8:      $maxT \leftarrow 0$
9:      **for** $f \in (C - R)$ **do**
10:          $val = \alpha * \gamma_{R \cup \{f\}}(D) + (1 - \alpha) * (1 - sum(q(R \cup \{f\})))$
11:          **if** $val > maxT$ **then**
12:              $T \leftarrow R \bigcup \{f\}$
13:              $maxT \leftarrow val$
14:      $R \leftarrow T$
15: **until** $\gamma_R(D) == \gamma_C(D)$
16:      $MR \leftarrow$ sort attributes in R according to test costs in descending order
17: **repeat**
18:      $MR \leftarrow MR - \{f\}$
19:      **if** $\gamma_{R - \{f\}}(D) == \gamma_C(D)$ **then**
20:          $R \leftarrow R - \{f\}$
21: **until** $MR \neq \emptyset$
22: **return** R

Uniform. Being "uniform" means that all values have equal probability of being drawn. Let c_u denote a test cost under the discrete uniform distribution and u be a uniform distributed random variable in $[0, 1]$, then

$$c_u = M + \lfloor (N - M + 1)u \rfloor \tag{8}$$

is distributed as a discrete bounded uniform distribution in $[M, N]$.

Gaussian. A Gaussian (or normal) distribution is the infamous and ubiquitous bell-shaped probability distribution towards which the sum of all other distributions tends under mild conditions due to the central limit theorem. It is symmetric and unimodal, it is continuous, and it is described by the following probability density function:

$$f(g) = \frac{1}{\sqrt{2\pi\sigma^2}} e^{-\frac{(g-\mu)^2}{2\sigma^2}} \tag{9}$$

where the the mean and variance, μ and σ^2, are the only parameters.

Let c_g denote a test cost under the discrete normal distribution and g be standard normal distributed ($\mu = 0$ and $\sigma = 1$), then

$$c_g = \begin{cases} M & g < M \\ N & g > N \\ \lfloor \frac{N+M+1}{2} + \beta y \rfloor & M < g < N \end{cases} \tag{10}$$

is distributed as a discrete bounded normal distribution in $[M, N]$ with an approximate mean of $(M + N)/2$ ($\beta = 8$ in the experiments).

Pareto. The Pareto distribution is a popular model for unequal exploitation of resources. If u is uniformly distributed in $[0, 1]$, then

$$c_p = \left\lfloor \left(-\left(\frac{u(N+1)^\delta - uM^\delta - (N+1)^\delta}{M^\delta(N+1)^\delta} \right) \right)^{-1/\delta} \right\rfloor \tag{11}$$

is distributed as a discrete bounded Pareto-distribution in $[M, N]$, where δ determines the shape of the distribution ($\delta = 2$ in the experiments).

Table 1. Test costs for dataset Zoo.

ID	Distribution	a	b	c	d	e	f	g	h	i	j	k	l	m	n	o	p
1	Uniform	9	42	97	77	10	63	70	84	42	97	62	65	69	55	86	53
2		35	28	58	18	50	18	53	56	31	9	10	46	28	6	43	1
3		98	72	73	4	4	63	28	58	25	48	68	53	26	82	37	26
4		46	52	59	91	54	52	48	74	17	62	43	77	74	55	49	28
5		59	19	85	11	75	12	63	30	93	16	98	58	20	93	99	52
6		13	34	33	35	77	51	85	10	27	6	19	4	40	53	86	16
7		52	87	21	28	6	2	51	38	91	71	90	99	45	42	24	57
8		42	60	22	33	87	96	65	48	32	22	86	38	3	58	17	96
9		96	75	71	25	21	35	87	29	98	44	40	84	22	17	5	28
10		83	9	12	5	99	7	72	5	23	91	85	70	44	80	55	35
11	Normal	44	64	53	57	47	50	55	45	44	59	53	44	33	43	58	51
12		54	68	53	38	49	51	65	56	67	54	45	56	49	49	51	45
13		48	57	62	55	51	41	63	70	70	30	39	48	44	35	54	40
14		46	48	54	42	66	47	44	57	62	39	59	45	40	51	51	53
15		58	53	55	58	53	50	44	50	53	52	54	47	41	55	51	47
16		62	50	53	63	54	53	70	51	50	57	51	54	54	63	36	60
17		49	52	36	44	52	49	39	62	47	51	69	47	39	42	47	53
18		62	61	45	47	54	57	52	59	41	47	57	65	37	37	35	51
19		53	55	44	61	50	54	57	42	40	40	56	43	57	46	55	46
20		65	45	46	39	38	53	46	54	54	51	40	39	54	59	47	54
21	Pareto	1	2	84	4	42	1	1	1	1	1	30	2	1	1	19	3
22		13	1	1	1	1	3	1	10	54	19	4	1	1	1	3	2
23		33	14	2	6	1	11	8	1	1	2	3	1	3	1	1	62
24		10	11	3	1	5	62	4	1	1	11	1	1	2	1	13	1
25		2	5	1	3	1	6	1	3	1	1	4	2	1	37	1	1
26		1	1	2	1	37	12	1	2	19	2	12	8	2	1	1	13
27		3	21	1	1	1	1	3	2	30	7	27	84	2	2	1	4
28		5	4	2	6	2	1	33	62	1	1	1	11	23	3	8	1
29		4	1	18	1	9	1	5	1	37	1	71	4	1	37	84	3
30		21	3	8	5	6	1	1	2	2	47	1	1	12	2	1	6

4.3   Comparison Metrics

In order to evaluate the performance of test-cost-sensitive algorithms, in [11] three metrics are proposed: Finding Optimal Factor (FOF from now on), Maximal Exceeding Factor (MEF from now on), and Average Exceeding Factor (AEF from now on).

FOF. Let the number of experiments be K and the number of successful searches of an optimal reduct be k. The FOF metric is defined as follows:

$$FOF = \frac{k}{K}. \tag{12}$$

This metric is both qualitative and quantitative, because it only counts optimal solutions and it is based on a number of searches. In the experiments, different test cost settings have been generated, obtaining many values for the FOF and allowing the computation of simple statistics on it.

FEF. For a dataset with a particular test cost setting, let R' be an optimal reduct. The Exceeding Factor (EF) of a reduct R is:

$$EF(R) = \frac{c^*(R) - c^*(R')}{c^*(R')} \tag{13}$$

EF provides a quantitative metric to evaluate the performance of a reduct. It indicates how far from optimality a reduct is: if R is an optimal reduct, then the exceeding factor is 0.

MEF. To measure the general performance of an algorithm, statistics are needed. Let the number of experiments be K. In the ith experiment ($1 \leq i \leq K$), the reduct computed by the algorithm is denoted R_i. The maximal exceeding factor (MEF) is defined as

$$MEF = \max_{1 \leq i \leq K} EF(R_i) \tag{14}$$

This shows the worst case of the algorithm given some data sets. Although it relates to the performance of one particular reduct, it should be viewed as a statistical rather than an individual metric.

AEF. The average exceeding factor (AEF) is defined as

$$AEF = \frac{\sum_{i=1}^{K} ef(R_i)}{K} \tag{15}$$

Table 2. Optimal reducts with associated test cost, reduct and exceeding factor of TCSQR.

ID	Optimal reduct	Minimal test cost	Constructed reduct	Exceeding factor
1	{d,f,i,l,m}	316	{a,d,f,i,l,m}	0.028
2	{d,f,j,k,m,n,p}	90	{d,f,j,k,m,n,p}	0.0
3	{d,f,i,l,m}	171	{d,e,f,i,l,m}	0.023
4	{c,f,i,m,p}	230	{c,f,i,m,p}	0.0
5	{d,f,h,l,m}	131	{d,f,h,l,m}	0.0
6	{a,f,h,j,l,m}	124	{a,f,h,j,l,m}	0.0
7	{c,d,f,h,m}	134	{c,d,f,h,m}	0.0
8	{c,d,f,i,m}	186	{c,d,f,i,m}	0.0
9	{d,f,h,k,m,p}	179	{d,f,h,k,m,o,p}	0.027
10	{c,d,f,h,m}	73	{c,d,f,h,m}	0.0
11	{d,f,i,l,m}	228	{d,f,i,l,m}	0.0
12	{c,d,f,h,m}	247	{d,f,h,l,m}	0.012
13	{a,f,j,l,m,n}	246	{d,f,h,l,m}	0.048
14	{d,f,h,l,m}	231	{d,f,h,l,m}	0.0
15	{c,f,h,m,p}	243	{d,f,h,l,m}	0.012
16	{c,f,h,j,m}	268	{d,f,i,l,m}	0.022
17	{c,d,f,i,m}	215	{d,f,i,l,m}	0.051
18	{c,d,f,i,m}	227	{d,f,i,l,m}	0.088
19	{c,f,h,j,m}	237	{d,f,i,l,m}	0.075
20	{d,f,h,l,m}	239	{d,f,h,l,m}	0.0
21	{a,f,i,j,l,m}	7	{a,f,j,l,m,n}	0.0
22	{d,f,h,l,m}	16	{d,f,h,l,m,n}	0.062
23	{c,f,h,l,m,n}	19	{c,f,h,l,m,n,o}	0.052
24	{d,f,h,l,m}	67	{d,f,i,l,m,p}	0.014
25	{c,f,i,m,p}	10	{c,f,i,m,p}	0.0
26	{c,d,f,h,m}	19	{c,d,f,j,m,n,o}	0.10
27	{c,d,f,h,m}	7	{c,d,f,h,m,o}	0.14
28	{c,f,i,m,p}	28	{c,f,i,m,p}	0.0
29	{d,f,h,l,m}	8	{d,f,h,l,m}	0.0
30	{d,f,h,l,m}	21	{d,f,i,l,m,o}	0.047

4.4 Results

Table 2 shows the optimal reduct for each test cost configuration with its associated total test cost and the reducts obtained with the proposed algorithm with the corresponding exceeding factor.

Results showed in Tables 2 and 3 are obtained with the following values of α for the uniform, normal and Pareto distributions respectively: 0.15, 0.7 and 0.05.

Table 3. Comparison between [11] and TCSQuickreduct considering: FOF, MEF and AEF.

Distribution	FOF		MEF		AEF	
	[11]	TCSQuickreduct	[11]	TCSQuickreduct	[11]	TCSQuickreduct
Uniform	0.9	0.7	0.025	0.028	0.0025	0.0081
Normal	0.8	0.3	0.189	0.088	0.0364	0.0312
Pareto	0.9	0.4	0.053	0.143	0.0053	0.0433

It can be noted that, although the proposed method is not able to always find the optimal reduct, the performance in terms of exceeding factor, both maximal and average, are comparable with published literature [11]. More in detail, with uniform test costs MEF and AEF are very close; with normal test costs TCSQR performs slightly better, while with test costs based on Pareto distribution the algorithm proposed in [11] performs better than TCSQR.

Table 4 shows comparison of the proposed approach with [11] in the case of the optimal choice of the α parameter. The test has been performed on the same data as the previous test. Also in this case it can be noted how TCSQR tends to find super-reducts, nevertheless the MEF is lower with all costs distributions while the AEF is slightly higher due the further selected features.

Table 4. Comparison between optimal choice of parameter in [11] and TCSQuickreduct considering: FOF, MEF and AEF.

Distribution	FOF		MEF		AEF	
	[11]	TCSQuickreduct	[11]	TCSQuickreduct	[11]	TCSQuickreduct
Uniform	0.7	0.7	0.391	0.028	0.0164	0.0081
Normal	0.7	0.3	0.127	0.088	0.0107	0.0312
Pareto	0.9	0.4	0.167	0.143	0.0011	0.0433

TCSQR has been also compared, on *Zoo*, with the three approaches proposed in [12], where only the uniform distribution and FOF metric have been tested. The FOF for the three strategies are: 0.18, 0.52 and 0.7. Even in this case, the performance of the proposed approach are comparable and in two cases out of three better than competitors.

For what concerns the choice of parameter α, Fig. 1 show the performance in terms of FOF (Fig. 1(a)), MEF (Fig. 1(b)) and AEF (Fig. 1(c)) while varying parameter α. It can be noted that, overall, better performance are obtained with values in the range $(0.3, 0.4)$.

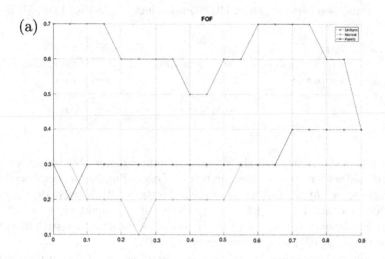

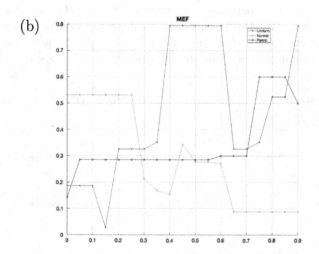

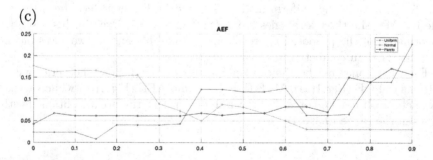

Fig. 1. (a) FOF (b) MEF and (c) AEF metrics with varying α. Uniform: red line; Normal: green line; Pareto: blue line. (Color figure online)

5   Conclusion

A new algorithm for solving the test-cost-sensitive reduct problem has been presented. The proposed approach is a modified version of the quick reduct algorithm in which test costs are combined with the rough degree of dependency in the objective function and a deletion step is added at the end. The resulting algorithm shares the same limitation as the original Quick Reduct, i.e. it can produce a super reduct and hence it is no guaranteed to find the optimal solution. Nevertheless, its performance are promising, in some cases outperforming state-of-the art competitors. Further tests and data with real costs are required to confirm its generalization capabilities.

Acknowledgement. The authors acknowledge the financial support for this research through "FFABR", granted by MIUR.

References

1. Bellman, R.: Adaptive Control Processes: A Guided Tour. Princeton University Press, Princeton (1961)
2. Camastra, F.: Data dimensionality estimation methods: a survey. Pattern Recogn. **36**(12), 2945–2954 (2003)
3. Chandrashekar, G., Sahin, F.: A survey on feature selection methods. Comput. Electr. Eng. **40**(1), 16–28 (2014). https://doi.org/10.1016/j.compeleceng.2013.11.024. http://www.sciencedirect.com/science/article/pii/S0045790613003066. 40th-year commemorative issue
4. Ding, S., Zhu, H., Jia, W., Su, C.: A survey on feature extraction for pattern recognition. Artif. Intell. Rev. **37**(3), 169–180 (2012)
5. Ferone, A., Petrosino, A.: A rough fuzzy perspective to dimensionality reduction. In: Masulli, F., Petrosino, A., Rovetta, S. (eds.) CHDD 2012. LNCS, vol. 7627, pp. 134–147. Springer, Heidelberg (2015). https://doi.org/10.1007/978-3-662-48577-4_9
6. He, H., Min, F.: Accumulated cost based test-cost-sensitive attribute reduction. In: Kuznetsov, S.O., Ślęzak, D., Hepting, D.H., Mirkin, B.G. (eds.) RSFDGrC 2011. LNCS (LNAI), vol. 6743, pp. 244–247. Springer, Heidelberg (2011). https://doi.org/10.1007/978-3-642-21881-1_39
7. Jensen, R., Tuson, A., Shen, Q.: Finding rough and fuzzy-rough set reducts with SAT. Inf. Sci. **255**, 100–120 (2014)
8. Lichman, M.: UCI machine learning repository (2013). http://archive.ics.uci.edu/ml
9. Ling, C.X., Sheng, V.S.: Cost-sensitive learning and the class imbalanced problem. In: Sammut, C. (eds.) Encyclopedia of Machine Learning, pp. 171–179. Springer (2007)
10. Maratea, A., Petrosino, A., Manzo, M.: Adjusted f-measure and kernel scaling for imbalanced data learning. Inf. Sci. **257**, 331–341 (2014)
11. Min, F., He, H., Qian, Y., Zhu, W.: Test-cost-sensitive attribute reduction. Inf. Sci. **181**(22), 4928–4942 (2011)
12. Min, F., Hu, Q., Zhu, W.: Feature selection with test cost constraint. Int. J. Approx. Reason. **55**(1, Part 2), 167–179 (2014). Special issue on Decision-Theoretic Rough Sets

13. Min, F., Zhu, W.: Optimal sub-reducts with test cost constraint. In: Yao, J.T., Ramanna, S., Wang, G., Suraj, Z. (eds.) RSKT 2011. LNCS (LNAI), vol. 6954, pp. 57–62. Springer, Heidelberg (2011). https://doi.org/10.1007/978-3-642-24425-4_10
14. Pan, G., Min, F., Zhu, W.: A genetic algorithm to the minimal test cost reduct problem. In: 2011 IEEE International Conference on Granular Computing, pp. 539–544, November 2011
15. Pawlak, Z.: Rough sets. Int. J. Comput. Inf. Sci. **11**, 341–356 (1982)
16. Pawlak, Z.: Granularity of knowledge, indiscernibility and rough sets. In: Proceedings of IEEE International Conference on Fuzzy Systems, pp. 106–110 (1998)
17. Petrosino, A., Ferone, A.: Feature discovery through hierarchies of rough fuzzy sets. In: Pedrycz, W., Chen, S.M. (eds.) Granular Computing and Intelligent Systems: Design with Information Granules of Higher Order and Higher Type. ISRL, pp. 57–73. Springer, Heidelberg (2011). https://doi.org/10.1007/978-3-642-19820-5_4
18. Susmaga, R.: Computation of minimal cost reducts. In: Raś, Z.W., Skowron, A. (eds.) ISMIS 1999. LNCS, vol. 1609, pp. 448–456. Springer, Heidelberg (1999). https://doi.org/10.1007/BFb0095132
19. Thai-Nghe, N., Gantner, Z., Schmidt-Thieme, L.: Cost-sensitive learning methods for imbalanced data. In: The 2010 International Joint Conference on Neural Networks (IJCNN), pp. 1–8. IEEE (2010)
20. Vapnik, V.: Statistical Learning Theory. Wiley, Hoboken (1998)
21. Xu, C., Min, F.: Weighted reduction for decision tables. In: Wang, L., Jiao, L., Shi, G., Li, X., Liu, J. (eds.) FSKD 2006. LNCS (LNAI), vol. 4223, pp. 246–255. Springer, Heidelberg (2006). https://doi.org/10.1007/11881599_28
22. Yao, J.T., Vasilakos, A.V., Pedrycz, W.: Granular computing: perspectives and challenges. IEEE Trans. Cybern. **43**(6), 1977–1989 (2013)
23. Yao, J.: A ten-year review of granular computing. In: IEEE International Conference on Granular Computing, GRC 2007, pp. 734–734. IEEE (2007)
24. Yao, Y., Zhao, Y., Wang, J.: On reduct construction algorithms. In: Wang, G.-Y., Peters, J.F., Skowron, A., Yao, Y. (eds.) RSKT 2006. LNCS (LNAI), vol. 4062, pp. 297–304. Springer, Heidelberg (2006). https://doi.org/10.1007/11795131_43

On the Problem
of Possibilistic-Probabilistic
Optimization with Constraints
on Possibility/Probability

Alexander Yazenin and Ilia Soldatenko[(✉)] [iD]

Tver State University, 33 Zhelyabova str., Tver, Russia
{Yazenin.AV,soldis}@tversu.ru

Abstract. The problem of possibilistic-probabilistic linear programming with constraints on possibility and probability is investigated. For the case of normally distributed random parameters of the model and under the most general assumptions concerning the properties of probability distributions, its equivalent determinate analogue is constructed which is the quadratic programming model.

Keywords: Possibilistic-probabilistic optimization ·
Constraints on possibility/probability ·
Equivalent determinate analogue · Fuzzy random variable ·
Strongest t-norm

1 Introduction

The problem of optimization under conditions of hybrid uncertainty of the possibilistic-probabilistic type is currently being actively developed. Its methods are used in the construction and study of generalized models of portfolio analysis, in economic and mathematical planning and other fields.

When indirect methods for solving such problems are developed that are based on the construction of equivalent determinate analogs, two-level procedures are used to remove the uncertainty of the probabilistic and fuzzy types. These methods are based on the principles of the expected possibility [1–8] and the implementation of constraints on the possibility and probability.

In this paper we investigate the problem of possibilistic-probabilistic linear programming with constraints on possibility and probability. In the class of quasi-concave upper semicontinuous probability distributions with finite support that characterize fuzzy parameters of the model and for normally distributed random parameters, an equivalent determinate analogue of the model is constructed. The properties of the constructed equivalent model are investigated.

R. Fullér et al. (Eds.): WILF 2018, LNAI 11291, pp. 43–54, 2019.
https://doi.org/10.1007/978-3-030-12544-8_4

2   Necessary Concepts and Notations

We introduce a number of definitions and concepts from the possibility theory following [9–11]. Let $(\Gamma, \mathrm{P}(\Gamma), \tau)$ and $(\Omega, \mathrm{B}, \mathrm{P})$ be possibility and probability spaces, where Ω is a sample space with $\omega \in \Omega$, Γ – a pattern space with elements $\gamma \in \Gamma$, B – σ-algebra of events, $\mathrm{P}(\Gamma)$ is the discrete topology on Γ, $\tau \in \{\pi, \nu\}$, π and ν – measures of possibility and necessity, respectively, and P – probability measure, E^1 – number line.

Definition 1. *Fuzzy random variable* $Y(\omega, \gamma)$ *is a real function* $Y : \Omega \times \Gamma \to E^1$ σ-*measurable for each fixed* γ, *where*

$$\mu_Y(\omega, t) = \pi\{\gamma \in \Gamma : Y(\omega, \gamma) = t\}$$

is called its distribution function.

It follows from Definition 1 that the distribution function of a fuzzy random variable depends on a random parameter, that is, it is a random function.

Definition 2. *Let* $Y(\omega, \gamma)$ *be a fuzzy random variable. Its expected value* $E[Y]$ *is a fuzzy variable with possibility distribution function*

$$\mu_{E[Y]}(t) = \pi\{\gamma \in \Gamma : E[Y(\omega, \gamma)] = t\},$$

where E *is the expectation operator*

$$E[Y(\omega, \gamma)] = \int_\Omega Y(\omega, \gamma) P(d\omega).$$

In this case, the distribution function of the expected value of a fuzzy random variable is no longer dependent on a random parameter and is therefore deterministic.

We use triangular norms and conorms (t-norms and t-conorms) as an instrument for aggregation of fuzzy information that extends min and max operations, laid in actions on fuzzy sets and fuzzy variables [12,13].

In particular, in this work we consider two extremal t-norms: $T_M(x, y) = \min(x, y)$ and

$$T_W(x, y) = \begin{cases} \min\{x, y\}, & if \max\{x, y\} = 1, \\ 0, & otherwise. \end{cases}$$

T_M is called the strongest t-norm and T_W – the weakest t-norm, since for any arbitrary t-norm T and $\forall x, y \in [0, 1]$, the inequality holds:

$$T_W(x, y) \le T(x, y) \le T_M(x, y).$$

One of the main properties of t-norms is their ability to control uncertainty ("fuzziness") growth, which is obvious, for example, when performing arithmetic operations on fuzzy numbers: when adding two fuzzy numbers of LR-type using the strongest t-norm T_M corresponding coefficients of fuzziness are summed, therefore uncertainty is growing. With the help of t-norms other than T_M we can have slower growth of fuzziness. The extreme cases of triangular norms which are considered in the work give us boundaries for control of fuzziness in our minimum risk portfolio model.

Following [14], we introduce the notion of mutual t-relatedness of fuzzy variables. It is used as an instrument for constructing joint possibility distribution functions.

Definition 3. *Fuzzy sets* $A_1, \ldots, A_n \in P(\Gamma)$ *are called mutually T-related, if for any index set* $\{i_1, \ldots, i_k\} \subset \{1, \ldots, n\}$, $k = 1, \ldots, n$, *we have*

$$\pi\left(A_{i_1} \cap \ldots \cap A_{i_k}\right) = T\left(\pi\left(A_{i_1}\right), \ldots, \pi\left(A_{i_k}\right)\right),$$

where

$$T\left(\pi\left(A_{i_1}\right), \ldots, \pi\left(A_{i_k}\right)\right) = T\left(T\left(\ldots T\left(T\left(\pi\left(A_{i_1}\right), \pi\left(A_{i_2}\right)\right), \pi\left(A_{i_3}\right)\right), \ldots\right), \pi\left(A_{i_k}\right)\right).$$

We can transfer the notion of mutual T-relatedness of fuzzy sets on fuzzy variables.

Definition 4. *Fuzzy variables* $Z_1(\gamma), \ldots, Z_n(\gamma)$ *are called mutually T-related, if for any index set* $\{i_1, \ldots, i_k\} \subset \{1, \ldots, n\}$, $k = 1, \ldots, n$, *we have*

$$\mu_{Z_{i_1}, \ldots, Z_{i_k}}\left(t_{i_1}, \ldots, t_{i_k}\right) = \pi\left\{\gamma \in \Gamma : Z_{i_1}(\gamma) = t_{i_1}, \ldots, Z_{i_k}(\gamma) = t_{i_k}\right\} =$$
$$\pi\left\{Z_{i_1}^{-1}\{t_{i_1}\} \cap \ldots \cap Z_{i_k}^{-1}\{t_{i_k}\}\right\} = T\left\{\pi\left(Z_{i_1}^{-1}\{t_{i_1}\}\right), \ldots, \pi\left(Z_{i_k}^{-1}\{t_{i_k}\}\right)\right\}, t_{i_j} \in E^1.$$

We use shift-scale representation [11] for fuzzy random variables:

$$R(\omega, \gamma) = a(\omega) + \sigma(\omega) Z(\gamma).$$

Further in this work we assume that in this representation fuzzy variables $Z(\gamma)$ are mutually T_M-related (min-related), and $a(\omega)$, $\sigma(\omega)$ are random offset and scale factors.

3 Construction of a Possibilistic-Probabilistic Programming Model with Constraints on Possibility and Probability

In [15] possibilistic-probabilistic optimization problems that use the principle of the expected possibility which allows to remove the stochastic uncertainty under the hybrid uncertainty of the possibilistic-probabilistic type are investigated. In this paper, we use probability constraints to remove it. This approach was previously applied for portfolio analysis problems in [5,6].

Consider a model of possibilistic-probabilistic programming of the following form:

$$k \to \min,$$
$$\tau\{P\{f_0(x, \omega, \gamma) \leq k\} \geq p_0\} \geq \alpha_0, \tag{1}$$

$$\begin{cases} \tau\{P\{f_i(x,\omega,\gamma) \le 0\} \ge p_i\} \ge \alpha_i, \ i = 1,\ldots,m, \\ x \in X. \end{cases} \tag{2}$$

In this model $f_i(x,\omega,\gamma)$, $i = 0,\ldots,m$ are possibilistic-probabilistic functions that have the meaning of mappings

$$f_i(\cdot,\cdot,\cdot) : X \times \Omega \times \Gamma \to E^1,$$
$$X \subset E_+^n = \{x \in E^n : x \ge 0\}, \ \tau \in \{\pi,\nu\}, \ p_i,\alpha_i, \ i = 0,\ldots,n$$

are given levels of probability and possibility, $p_i, \alpha_i \in (0,1]$, k—additional scalar variable.

Let us consider concrete representations of functions $f_i(x,\omega,\gamma)$—linear possibilistic-probabilistic functions. In this case

$$f_i(x,\omega,\gamma) = \sum_{j=1}^{n} A_{ij}(\omega,\gamma)x_j - B_i(\omega,\gamma) \tag{3}$$

with $i = 0,\ldots,m$ and $B_0(\omega,\gamma) \equiv 0$.

We assume that $A_{ij}(\omega,\gamma)$ and $B_i(\omega,\gamma)$ have a shift-scale representation:

$$A_{ij}(\omega,\gamma) = a_{ij}(\omega) + \sigma_{ij}(\omega)Y_{ij}(\gamma), \ B_i(\omega,\gamma) = b_i(\omega) + \sigma_i(\omega)Y_i(\gamma),$$

and

$$a_{ij}(\omega) \in \mathcal{N}_p(a_{ij}^0, d_{ij}^a), \sigma_{ij}(\omega) \in \mathcal{N}_p(\sigma_{ij}^0, d_{ij}^\sigma),$$
$$b_i(\omega) \in \mathcal{N}_p(b_i^0, d_i^b), \sigma_i(\omega) \in \mathcal{N}_p(\sigma_i^0, d_i^\sigma),$$

$\mathcal{N}_p$—is a class of normal probability distributions.

Let the interaction of fuzzy parameters of the model is described by the strongest t-norm T_M, $t^i = (t_{i_1}, t_{i_2}, \ldots, t_{i_n}, t_i)$,

$$f_i(x,\omega,t^i) = \sum_{j=1}^{n}(a_{ij}(\omega) + \sigma_{ij}(\omega)t_{ij})x_j - (b_i(\omega) + \sigma_i(\omega)t_i).$$

It is clear that for fixed x and t^i, $i = 1,\ldots,m$ the function $f_i(x,\omega,t^i)$ is a normal random variable. Then [11] with possibility

$$\mu(t^i) = \min\{\min_{1 \le j \le n}\{\mu_{Y_{ij}}(t_{ij})\}, \mu_{Y_i}(t_i)\}$$

its mathematical expectation is defined by formula

$$m_i(x,t^i) = E\{f_i(x,\omega,t^i)\} = \sum_{j=1}^{n}(a_{ij}^0 + d_{ij}^0 t_{ij})x_j - (b_i^0 + \sigma_i^0 t_i).$$

For what follows we also need the variance of the function $f_i(x,\omega,t^i)$. In accordance with the classical approach, it can be defined as follows:

$$d_i(x,t^i) = E\{(f_i(x,\omega,t^i) - m_i(x,t_i))^2\}.$$

We will specify the formula. We have after appropriate substitutions

$$d_i(x, t^i) = E\{(\sum_{j=1}^{n}(a_{ij}(\omega) + \sigma_{ij}(\omega)t_{ij} - (a_{ij}^0 + d_{ij}^0 t_{ij}))x_j -$$

$$(b_i(\omega) - b_i^0 + (\sigma_i(\omega) - \sigma_i^0)t_i)^2\} = E\{\sum_{j=1}^{n}\sum_{k=1}^{n}(a_{ij}(\omega) + \sigma_{ij}(\omega)t_{ij}$$

$$-(a_{ij}^0 + d_{ij}^0)t_{ij})(a_{ik}(\omega) + \sigma_{ik}(\omega)t_{ik} - (a_{ik}^0 + d_{ik}^0)t_{ik})x_j x_k - 2\sum_{j=1}^{n}(a_{ij}(\omega)$$

$$+\sigma_{ij}(\omega)t_{ij} - (a_{ij}^0 + d_{ij}^0)t_{ij})(b_i(\omega) - b_i^0 + (\sigma(\omega) - \sigma_i^0)t_i)x_j$$

$$+(b_i(\omega) - b_i^0 - (\sigma_i(\omega) - \sigma_i^0)t_i)^2 +\}$$

$$= \sum_{j=1}^{n}\sum_{k=1}^{n}E\{((a_{ij}(\omega) - a_{ij}^0) + (\sigma_{ij}(\omega) - \sigma_{ij}^0)t_{ij}) \cdot$$

$$((a_{ik}(\omega) - a_{ik}^0) + (\sigma_{ik}(\omega) - \sigma_{ik}^0)t_{ik})x_j x_k\}$$

$$-2\sum_{j=1}^{n}E\{((a_{ij}(\omega) - a_{ij}^0) + (\sigma_{ij}(\omega) - \sigma_{ij}^0)t_{ij})((b_i(\omega) - b_i^0) + (\sigma_i(\omega) - \sigma_i^0)t_i)x_j\}$$

$$+E\{((b_i(\omega) - b_i^0) + (\sigma_i(\omega) - \sigma_i^0)t_i)^2\}.$$

We introduce the notation for the covariance coefficients of the random variables participating in the shift-scale representation of fuzzy random variables:

$$C_{a_{ij}a_{ik}} = cov(a_{ij}, a_{ik}); \quad C_{\sigma_{ij}\sigma_{ik}} = cov(\sigma_{ij}, \sigma_{ik});$$

$$C_{a_{ij}\sigma_{ik}} = cov(a_{ij}, \sigma_{ik}); \quad C_{\sigma_{ij}a_{ik}} = cov(\sigma_{ij}, a_{ik});$$

$$C_{b_i\sigma_i} = cov(b_i, \sigma_i); \quad C_{a_{ij}b_i} = cov(a_{ij}, b_i);$$

$$C_{a_{ij}\sigma_i} = cov(a_{ij}, \sigma_i); \quad C_{\sigma_{ij}b_i} = cov(\sigma_{ij}, b_i);$$

$$C_{\sigma_{ij}\sigma_i} = cov(\sigma_{ij}, \sigma_i).$$

Suppose further that

$$\mathbb{C}_{ijk} = C_{a_{ij}a_{ik}} + C_{a_{ij}\sigma_{ik}}t_{ik} + C_{\sigma_{ij}a_{ik}}t_{ij} + C_{\sigma_{ij}\sigma_{ik}}t_{ij}t_{ik}.$$

Using these notations, we have

$$d_i(x, t^i) = \sum_{j=1}^{n}\sum_{k=1}^{n}\mathbb{C}_{ijk}x_j x_k - 2\sum_{j=1}^{n}(C_{a_{ij}b_i} + C_{a_{ij}\sigma_i}t_i + C_{\sigma_{ij}b_i}t_{ij}$$

$$+C_{\sigma_{ij}\sigma_i}t_{ij}t_i)x_j + d_{b_i} + d_{\sigma_i}t_i^2 + 2C_{\sigma_i b_i}t_i, \tag{4}$$

where d_{b_i}, d_{σ_i}—variances of random variables $b_i(\omega)$ and $\sigma_i(\omega)$.

In the case when the random variables are uncorrelated, the resulting formula (4) takes the form

$$d_i(x, t^i) = \sum_{j=1}^{n} (C_{a_{ij}a_{ij}} + C_{\sigma_{ij}\sigma_{ij}} t_{ij}^2) x_j^2 + d_{b_i} + d_{\sigma_i} t_i^2. \tag{5}$$

Function $d_i(x, t^i)$ possesses the following properties due to the covariance matrix $\mathbb{C}^i = \{C_{ijk}\}_{j,k=1}^{n}$:

- $d_i(x, t^i)$—convex function with respect to x for fixed t^i;
- for any vectors x and t^i the function $d_i(x, t^i)$ is nonnegative;
- function $d_i(x, t^i)$ is convex with respect to t^i for fixed x.

4   Equivalent Crisp Analog

We now turn to the construction of an equivalent crisp analogue of problem (1)–(2). In order to do this we need to construct an equivalent deterministic system of constraints for (2). After that the model of criterion (1) can be reduced to an equivalent deterministic one in a similar way.

It is clear that for a fixed vector t^i

$$f(x, \omega, t^i) \in \mathcal{N}_p(m_i(x, t^i), \sqrt{d_i(x, t^i)}).$$

Then, in accordance with the classical results of stochastic programming [16] we have

$$P\{f_i(x, \omega, t^i) \le 0\} = P\{\frac{f_i(x, \omega, t^i) - m_i(x, t^i) + m_i(x, t^i)}{\sqrt{d_i(x, t^i)}} \le 0\}$$

$$= P\{\frac{f_i(x, \omega, t^i) - m_i(x, t^i)}{\sqrt{d_i(x, t^i)}} + \frac{m_i(x, t^i)}{\sqrt{d_i(x, t^i)}} \le 0\}$$

$$= 1 - \mathcal{F}_i(\frac{m_i(x, t^i)}{\sqrt{d_i(x, t^i)}}) \ge p_i,$$

where $\mathcal{F}_i$—is standard normal distribution function.

The last inequality is equivalent to the inequality

$$\frac{m_i(x, t^i)}{\sqrt{d_i(x, t^i)}} \ge \beta_i,$$

in which β_i is a solution of equation

$$\mathcal{F}_i(t) = 1 - p_i.$$

As a result, we obtain the inequality

$$m_i(x, t^i) - \beta_i \sqrt{d_i(x, t^i)} \le 0 \tag{6}$$

which is equivalent to the i-th constraint of (2) with the possibility of $\mu(t^i)$.

After the specification (6) takes the form

$$\sum_{j=1}^{n}(a_{ij}^0 + d_{ij}^0 t_{ij})x_j - \beta_i\sqrt{d_i(x, t^i)} \leq b_i^0 + \sigma_i^0 t_i. \tag{7}$$

If $p_i > 0.5$, then $\beta_i < 0$ and the function on the left-hand side of (7) is convex and monotonic with respect to fuzzy parameters.

After substituting the corresponding fuzzy values A_{ij} and B_i into the inequality (7) instead of the parameters t_{ij}, t^i, t_i and demanding the fulfillment of the resulting possibilistic inequality with the possibility of α_i we obtain an equivalent restriction for the i-th constraint of the system (2) which contains only fuzzy parameters. We have

$$\pi\Big\{\sum_{j=1}^{n}(a_{ij}^0 + d_{ij}^0 A_{ij}(\gamma))x_j$$
$$-\beta_i\sqrt{d_i(x, (A_{i1}(\gamma), \ldots, A_{in}(\gamma), B_i(\gamma)))} \leq b_i^0 + \sigma_i^0 B_i(\gamma)\Big\} \geq \alpha_i, \tag{8}$$

where

$$d_i(x, (A_{i1}(\gamma), \ldots, A_{in}(\gamma), B_i(\gamma)))$$
$$= \sum_{j=1}^{n}\sum_{k=1}^{n}\tilde{C}_{ijk}(A_{ij}(\gamma), A_{ik}(\gamma))x_j x_k$$
$$-2\sum_{j=1}^{n}(C_{a_{ij}b_i} + C_{a_{ij}\sigma_i}B_i(\gamma) + C_{\sigma_{ij}b_i}A_{ij}(\gamma)$$
$$+C_{\sigma_{ij}\sigma_i}A_{ij}(\gamma)B_i(\gamma))x_j + d_{b_i} + d_{\sigma_i}B_i^2(\gamma) + 2C_{\sigma_i b_i}B_i(\gamma),$$
$$\tilde{C}_{ijk}(A_{ij}(\gamma), A_{ik}(\gamma)) = C_{a_{ij}a_{ik}} + C_{a_{ij}\sigma_{ik}}A_{ik}(\gamma) + C_{\sigma_{ij}a_{ik}}A_{ij}(\gamma)$$
$$+C_{\sigma_{ij}\sigma_{ik}}A_{ij}(\gamma)A_{ik}(\gamma).$$

Now we are ready to formulate and prove the following theorem.

Theorem 1. *Let in the constraint model (2)* $\tau = \pi$, $p_i > 0.5$ $(i = 1, \ldots, m)$, *fuzzy variables* $Y_{ij}(\gamma)$, $Y_i(\gamma)$ *have shift-scale representation and are* T_M-*related, characterized by quasi-concave, semicontinuous distributions with finite supports, and the shift and scale coefficients are normally distributed random variables with the corresponding parameters. Then the constraint system (2) is equivalent to the determinate constraint model*

$$\begin{cases} \sum_{j=1}^{n}(a_{ij}^0 + d_{ij}^0 A_{ij}^-)x_j - \beta_i\sqrt{d_i(x, A_{i1}^{\{\pm\}}, \ldots, A_{in}^{\{\pm\}}, B_i^{\{\pm\}})} \leq b_i^0 + \sigma_i^0 B_i^+, \\ x \in X, i = 1, \ldots, m, \end{cases} \tag{9}$$

where $A_{ij}^{\{\pm\}}$, $B_i^{\{\pm\}}$ are right and left borders of α_i-level sets of fuzzy quantities $A_{ij}(\gamma)$, $B_i(\gamma)$,

$$d_i(x, A_{i1}^{\{\pm\}}, \ldots, A_{in}^{\{\pm\}}, B_i^{\{\pm\}}) = \sum_{j=1}^{n}\sum_{k=1}^{n} C_{ijk}(A_{ij}^{\{\pm\}}, A_{ik}^{\{\pm\}}) x_j x_k$$

$$+2\sum_{j=1}^{n}(-C_{a_{ij}b_i} - C_{a_{ij}\sigma_i}B_i^- - C_{\sigma_{ij}b_i}A_{ij}^- - C_{\sigma_{ij}\sigma_i}(B_i A_{ij})^-)x_j$$

$$+d_{b_i} + d_{\sigma_i}(B_i^2)^- + 2C_{\sigma_i b_i}(B_i)^-,$$

$$C_{ijk}(A_{ij}^{\{\pm\}}, A_{ik}^{\{\pm\}}) = C_{a_{ij}a_{ik}} + C_{a_{ij}\sigma_{ik}}A_{ik}^- + C_{\sigma_{ij}a_{ik}}A_{ij}^- + C_{\sigma_{ij}\sigma_{ik}}(A_{ij}A_{ik})^-.$$

Proof. The possibilistic function on the left-hand side of the inequality under the sign of the probability measure in (8) is a monotonic function of fuzzy parameters. Hence, on the basis of [17, 18], the i-th restriction of system (8) is equivalent to the i-th constraint of system (9).

Remark 1. The boundaries of α-level sets of fuzzy variables and their products are found using the results from [11, 19–21]. For example:

$$(B_i A_{ij})^- = \min\{B_i^- A_{ij}^-, \ B_i^- A_{ij}^+, \ B_i^+ A_{ij}^-, \ B_i^+ A_{ij}^+\}.$$

Remark 2. The boundary $(B_i^2)^-$ can be calculated for a number of parametrized distributions of LR-type [11].

The following theorem can be proved by a similar scheme.

Theorem 2. *Let in the criterion model (1) $\tau = \pi$, $p_0 > 0.5$, fuzzy variables Y_{0j} ($j = 1, \ldots, n$) have shift-scale representation and are T_M-related, characterized by quasi-concave, semicontinuous distributions with finite supports, and the shift and scale coefficients are normally distributed random variables.*
Then the equivalent for the criterion model (1) is the following model:

$$\sum_{j=1}^{n}(a_{0j}^0 + d_{0j}^0 A_{0j}^-) - \beta_0 \sqrt{\sum_{j=1}^{n}\sum_{k=1}^{n} C_{0jk}(A_{0j}^{\{\pm\}}, A_{0k}^{\{\pm\}}) x_j x_k} \to \min_{x \in X}.$$

The objective function in the latter problem is convex, and in the case when $p_0 < 0.5$, the problem goes to the class of nonconvex multiextremal problems. However, the requirement $p_0 > 0.5$ is consistent with the requirements of practical problems.

The following theorem is also important for applications.

Theorem 3. *Let in the model (1)–(2) $\tau = \pi$, $p_i = 0.5$, $i = 0, \ldots, m$, fuzzy variables Y_{ij} ($i = 0, \ldots, m$, $j = 1, \ldots, n$) have shift-scale representation and are T_M-related, characterized by quasi-concave, semicontinuous distributions with finite supports, and the shift and scale coefficients are normally distributed random variables.*

Then the equivalent deterministic analogue for the model (1)–(2) is the following model:

$$\sum_{j=1}^{n}(a_{0j}^0 + d_{0j}^0 A_{0j}^-) \to \min, \tag{10}$$

$$\begin{cases} \sum_{j=1}^{n}(a_{ij}^0 + d_{ij}^0 A_{ij}^-)x_j \le b_i^0 + \sigma_i^0 B_i^+, \\ x \in X, i = 1, \dots, m. \end{cases} \tag{11}$$

Proof. If probabilities $p_i = 0.5$, $i = 0, \dots, m$, then $\beta_i = 0$ and the quadratic components in the goal function of the problem and in the constraints system vanish. This concludes the proof.

Remark 3. In the case $\tau = \nu$ the equivalent problem for the model (1)–(2) takes the form:

$$\sum_{j=1}^{n}(a_{0j}^0 + d_{0j}^0 A_{0j}^-) \to \min, \tag{12}$$

$$\begin{cases} \sum_{j=1}^{n}(a_{ij}^0 + d_{ij}^0 A_{ij}^+)x_j \le b_i^0 + \sigma_i^0 B_i^-, \\ x \in X, i = 1, \dots, m, \end{cases} \tag{13}$$

where A_{ij}^+ and B_i^- are borders of α-level sets for levels $1 - \alpha_i$.

Now we will establish connection between the model (1)–(2) and the model in which the principle of the expected possibility is used in order to remove the uncertainty of the probabilistic type:

$$k \to \min,$$
$$\tau\{E[f_0(x, \omega, \gamma)] \le k\} \ge \alpha_0, \tag{14}$$

$$\begin{cases} \tau\{E[f_i(x, \omega, \gamma)] \le 0\} \ge \alpha_i, i = 1, \dots, m, \\ x \in X. \end{cases} \tag{15}$$

This connection is characterized by Theorem 4.

Theorem 4. *Let in the model (14)–(15) $\tau = \pi$, $p_i = 0.5$, $i = 0, \dots, m$, fuzzy variables Y_{ij} $(i = 0, \dots, m, j = 1, \dots, n)$ have shift-scale representation and are T_M-related, characterized by quasi-concave, semicontinuous distributions with finite supports, and the shift and scale coefficients are normally distributed random variables.*
Then the equivalent for the model (14)–(15) has the form (10)–(11).

Proof. Really,

$$E[f_i(x, \omega, \gamma)] = \sum_{j=1}^{n} E[A_{ij}(\omega, \gamma)]x_j - E[B_i(\omega, \gamma)].$$

Taking into account the shift-scale representation of fuzzy random variables A_{ij} and B_i we have:

$$E[f_i(x, \omega, \gamma)] = \sum_{j=1}^{n}(a_{ij}^0 + \sigma_{ij}^0 Y_{ij}(\gamma))x_j - (b_i^0 + \sigma_i^0 Y_i(\gamma)).$$

Then the model (14)–(15) takes the form:

$$k \to \min,$$

$$\tau\left\{\sum_{j=1}^{n}(a_{0j}^0 + \sigma_{0j}^0 Y_{0j}(\gamma))x_j \leq k\right\} \geq \alpha_0, \tag{16}$$

$$\begin{cases} \tau\left\{\sum_{j=1}^{n}(a_{ij}^0 + \sigma_{ij}^0 Y_{ij}(\gamma))x_j \leq b_i^0 + \sigma_i^0 Y_i(\gamma)\right\} \geq \alpha_i, \ i = 1, \dots, m, \\ x \in X. \end{cases} \tag{17}$$

Earlier we made the assumption that the distribution functions of possiblistic components in the shift-scale representation of fuzzy random variables are characterized by quasi-concave upper semicontinuous distributions and their interaction is described by the strongest t-norm. Thus we are in the conditions of applicability of the corresponding theorem from [23], on which the proof of the statement presented here can be based. In the case $\tau = \pi$ the model (10)–(11) is equivalent to (16)–(17).

Remark 4. For $\tau = \nu$ the model (14)–(15) has (12)–(13) as an equivalent.

5 Conclusion

The problem of possibilistic-probabilistic linear programming with constraints on possibility and probability is investigated. For the case of normally distributed random parameters of the model and under the most general assumptions concerning the properties of probability distributions, its equivalent determinate analogue is constructed which is the quadratic programming model. Its properties (convexity-concavity), as in classical stochastic programming [16], essentially depend on the levels of probability that are given in the initial model. For the case of necessity measure ($\tau = \nu$) in an equivalent model (we can prove this by using [18]) the right boundaries of $(1 - \alpha_i)$-level sets will be used in left-hand side of inequality (9), and left boundaries will be used in right-hand side of the inequality.

It is clear that for the probability values $p_i = 0$ moments of the second order (variance) are excluded and the model (1)–(2) is transformed into the problem of possibility-necessity linear programming.

In terms of prospective studies it seems appropriate to generalize the obtained results to other probability distributions and to the case of the weakest t-norm describing the interaction of fuzzy parameters [22].

References

1. Yazenin, A.V.: Linear programming with random fuzzy data. Izv. AN SSSR. Techn. Cybern. **3**, 52–58 (1991). (in Russian)
2. Yazenin, A.V.: On a method of solving a problem of linear programming with random fuzzy data. J. Comput. Syst. Sci. Int. **36**(5), 737–741 (1997)
3. Luhandjula, M.K.: Fuzziness and randomness in an optimization framework. Fuzzy Sets Syst. **77**(3), 291–297 (1996)
4. Luhandjula, M.K.: Linear programming under randomness and fuzziness. Fuzzy Sets Syst. **10**(1–3), 45–55 (1983)
5. Yazenin, A.V., Shefova, N.A.: On one possibilistic-probabilistic model of a portfolio of minimum risk. Vestnik TvGU. Seriya: Prikladnaya matematika **17**, 85–96 (2010). (in Russian)
6. Xu, J., Zhou, X.: Fuzzy-like multiple objective decision making. Studies in Fuzziness and Soft Computing, vol. 263. Springer, Berlin (2011). https://doi.org/10. 1007/978-3-642-16895-6
7. Egorova, Y.E., Yazenin, A.V.: The problem of possibilistic-probabilistic optimization. J. Comput. Syst. Sci. Int. **56**(4), 652–667 (2017). https://doi.org/10.7868/ S0002338817040096
8. Yazenin, A., Soldatenko, I.: A portfolio of minimum risk in a hybrid uncertainty of a possibilistic-probabilistic type: comparative study. In: Kacprzyk, J., Szmidt, E., Zadrożny, S., Atanassov, K.T., Krawczak, M. (eds.) IWIFSGN/EUSFLAT -2017. AISC, vol. 643, pp. 551–563. Springer, Cham (2018). https://doi.org/10.1007/978-3-319-66827-7_51
9. Nahmias, S.: Fuzzy variables in a random environment. In: Gupta, M.M., Ragade, R.K., Yager, R.R. (eds.) Advances in Fuzzy Sets Theory and Applications, pp. 165–180. NHCP, Amsterdam (1979)
10. Puri, M.L., Ralescu, D.A.: Fuzzy random variables. J. Math. Anal. Appl. **114**, 409–422 (1986)
11. Yazenin, A.V.: Basic concepts of possibility theory: a mathematical apparatus for decision-making under hybrid uncertainty conditions. Fizmatlit Publ., Moscow (2016). (in Russian)
12. Nguyen, H.T., Walker, E.A.: A First Course in Fuzzy Logic. CRC Press, Boca Raton (1997)
13. Mesiar, R.: Triangular-norm-based addition of fuzzy intervals. Fuzzy Sets Syst. **91**, 231–237 (1997)
14. Hong, D.H.: Parameter estimations of mutually T-related fuzzy variables. Fuzzy Sets Syst. **123**, 63–71 (2001)
15. Yazenin, A.V.: Possibilistic-probabilistic models and methods of portfolio optimization. In: Batyrshin, I., Kacprzyk, J., et al. (eds.) Perception-based Data Mining and Decision Making in Economics and Finance. SCI, vol. 36, pp. 241–259. Springer, Heidelberg (2007). https://doi.org/10.1007/978-3-540-36247-0_9
16. Ermolyev, Y.M.: Methods of stochastic programming. Nauka Publ., Moscow (1976). (in Russian)
17. Alefeld, G., Herzberger, J.: Introduction to Interval Computation. Academic Press, Cambridge (2012)
18. Yazenin, A., Wagenknecht, M.: Possibilistic optimization. Brandenburgische Technische Universitat, Cottbus (1996)
19. Fuller, R., Keresztfalvi, T.: On generalization of Nguyen's theorem. Fuzzy Sets Syst. **41**(3), 371–374 (1991). https://doi.org/10.1016/0165-0114(91)90139-H

20. Fuller, R., Keresztfalvi, T.: t-Norm-based addition of fuzzy intervals. Fuzzy Sets Syst. **51**(2), 155–159 (1992). https://doi.org/10.1016/0165-0114(92)90188-A
21. Dubois, D., Prade, H.: Théorie des Possibilités, Application à la Représentation des Connaissances en Informatique. Masson, Paris (1988)
22. Yazenin, A.V., Soldatenko, I.S.: Possibilistic optimization tasks with mutually t-related parameters: solution methods and comparative analysis. In: Lodwick, W.A., Kacprzyk, J. (eds.) Fuzzy Optimization. STUDFUZZ, vol. 254, pp. 163–192. Springer, Heidelberg (2010). https://doi.org/10.1007/978-3-642-13935-2_8
23. Yazenin, A.V.: On the problem of possibilistic optimization. Fuzzy Sets Syst. **81**(1), 133–140 (1996)

Any F-Transform Is Defined by a Powerset Theory

Jiří Močkoř[(✉)]

Institute for Research and Applications of Fuzzy Modeling, University of Ostrava,
Centre of Excellence IT4Innovations, 30. dubna 22, 701 03 Ostrava 1, Czech Republic
Jiri.Mockor@osu.cz
http://irafm.osu.cz/

Abstract. Relationships between powerset theories and F-transforms are investigated. Both these methods represent strong tools in fuzzy sets theory and applications. Although both methods deal with similar objects, both these methods use different tools and, so far, the relationship between the two methods has not been investigated. The aim of this paper is to show that there is a strong relationship between the two methods. Namely, arbitrary lower or upper F-transform of lattice-valued fuzzy sets can be derived from a special powerset theory and, conversely, there exists a special class of powerset theories, such that maps defined by these powerset theories are lower or upper F-transforms. These results allow, among other things, to extend the range of methods and tools that are used in both theories.

1 Introduction

In fuzzy set theory there are two important methods which are frequently used both in theoretical research and applications. These methods are the *powerset theory and the F-transform*. Both these methods were, in full details and theoretical backgrounds, introduced relatively recently and, in the present, both methods represent very strong tools in the theory and applications.

The powerset structures are widely used in algebra, logic, topology and also in computer science. The standard example of a powerset structure $P(X) = \{A : A \subseteq X\}$ and the corresponding extension of a mapping $f : X \to Y$ to the map $f_P^\to : P(X) \to P(Y)$ is widely used in almost all branches of mathematics and their applications, including computer science. For illustrative examples of possible applications see, e.g., the introductory part of the paper of [24]. Because the classical set theory can be considered to be a special part of the fuzzy set theory, introduced by [26], it is natural that powerset objects associated with fuzzy sets were soon investigated as generalizations of classical powerset objects. The first approach was done again by Zadeh [26], who defined $[0,1]^X$ to be a new powerset object $Z(X)$ instead of $P(X)$ and introduced the new powerset operator $f_Z^\to : Z(X) \to Z(Y)$, such that for $s \in Z(X), y \in Y$,

This research was partially supported by the project 18-06915S provided by the Grant Agency of the Czech Republic.

$$f_{\overrightarrow{Z}}(s)(y) = \bigvee_{x, f(x)=y} s(x).$$

A lot of papers were published about Zadeh's extension and its generalizations, see, e.g., [5,10,11,14,21–24]. Zadeh's extension was for the first time intensively studied by Rodabaugh in [21] for lattice-valued fuzzy sets. This paper was, in fact, the first real attempt to uniquely derive the powerset operator $f_{\overrightarrow{Z}}$ from $f_{\overrightarrow{P}}$ and not only explicitly stipulate them. The works of Rodabaugh gave very serious basis for further research of powerset objects and operators. That new approach to the powerset theory was based on application of the *theory of monads in clone form*, introduced by Manes [9]. A special example of monads in clone form was introduced by Rodabaugh [23] as a special structure describing powerset objects. In the papers [10] and [11] we presented some examples of powerset theories based on fuzzy sets which are generated by monads in clone form.

Another important method which was recently introduced in the fuzzy set theory is the F-transform. This theory was in lattice-valued form introduced by Perfilieva [19] and elaborated in many other papers (see, e.g., [16–18,20]). Analogically as the powerset operator $f_{\overrightarrow{P}} : P(X) \to P(Y)$, F-transform is a special transformation map $F : \mathcal{L}^X \to \mathcal{L}^Y$, that transforms $\mathcal{L}$-valued fuzzy sets defined in the set X to $\mathcal{L}$-valued fuzzy sets defined in another set Y.

Fuzzy transforms represent new methods that have been successfully used in signal and image processing [1,2,5], signal compressions [16], numerical solutions of ordinary and partial differential equations [7,25], data analysis [3,4,18] and many other applications.

Although both methods deal also with the same object, i.e. $\mathcal{L}$-valued fuzzy sets, in general, both these methods use different tools and, so far, the relationship between the two methods has not been investigated. The aim of this paper is to show that, in fact, there is a very strong relationship between the two methods. We show, that arbitrary F-transform of $\mathcal{L}$-valued fuzzy sets defined by a fuzzy partition can be derived from a special powerset theory and, conversely, there exists a special class of powerset theories, such that maps defined by these powerset theories are F-transforms. This result allows, among other things, to extend the range of methods and tools that are used in both theories.

2　Preliminaries

A principal structure used in the paper is a *complete residuated lattice* (see e.g. [9,15]), i.e. a structure $\mathcal{L} = (L, \wedge, \vee, \otimes, \to, 0_L, 1_L)$ such that $(L, \wedge, \vee)$ is a complete lattice, $(L, \otimes, 1_L)$ is a commutative monoid with operation $\otimes$ isotone in both arguments and $\to$ is a binary operation which is residuated with respect to $\otimes$, i.e.

$$\alpha \otimes \beta \leq \gamma \quad \text{iff} \quad \alpha \leq \beta \to \gamma.$$

Recall that a negation of an element a in $\mathcal{L}$ is defined by $\neg a = a \to 0_L$.

A special example of a residuated lattice $\mathcal{L}$ is a MV-algebra, i.e., a structure $\mathcal{L} = (L, \oplus, \otimes, \neg, 0_L, 1_L)$ satisfying the following axioms:

(i) $(L, \otimes, 1_L)$ is a commutative monoid,

(ii) $(L, \oplus, 0_L)$ is a commutative monoid,

(iii) $\neg\neg x = x$, $\neg 0_L = 1_L$,

(iv) $x \oplus 1_L = 1_L$, $x \oplus 0_L = x$, $x \otimes 0_L = 0_L$,

(v) $x \oplus \neg x = 1_L$, $x \otimes \neg x = 0_L$,

(vi) $\neg(x \oplus y) = \neg x \otimes \neg y$, $\neg(x \otimes y) = \neg x \oplus \neg y$,

(vii) $\neg(\neg x \oplus y) \oplus y = \neg(\neg y \oplus x) \oplus x$,

for all $x, y \in X$.

If we put

$$x \vee y = (x \oplus \neg y) \otimes y, \quad x \wedge y = (x \otimes \neg y) \oplus y, \quad x \to y = \neg x \oplus y,$$

then, $(L, \wedge, \vee, 0_L, 1_L)$ is a distributive lattice and $(L, \wedge, \vee, \otimes, \to, 0_L, 1_L)$ is a residuated lattice. MV-algebra is called a complete algebra, if that lattice is a complete lattice.

MV-algebras have their origin in algebraic analysis of Łukasiewicz logic by Chang in [6] and represent a generalization of Boolean algebras. A standard example of a MV-algebra is the *Łukasiewicz algebra* $\mathcal{L}_{\mathbf{L}} = ([0,1], \oplus, \otimes, \neg, 0, 1)$, where

$$x \otimes y = 0 \vee (x + y - 1), \quad \neg x = 1 - x, \quad x \oplus y = 1 \wedge (x + y).$$

If $\mathcal{L}$ is a complete residuated lattice, a $\mathcal{L}$-fuzzy set in a crisp set X is a map $f : X \to L$. f is a non-trivial $\mathcal{L}$-fuzzy set, if f is not identical to the zero function. The core of a $\mathcal{L}$-fuzzy set f in a set X is defined by $core(f) = \{x \in X : f(x) = 1_L\}$.

We recall some basic facts about F-transforms. An F-transform in a form introduced by Perfilieva [20] is based on the so called fuzzy partitions on the crisp set. Unless otherwise stated, by $\mathcal{L}$ we denote the complete residuated lattice $\mathcal{L} = (L, \wedge, \vee, \otimes, \to, 0_L, 1_L)$.

Definition 1. *Let X be a set. A system $\mathcal{A} = \{A_\lambda : \lambda \in \Lambda\}$ of normal $\mathcal{L}$-fuzzy sets in X is a fuzzy partition of X, if $\{core(A_\lambda) : \lambda \in \Lambda\}$ is a partition of X. A pair $(X, \mathcal{A})$ is called a space with a fuzzy partition. The index set of $\mathcal{A}$ will be denoted by $|\mathcal{A}|$.*

In [12,13] we introduced the category **SpaceFP** of spaces with fuzzy partitions. In the paper we consider the modified version of this category.

Definition 2. *The category* **SpaceFP** *is defined by*

1. *Fuzzy partitions $(X, \mathcal{A})$, as objects,*
2. *Morphisms $(g, \sigma) : (X, \{A_\lambda : \lambda \in \Lambda\}) \to (Y, \{B_\omega : \omega \in \Omega\})$, such that*
   (a) *$g : X \twoheadrightarrow Y$ and is $\sigma : \Lambda \twoheadrightarrow \Omega$ are surjective mappings,*
   (b) *$\forall \lambda \in \Lambda$, $A_\lambda(x) = B_{\sigma(\lambda)}(g(x))$, for each $x \in X$.*
3. *The composition of morphisms in* **SpaceFP** *is defined by $(h, \tau) \circ (g, \sigma) = (h \circ g, \tau \circ \sigma)$.*

Objects of the category **SpaceFP** represent ground structures for a fuzzy transform, firstly proposed by Perfilieva [19] and, in the case where it is applied to $\mathcal{L}$-fuzzy sets with $\mathcal{L}$-valued partitions, in [20].

Definition 3. *Let $(X, \mathcal{A})$ be a space with a fuzzy partition $\mathcal{A} = \{A_\lambda : \lambda \in |\mathcal{A}|\}$.*

1. *An upper F-transform with respect to the space $(X, \mathcal{A})$ is a function $F^{\uparrow}_{X,\mathcal{A}} :$ $\mathcal{L}^X \to \mathcal{L}^{|\mathcal{A}|}$, defined by*

$$f \in \mathcal{L}^X, \lambda \in |\mathcal{A}|, \quad F^{\uparrow}_{X,\mathcal{A}}(f)(\lambda) = \bigvee_{x \in X} (f(x) \otimes A_\lambda(x)).$$

2. *A lower F-transform with respect to the space $(X, \mathcal{A})$ is a function $F^{\downarrow}_{X,\mathcal{A}} :$ $\mathcal{L}^X \to \mathcal{L}^{|\mathcal{A}|}$, defined by*

$$f \in \mathcal{L}^X, \lambda \in |\mathcal{A}|, \quad F^{\downarrow}_{X,\mathcal{A}}(f)(\lambda) = \bigwedge_{x \in X} (A_\lambda(x) \to f(x)).$$

3   Powerset Theories in the Category SpaceFP

In what follows, by $CSLAT(\vee)$ or $CSLAT(\wedge)$ we denote the category of complete $\vee$- or $\wedge$-semilattices as objects, respectively, with $\vee$- or $\wedge$-preserving maps as morphisms. If there is no need to distinguish between $\vee$ and $\wedge$, we will only write $CSLAT$. The standard definition of powerset theories was presented by Rodabaugh [23].

Definition 4. *Let $\mathbf{K}$ be a ground category. Then $\mathbf{T} = (T, \to, V, \eta)$ is called $CSLAT$-powerset theory in $\mathbf{K}$, if*

1. *$T : \mathbf{K} \to CSLAT$ is an object-mapping,*
2. *for each morphism $f : A \to B$ in $\mathbf{K}$, there exists $f^{\to}_T : T(A) \to T(B)$ in $CSLAT$,*
3. *There exists a concrete functor $V : \mathbf{K} \to Set$, such that η determines in Set for each $A \in \mathbf{K}$ a mapping $\eta_A : V(A) \to T(A)$,*
4. *For each $f : A \to B$ in $\mathbf{K}$, $f^{\to}_T \circ \eta_A = \eta_B \circ V(f)$.*

In the paper we deal with powerset theories in the category **SpaceFP** which satisfy additional properties, typical for fuzzy sets structures. Two types of these powerset theories are introduced in the following definitions.

Definition 5. *A structure $\mathbf{T} = (T, \to, V, \eta)$ is called a $\mathcal{L}^{\vee}$-powerset theory in the category **SpaceFP**, if*

1. *$\mathbf{T}$ is a $CSLAT(\vee)$-powerset theory in the category **SpaceFP**,*
2. *For each object $(X, \mathcal{A}) \in \mathbf{SpaceFP}$,*
   (a) *there exists a $\bigvee$-preserving embedding $i_{(X,\mathcal{A})} : T(X, \mathcal{A}) \hookrightarrow \mathcal{L}^{|\mathcal{A}|}$,*
   (b) *for each $x \in V(X, \mathcal{A})$ there exists $\alpha \in |\mathcal{A}|$, such that $core(i_{(X,\mathcal{A})}$ $(\eta_{(X,\mathcal{A})}(x))) = \{\alpha\}$,*

(c) there exists an external operation $\star : \mathcal{L} \times T(X,\mathcal{A}) \to T(X,\mathcal{A})$, such that $i_{(X,\mathcal{A})}(\alpha \star f) = \alpha \otimes i_{(X,\mathcal{A})}(f)$, for each $f \in T(X,\mathcal{A}), \alpha \in \mathcal{L}$.

If $\mathcal{L} = (L, \oplus, \otimes, \neg, 0_L, 1_L)$ is a complete MV-algebra, we can also define the $\mathcal{L}^\wedge$-powerset theory in the category **SpaceFP**.

Definition 6. *A structure* $\mathbf{S} = (S, \to, W, \mu)$ *is called a* $\mathcal{L}^\wedge$-*powerset theory in the category* **SpaceFP**, *if*

1. *$\mathbf{S}$ is a $CSLAT(\wedge)$-powerset theory in the category* **SpaceFP**,
2. *For each object $(X,\mathcal{A}) \in$ **SpaceFP**,*
   (a) there exists a $\bigwedge$-preserving embedding $j_{(X,\mathcal{A})} : S(X,\mathcal{A}) \hookrightarrow \mathcal{L}^{|\mathcal{A}|}$,
   (b) for each $x \in W(X,\mathcal{A})$ there exists $\alpha \in |\mathcal{A}|$, such that $core(j_{(X,\mathcal{A})}(\mu_{(X,\mathcal{A})}(x))) = \{\alpha\}$,
   (c) there exists an external operation $+ : \mathcal{L} \times S(X,\mathcal{A}) \to S(X,\mathcal{A})$, such that $j_{(X,\mathcal{A})}(\alpha + f) = \alpha \oplus j_{(X,\mathcal{A})}(f)$, for each $f \in S(X,\mathcal{A}), \alpha \in \mathcal{L}$.

Let us consider the following examples of the $\mathcal{L}^\vee$-and $\mathcal{L}^\wedge$-powerset theory.

Example 1. Let $\mathcal{U} = \{\tau_{(X,\mathcal{A})} : (X,\mathcal{A}) \in$ **SpaceFP**$\}$ be a system of $\mathcal{L}$-valued similarity relations defined on sets $|\mathcal{A}|$, such that for arbitrary morphism $(f, \sigma) : (X,\mathcal{A}) \to (Y,\mathcal{B})$ in the category **SpaceFP**, $\tau_{(X,\mathcal{A})}(\alpha, \beta) = \tau_{(Y,\mathcal{B})}(\sigma(\alpha), \sigma(\beta))$ holds for arbitrary $\alpha, \beta \in |\mathcal{A}|$. Moreover, let the following condition holds for arbitrary $(X,\mathcal{A})$:

$$\alpha, \beta \in \mathcal{L}, \quad \tau_{(X,\mathcal{A})}(\alpha, \beta) = 1_L \Leftrightarrow \alpha = \beta.$$

For arbitrary morphism $(f, \sigma) : (X,\mathcal{A}) \to (Y,\mathcal{B})$ in **SpaceFP**, we set

$$V(X,\mathcal{A}) = |\mathcal{A}|, \quad V(f, \sigma) = \sigma,$$

$$T(X,\mathcal{A}) = \{g \in \mathcal{L}^{|\mathcal{A}|} : g \text{ is extensional with respect to } \tau_{(X,\mathcal{A})}\} \hookrightarrow \mathcal{L}^{|\mathcal{A}|},$$

$$(f, \sigma)_{\overrightarrow{T}} : T(X,\mathcal{A}) \to T(Y,\mathcal{B}), \quad (f, \sigma)_{\overrightarrow{T}}(g)(\beta) = \bigvee_{\alpha \in |\mathcal{A}|} g(\alpha) \otimes \tau_{(Y,\mathcal{B})}(\beta, \sigma(\alpha)),$$

$$\eta_{(X,\mathcal{A})} : V(X,\mathcal{A}) = |\mathcal{A}| \to T(X,\mathcal{A}), \quad \eta_{(X,\mathcal{A})}(\alpha)(\beta) = \tau_{(X,\mathcal{A})}(\alpha, \beta).$$

It is clear that $T(X,\mathcal{A})$ is a complete $\bigvee$-semilattice and $(f, \sigma)_{\overrightarrow{T}}(g)$ is also extensional with respect to $\tau_{(Y,\mathcal{B})}$. Then, $\mathbf{T} = (T, \to, V, \eta)$ is the $\mathcal{L}^\vee$-powerset theory called *powerset theory defined by* $\mathcal{U}$. In fact, we set

$$\alpha \in \mathcal{L}, g \in T(X,\mathcal{A}), \quad \alpha \star g = \alpha \otimes g.$$

It can be proven simply that $\alpha \otimes g$ are elements of $T(X,\mathcal{A})$ and the following diagram commutes,

$$
\begin{array}{ccc}
|\mathcal{A}| & \xrightarrow{\sigma} & |\mathcal{B}| \\
\downarrow{\scriptstyle \eta_{(X,\mathcal{A})}} & & \downarrow{\scriptstyle \eta_{(Y,\mathcal{B})}} \\
T(X,\mathcal{A}) & \xrightarrow{(f,\sigma)_{\overrightarrow{T}}} & T(Y,\mathcal{B})
\end{array}
$$

Hence, $(T, \to, V, \eta)$ is a $\mathcal{L}^\vee$-powerset theory.   □

Example 2. Let $\mathcal{L}$ be a complete MV-algebra. Let $\mathcal{U} = \{\tau_{(X,\mathcal{A})} : (X,\mathcal{A}) \in$ **SpaceFP**$\}$ be the same sets of similarity relation as in the Example 1.

For arbitrary morphism $(f,\sigma) : (X,\mathcal{A}) \to (Y,\mathcal{B})$ we set

$$S(X,\mathcal{A}) = T(X,\mathcal{A}),$$

$$(f,\sigma)_{\overrightarrow{S}} : S(X,\mathcal{A}) \to S(Y,\mathcal{B}), \quad (f,\sigma)_{\overrightarrow{S}}(g)(\beta) = \bigwedge_{\alpha \in |\mathcal{A}|} \neg\tau_{(Y,\mathcal{B})}(\sigma(\alpha),\beta) \oplus g(\alpha),$$

$$\alpha \in \mathcal{L}, g \in S(X,\mathcal{L}), \quad \alpha + g := \alpha \oplus g.$$

It can be proven that $(f,\sigma)_{\overrightarrow{S}}$ is defined correctly, i.e., $(f,\sigma)_{\overrightarrow{S}}(g) \in S(Y,\mathcal{B})$, for arbitrary $g \in S(X,\mathcal{A})$. In fact, for $\beta, \omega \in |\mathcal{B}|$, we have

$$\tau_{(Y,\mathcal{B})}(\sigma(\alpha),\beta) \geq \tau_{(Y,\mathcal{B})}(\beta,\omega) \otimes \tau_{(Y,\mathcal{B})}(\sigma(\alpha),\omega) \quad \Rightarrow$$
$$\tau_{(Y,\mathcal{B})}(\sigma(\alpha),\beta) \to g(\alpha) \leq \tau_{(Y,\mathcal{B})}(\beta,\omega) \otimes \tau_{(Y,\mathcal{B})}(\sigma(\alpha),\omega) \to g(\alpha) =$$
$$\tau_{(Y,\mathcal{B})}(\beta,\omega) \to (\tau_{(Y,\mathcal{B})}(\sigma(\alpha),\omega) \to g(\alpha)) \quad \Rightarrow$$
$$(\tau_{(Y,\mathcal{B})}(\sigma(\alpha),\beta) \to g(\alpha)) \otimes \tau_{(Y,\mathcal{B})}(\beta,\omega) \leq \tau_{(Y,\mathcal{B})}(\sigma(\alpha),\omega) \to g(\alpha) \quad \Rightarrow$$
$$(f,\sigma)_{\overrightarrow{S}}(g)(\beta) \otimes \tau_{(Y,\mathcal{B})}(\beta,\omega) \leq (f,\sigma)_{\overrightarrow{S}}(g)(\omega),$$

and $(f,\sigma)_{\overrightarrow{S}}(g)$ is extensional with respect to $\tau_{(Y,\mathcal{B})}$ and $\alpha \oplus g \in S(X,\mathcal{A})$. In fact, for arbitrary $\beta, \omega \in \mathcal{L}$, we have

$$(\neg\alpha \to g(\beta)) \otimes \neg\alpha \otimes \tau_{(X,\mathcal{A})}(\beta,\omega) \leq g(\beta) \otimes \tau_{(X,\mathcal{A})}(\beta,\omega) \leq g(\omega),$$

and it follows that

$$(\alpha \oplus g(\beta)) \otimes \tau_{(X,\mathcal{A})}(\beta,\omega) = (\neg\alpha \to g(\beta)) \otimes \tau_{(X,\mathcal{A})}(\beta,\omega) \leq$$
$$\neg\alpha \to (\omega) = \alpha \oplus g(\omega).$$

Therefore, $\mathbf{S} = (S,\to,V,\eta)$ is the $\mathcal{L}^\wedge$-powerset theory, where η is the same as in the previous Example. $\qquad\square$

As we mentioned in the Introduction, our goal is to show that the classical F-transform $F_{X,\mathcal{A}} : \mathcal{L}^X \to \mathcal{L}^{|\mathcal{A}|}$ defined by the space with a fuzzy partition $(X,\mathcal{A})$ can be derived from a powerset theory and, vice versa, that each suitable powerset theory $\mathbf{T}$ in the category **SpaceFP**, defines for arbitrary $(X,\mathcal{A}) \in$ **SpaceFP** the map $T_{[X,\mathcal{A}]} : \mathcal{L}^{V(X,\mathcal{A})} \to T(X,\mathcal{A})$, which can be represented by the F-transform $F_{V(X,\mathcal{A}),\mathcal{B}}$ defined by (possible different) space with a fuzzy partition $(V(X,\mathcal{A}),\mathcal{B})$.

Let us introduce the definition of the map defined by a $\mathcal{L}^\vee$- or $\mathcal{L}^\wedge$-powerset theories.

Definition 7. *1. Let* $\mathbf{T} = (T,\to,V,\eta)$ *be a* $\mathcal{L}^\vee$*-powerset theory in the category* **SpaceFP***. For* $(X,\mathcal{A}) \in$ **SpaceFP***, the map defined by* $\mathbf{T}$ *is*

$$T^{[X,\mathcal{A}]} : \mathcal{L}^{V(X,\mathcal{A})} \to T(X,\mathcal{A}),$$

$$f \in \mathcal{L}^{V(X,\mathcal{A})}, \quad T^{[X,\mathcal{A}]}(f) := \bigvee_{x \in V(X,\mathcal{A})} \eta_{(X,\mathcal{A})}(x) \star f(x) \in T(X,\mathcal{A}).$$

2. *Let $\mathcal{L}$ be a complete MV-algebra and let $\mathbf{S} = (S, \rightarrow, W, \mu)$ be a $\mathcal{L}^\wedge$-powerset theory in the category $\mathbf{SpaceFP}$. For $(X, \mathcal{A}) \in \mathbf{SpaceFP}$, the map defined by $\mathbf{S}$ is*

$$S_{[X,\mathcal{A}]} : \mathcal{L}^{W(X,\mathcal{A})} \rightarrow S(X, \mathcal{A}),$$

$$f \in \mathcal{L}^{W(X,\mathcal{A})}, \quad S_{[X,\mathcal{A}]}(f) := \bigwedge_{x \in W(X,\mathcal{A})} \neg \eta_{(X,\mathcal{A})}(x) + f(x) \in T(X, \mathcal{A}).$$

In the next theorem we prove that lower and upper F-transforms are derived from powerset theories. We show that for arbitrary space with a fuzzy partition $(X, \mathcal{A})$, the upper F-transform $F^\uparrow_{X,\mathcal{A}} : \mathcal{L}^X \rightarrow \mathcal{L}^{|\mathcal{A}|}$ is identical to the map $T^{[X,\mathcal{A}]}$ defined by a $\mathcal{L}^\vee$-powerset theory $\mathbf{T}$. An analogical result we can obtain for lower F-transform $F^\downarrow_{X,\mathcal{A}}$, which is identical to the map $T_{[X,\mathcal{A}]}$.

Theorem 1. *There exists the powerset theory $\mathbf{T} = (T, \rightarrow, V, \eta)$ of the category $\mathbf{SpaceFP}$, such that*

1. *$\mathbf{T}$ is $\mathcal{L}^\vee$-powerset theory.*
2. *If $\mathcal{L}$ is a complete MV-algebra, then $\mathbf{T}$ is also $\mathcal{L}^\wedge$-powerset theory,*
3. *For each $(X, \mathcal{A}) \in \mathbf{SpaceFP}$,*

$$T^{[X,\mathcal{A}]} = F^\uparrow_{X,\mathcal{A}}, \quad T_{[X,\mathcal{A}]} = F^\downarrow_{X,\mathcal{A}}.$$

Proof. Let $(f, \sigma) : (X, \mathcal{A}) \rightarrow (Y, \mathcal{B})$ be a morphism in the category $\mathbf{SpaceFP}$.

(1) We define

$$T : \mathbf{SpaceFP} \rightarrow CSLAT(\vee), \quad V : \mathbf{SpaceFP} \rightarrow Set,$$
$$T(X, \mathcal{A}) = \mathcal{L}^{|\mathcal{A}|}, \quad V(X, \mathcal{A}) = X,$$
$$(f, \sigma)^\rightarrow_T = T(f, \sigma) : T(X, \mathcal{A}) \rightarrow T(Y, \mathcal{B}), \quad V(f, \sigma) = f,$$
$$g \in T(X, \mathcal{A}), \quad (f, \sigma)^\rightarrow_T(g) = \sigma^\rightarrow_Z(g) \in T(Y, \mathcal{B}),$$

where $\sigma^\rightarrow_Z$ is the Zadeh's extension of the map $\sigma : |\mathcal{A}| \rightarrow |\mathcal{B}|$ to the map $\mathcal{L}^{|\mathcal{A}|} \rightarrow \mathcal{L}^{|\mathcal{B}|}$. The ordering on the set $T(X, \mathcal{A})$ is point-wise and it is clear that $T(X, \mathcal{A})$ is a complete $\vee$-semilattice and $\sigma^\rightarrow$ is $\vee$-preserving map.

We define the map $\eta_{(X,\mathcal{A})} : X \rightarrow T(X, \mathcal{A})$ by

$$x \in X, \alpha \in |\mathcal{A}| \quad \eta_{(X,\mathcal{A})}(x)(\alpha) = A_\alpha(x),$$

where $\mathcal{A} = \{A_\alpha : \alpha \in |\mathcal{A}|\}$. We show that the following diagram commutes.

$$\begin{array}{ccc} X & \xrightarrow{\quad f \quad} & Y \\ {\scriptstyle \eta_{(X,\mathcal{A})}}\downarrow & & \downarrow{\scriptstyle \eta_{(Y,\mathcal{B})}} \\ T(X, \mathcal{A}) & \xrightarrow{(f,\sigma)^\rightarrow_T} & T(Y, \mathcal{B}). \end{array}$$

In fact, for $x \in X, \beta = \sigma(\alpha) \in |\mathcal{B}|$, we have

$$(f,\sigma)_T^{\rightarrow}(\eta_{(X,\mathcal{A})}(x))(\beta) = \sigma_Z^{\rightarrow}(\eta_{(X,\mathcal{A})}(x))(\beta) = \bigvee_{\alpha,\sigma(\alpha)=\beta} \eta_{(X,\mathcal{A})}(x)(\alpha) =$$

$$\bigvee_{\alpha,\sigma(\alpha)=\beta} A_\alpha(x) = \bigvee_{\alpha,\sigma(\alpha)=\beta} B_{\sigma(\alpha)}(f(x)) = B_\beta(f(x)) = \eta_{(Y,\mathcal{B})}(f(x))(\beta).$$

Hence, $\mathbf{T} = (T,\rightarrow,V,\eta)$ is the $CSLAT(\vee)$-powerset theory. To prove that $\mathbf{T}$ is the $\mathcal{L}^\vee$-powerset theory, we define the external operation $\star$ by

$$\alpha \in \mathcal{L}, \omega \in |\mathcal{A}|, g \in \mathcal{L}^{|\mathcal{A}|}, \quad (\alpha \star g)(\omega) := \alpha \otimes g(\omega).$$

Moreover, we have

$$core(\eta_{(X,\mathcal{A})}(x)) = \{\alpha \in |\mathcal{A}| : A_\alpha(x) = 1_L\} = \{u_{\mathcal{A}}(x)\},$$

where $u_{\mathcal{A}} : X \to |\mathcal{A}|$ is the map defined by $u_{\mathcal{A}}(x) = \alpha \Leftrightarrow x \in core(A_\alpha)$. Hence, the condition (b) is also satisfied. Finally, for the map $T^{[X,\mathcal{A}]}$ defined by $\mathbf{T}$, for arbitrary $h \in \mathcal{L}^X, \alpha \in |\mathcal{A}|$ we have

$$T^{[X,\mathcal{A}]}(h)(\alpha) = (\bigvee_{x \in X} \eta_{(X,\mathcal{A})}(x) \star h(x))(\alpha) = \bigvee_{x \in X} \eta_{(X,\mathcal{A})}(x)(\alpha) \otimes h(x) =$$

$$\bigvee_{x \in X} A_\alpha(x) \otimes h(x) = F_{X,\mathcal{A}}^{\uparrow}(h)(\alpha).$$

Hence, $T^{[X,\mathcal{A}]} = F_{X,\mathcal{A}}^{\uparrow}$.

(2) Let $\mathcal{L}$ be the complete MV-algebra. For arbitrary morphism (f,σ) : $(X,\mathcal{A}) \to (Y,\mathcal{B})$, the set $T(X,\mathcal{A}) = \mathcal{L}^{|\mathcal{A}|}$ is also complete $\bigwedge$-semilattice. Since any complete MV-algebra is completely distributive ([8]), the map $\sigma_Z^{\rightarrow}$ is $\bigwedge$-preserving map, as follows from

$$\sigma_Z^{\rightarrow}(\bigwedge_{j \in J} h_j)(\beta) = \bigvee_{\alpha,\sigma(\alpha)=\beta} (\bigwedge_{j \in J} h_j(\alpha)) = \bigwedge_{j \in J} (\bigvee_{\alpha_j,\sigma(\alpha_j)=\beta} h_j(\alpha_j)) = \bigwedge_{j \in J} \sigma_Z^{\rightarrow}(h_j)(\beta).$$

Hence, the object function T from the previous case is also the object function $T : \mathbf{SpaceFP} \to CSLAT(\wedge)$ and $\mathbf{T} = (T,\rightarrow,V,\eta)$ can be consider to be also the $CSLAT(\wedge)$-powerset theory in the category $\mathbf{SpaceFP}$. To prove that $\mathbf{T}$ is also $\mathcal{L}^\wedge$-powerset theory, we need to change only the definition of the external operation $+$ as follows:

$$g \in T(X,\mathcal{A}), \alpha \in \mathcal{L}, \omega \in |\mathcal{A}|, \quad (\alpha + g)(\omega) := \alpha \oplus g(\omega).$$

Then, for the map $T_{[X,\mathcal{A}]}$ defined by $\mathbf{T}$, for arbitrary $h \in \mathcal{L}^X, \alpha \in |\mathcal{A}|$ we have

$$T_{[X,\mathcal{A}]}(h)(\alpha) = (\bigwedge_{x \in X} \neg\eta_{(X,\mathcal{A})}(x) + h(x))(\alpha) = \bigwedge_{x \in X} \neg\eta_{(X,\mathcal{A})}(x)(\alpha) \oplus h(x) =$$

$$\bigwedge_{x \in X} \neg A_\alpha(x) \oplus h(x) = \bigwedge_{x \in X} A_\alpha(x) \to h(x) = F_{X,\mathcal{A}}^{\downarrow}(h)(\alpha).$$

Hence, $T_{[X,\mathcal{A}]} = F^{\downarrow}_{X,\mathcal{A}}$. $\qquad\qquad\qquad\qquad\qquad\qquad\qquad\qquad\qquad\qquad\square$

In the next theorem we deal with the converse problem: Is it true that for arbitrary $\mathcal{L}^{\vee}$-powerset theory of the category **SpaceFP**, the map $T^{[X,\mathcal{A}]}$ defines an F-transform? The answer is "yes" and it allows to derive new types of F-transform maps $F : \mathcal{L}^X \to \mathcal{L}^{|\mathcal{A}|}$, where the transformed map $F(g)$ could have some additional properties.

Theorem 2. *Let* $\mathbf{T} = (T, \to, V, \eta)$ *be an arbitrary* $\mathcal{L}^{\vee}$-*powerset theory in the category* **SpaceFP**. *Then for arbitrary space with a fuzzy partition* $(X, \mathcal{A}) \in$ **SpaceFP** *there exists another space with a fuzzy partition* $(V(X, \mathcal{A}), \mathcal{B}) \in$ **SpaceFP**, *such that the following diagram commutes*

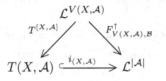

Proof. For arbitrary $\alpha \in |\mathcal{A}|, x \in V(X, \mathcal{A})$, we set $B_\alpha(x) = i_{(X,\mathcal{A})}$ $(\eta_{(X,\mathcal{A})}(x))(\alpha)$. Then $(V(X, \mathcal{A}), \mathcal{B})$ is a space with a fuzzy partition, where $\mathcal{B} = \{B_\alpha : \alpha \in |\mathcal{A}|\}$ is a fuzzy partition, as simply follows from the properties of η. Then, for $\alpha \in |\mathcal{A}|, g \in \mathcal{L}^{V(X,\mathcal{A})}$, we have

$$i_{(X,\mathcal{A})}.T^{[X,\mathcal{A}]}(g)(\alpha) = i_{(X,\mathcal{A})}\left(\bigvee_{x \in V(X,\mathcal{A})} \eta_{(X,\mathcal{A})}(x) \star g(x) \right)(\alpha) =$$

$$\bigvee_{x \in V(X,\mathcal{A})} i_{(X,\mathcal{A})}\eta_{(X,\mathcal{A})}(x)(\alpha) \otimes g(x) = \bigvee_{x \in V(X,\mathcal{A})} B_\alpha(x) \otimes g(x) = F^{\uparrow}_{V(X,\mathcal{A}),\mathcal{B}}(g)(\alpha).$$

$\qquad\qquad\qquad\qquad\qquad\qquad\qquad\qquad\qquad\qquad\qquad\qquad\qquad\qquad\square$

An analogical result we obtain for lower F-transform. The proof is similar and will be omitted.

Theorem 3. *Let* $\mathcal{L}$ *be a complete MV-algebra and let* $\mathbf{S} = (S, \to, W, \mu)$ *be an arbitrary* $\mathcal{L}^{\wedge}$-*powerset theory in the category* **SpaceFP**. *Then for arbitrary space with a fuzzy partition* $(X, \mathcal{A}) \in$ **SpaceFP** *there exists another space with a fuzzy partition* $(W(X, \mathcal{A}), \mathcal{B}) \in$ **SpaceFP**, *such that the following diagram commutes*

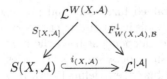

To illustrate the meaning of the preceding theorems, we show upper and lower F-transforms generated by the Theorems 2 and 3 from the $\mathcal{L}^{\vee}$- and $\mathcal{L}^{\wedge}$-powerset theories from the Examples 1 and 2, respectively.

Recall that for an arbitrary set X and an $\mathcal{L}$-valued similarity relation δ on the set X, a function $g \in \mathcal{L}^X$ is called the *extensional core* of a function $f \in \mathcal{L}^X$ with respect to δ, if

1. $\forall x \in X, \quad g(x) \le f(x)$,
2. g is extensional with respect to δ,
3. if $h \in \mathcal{L}^X$ is extensional with respect to δ, $h \le f$, then $g \ge h$.

Example 3. Let $\mathbf{T} = (T, \to, V, \eta)$ be the $\mathcal{L}^\vee$-powerset theory in the category **SpaceFP** from the Example 1. Then, according to the proof of the Theorem 2, for arbitrary $(X, \mathcal{A}) \in$ **SpaceFP**, the set $\mathcal{B} = \{\tau_{(X,\mathcal{A})}(\alpha, -) : \alpha \in |\mathcal{A}|\}$ is a fuzzy partition on $|\mathcal{A}|$, such that

$$T^{[X,\mathcal{A}]} = F^\uparrow_{|\mathcal{A}|,\mathcal{B}} : \mathcal{L}^{|\mathcal{A}|} \to \mathcal{L}^{|\mathcal{A}|},$$

$$g \in \mathcal{L}^{|\mathcal{A}|}, \quad F^\uparrow_{|\mathcal{A}|,\mathcal{B}}(g)(\omega) = \bigvee_{\alpha \in |\mathcal{A}|} g(\alpha) \otimes \tau_{(X,\mathcal{A})}(\alpha, \omega) = \widehat{g}(\omega).$$

It is clear that $\widehat{g}$ is the extensional hull of g with respect to $\tau_{(X,\mathcal{A})}$. Therefore, in that case, the upper F-transform $F^\uparrow_{|\mathcal{A}|,\mathcal{B}}$ represents the extensional hull transformation. □

Example 4. Let $\mathcal{L}$ be a complete MV-algebra and let $\mathbf{S} = (S, \to, V, \eta)$ be the $\mathcal{L}^\wedge$-powerset theory in the category **SpaceFP** from the Example 2. Then, according to the proof of the Theorem 3, for arbitrary $(X, \mathcal{A}) \in$ **SpaceFP**, the set $\mathcal{B} = \{\tau_{(X,\mathcal{A})}(\alpha, -) : \alpha \in |\mathcal{A}|\}$ is a fuzzy partition on the set $|\mathcal{A}|$, such that

$$T_{[X,\mathcal{A}]} = F^\downarrow_{|\mathcal{A}|,\mathcal{B}} : \mathcal{L}^{|\mathcal{A}|} \to \mathcal{L}^{|\mathcal{A}|},$$

$$g \in \mathcal{L}^{|\mathcal{A}|}, \quad F^\downarrow_{|\mathcal{A}|,\mathcal{B}}(g)(\beta) = \bigwedge_{\alpha \in |\mathcal{A}|} \tau_{(X,\mathcal{A})}(\alpha, \beta) \to g(\alpha) = \underline{g}(\beta).$$

It can be proven that $\underline{g}$ is the extensional core of g with respect to $\tau_{(X,\mathcal{A})}$. In fact, analogously as in the Example 2, we can prove that $\underline{g}$ is extensional with respect to $\tau_{(X,\mathcal{A})}$, $\underline{g} \le g$ and $\underline{g}$ is the largest extensional map with these properties. Therefore, in that case, the lower F-transform $F^\downarrow_{|\mathcal{A}|,\mathcal{B}}$ represents the extensional core transformation. □

4   Conclusions

F-transforms of lattice-valued fuzzy sets and powerset theories in fuzzy structures are frequently used tools in the fuzzy set theory and applications. Although these theories seem to be independent from the point of view of methods used, there exist deep relationships between these theories. We proved that arbitrary F-transform of $\mathcal{L}$-valued fuzzy sets defined by a fuzzy partition can be derived from a special powerset theory defined on the set of all $\mathcal{L}$-valued fuzzy sets and, conversely, there exists a special class of powerset theories, such that maps defined by these powerset theories are F-transforms. Using these relations, we can define new types of F-transforms and we can use, for example, new methods in the F-transform theory, including the theory of monads in special categories, which are typical tools in the powerset theories.

References

1. Di Martino, F., et al.: An image coding/decoding method based on direct and inverse fuzzy tranforms. Int. J. Approximation Reasoning **48**, 110–131 (2008)
2. Di Martino, F., et al.: A segmentation method for images compressed by fuzzy transforms. Fuzzy Sets Syst. **161**(1), 56–74 (2010)
3. Di Martino, F., et al.: Fuzzy transforms method and attribute dependency in data analysis. Inf. Sci. **180**(4), 493–505 (2010)
4. Di Martino, F., et al.: Fuzzy transforms method in prediction data analysis. Fuzzy Sets Syst. **180**(1), 146–163 (2011)
5. Gerla, G., Scarpati, L.: Extension principles for fuzzy set theory. J. Inf. Sci. **106**, 49–69 (1998)
6. Chang, C.C.: Algebraic analysis of many-valued logic. Trans. A.M.S. **93**, 74–80 (1958)
7. Khastan, A., Perfilieva, I., Alijani, Z.: A new fuzzy approximation method to Cauchy problem by fuzzy transform. Fuzzy Sets Syst. **288**, 75–95 (2016)
8. Jakubík, J.: Higher degrees of distributivity in MV-algebras. Czechoslovak Math. J. **53**(128), 641–653 (2003)
9. Manes, E.G.: Algebraic Theories. Springer, Berlin (1976). https://doi.org/10.1007/978-1-4612-9860-1
10. Močkoř, J.: Closure theories of powerset theories. Tatra Mt. Math. Publ. **64**, 101–126 (2015)
11. Močkoř, J.: Powerset operators of extensional fuzzy sets. Iran. J. Fuzzy Syst. **15**(2), 143–163 (2018)
12. Močkoř, J., Holčapek, M.: Fuzzy objects in spaces with fuzzy partitions. Soft Comput. **21**(24), 7269–7284 (2017)
13. Močkoř, J.: Spaces with fuzzy partitions and fuzzy transform. Soft Comput. **21**(13), 3479–3492 (2017)
14. Nguyen, H.T.: A note on the extension principle for fuzzy sets. J. Math. Anal. Appl. **64**, 369–380 (1978)
15. Novák, V., Perfilijeva, I., Močkoř, J.: Mathematical Principles of Fuzzy Logic. Kluwer Academic Publishers, Boston (1999)
16. Perfilieva, I.: Fuzzy transforms and their applications to image compression. In: Bloch, I., Petrosino, A., Tettamanzi, A.G.B. (eds.) WILF 2005. LNCS (LNAI), vol. 3849, pp. 19–31. Springer, Heidelberg (2006). https://doi.org/10.1007/11676935_3
17. Perfilieva, I.: Fuzzy transforms: a challange to conventional transform. In: Hawkes, P.W. (ed.) Advances in Image and Electron Physics, vol. 147, pp. 137–196. Elsevier Academic Press, San Diego (2007)
18. Perfilieva, I., Novak, V., Dvořak, A.: Fuzzy transforms in the analysis of data. Int. J. Approximate Reasoning **48**, 36–46 (2008)
19. Perfilieva, I.: Fuzzy transforms: theory and applications. Fuzzy Sets Syst. **157**, 993–1023 (2006)
20. Perfilieva, I., Singh, A.P., Tiwari, S.P.: On the relationship among F-transform, fuzzy rough set and fuzzy topology. In: Proceedings of IFSA-EUSFLAT 2015, Gijon, pp. 1324–1330. Atlantis Press, Amsterdam (2015)
21. Rodabaugh, S.E.: Powerset operator foundation for poslat fuzzy SST theories and topologies. In: Höhle, U., Rodabaugh, S.E. (eds.) Mathematics of Fuzzy Sets: Logic, Topology and Measure Theory, The Handbook of Fuzzy Sets Series, vol. 3, pp. 91–116. Kluwer Academic Publishers, Boston (1999)

22. Rodabaugh, S.E.: Powerset operator based foundation for point-set lattice theoretic (poslat) fuzzy set theories and topologies. Quaestiones Mathematicae **20**(3), 463–530 (1997)

23. Rodabaugh, S.E.: Relationship of algebraic theories to powerset theories and fuzzy topological theories for lattice-valued mathematics. Int. J. Math. Math. Sci. **2007**, 71 p. (2007). Article no. 43645

24. Solovyov, S.A.: Powerset operator foundations for catalg fuzzy set theories. Iran. J. Fuzzy Syst. **8**(2), 1–46 (2001)

25. Tomasiello, S.: An alternative use of fuzzy transform with application to a class of delay differential equations. Int. J. Comput. Math. **94**(9), 1719–1726 (2017)

26. Zadeh, L.A.: Fuzzy sets. Inf. Control **8**, 338–353 (1965)

Soft Clustering: Why and How-To

Stefano Rovetta[1] and Francesco Masulli[1,2]

[1] DIBRIS, University of Genova, Via Dodecaneso 35, 16146 Genoa, Italy
{stefano.rovetta,francesco.masulli}@unige.it
[2] Sbarro Institute for Cancer Research and Molecular Medicine, Temple University,
Philadelphia, PA, USA

Abstract. Despite the huge success of machine learning methods in the last decade, a crucial issue is to control the support of the data used in inference, so that data that are too far from the training set are given low confidence by default. The most important class that features this ability is that of prototype-based methods which are based on clustering or vector quantization as a representation learning model. This paper surveys a family of popular soft clustering methods, framing them in a unified formalism. It also discusses the peculiarities of each of them. A large fraction of the paper is devoted to clarifying the role of model parameters and to providing some guidelines on how to set up these parameters.

Keywords: Fuzzy clustering · Possibilistic clustering ·
Graded possibilistic clustering · Model parameters

1 Introduction

Despite the huge success of machine learning methods in the last decade, several issues remain unsolved. Machine learning usually focuses on black-box models which suffer from lack of explainability and, dually, from difficulty in using prior knowledge. In the specific case of deep learning [15] an additional issue is that theories of generalisation are apparently not applicable. In fact, adversarial machine learning techniques [16] seem to prove that generalisation ability is actually low in deep neural networks, and that bad quality outputs can easily be produced with high confidence. This is a very serious issue when machine learning is used in life-critical contexts like autonomous vehicle guidance or condition-based monitoring in predictive maintenance of sensitive plants.

In view of these problems, it is imperative to control the support of the data used in inference, so that data that are too far from the training set are given low confidence by default. The most important class that features this ability is that of prototype-based methods which are based on clustering or vector quantization as a representation learning model.

In the literature, these methods have been used extensively [2, 23, 26, 27] although they may appear to be less popular than other approaches (in particular deep learning and support vector machines). As noted, clustering, and specifically soft clustering, is a key component.

R. Fullér et al. (Eds.): WILF 2018, LNAI 11291, pp. 67–82, 2019.
https://doi.org/10.1007/978-3-030-12544-8_6

In this perspective, this paper surveys a family of popular soft clustering methods, framing them in a unified formalism. It also discusses the peculiarities of each of them. A large fraction of the paper is devoted to clarifying the role of model parameters and to providing some guidelines on how to set up these parameters.

2   Soft Clustering

The clustering problem is usually stated as the task of partitioning a set of data vectors or patterns $X = \{x_k\}$, $k \in \{1, \ldots, n\}$, $x_k \in \mathbb{R}^d$ by attributing each data point x_k to a subset $\omega_j \subset X$, $j \in \{1, \ldots, c\}$, defined by its *centroid* $y_j \in \mathbb{R}^d$. This attribution is made based on a given distance function that is used to measure the degree of centroid-observation closeness (in the following always assumed to be the Euclidean distance).

Some methods also employ a relational approach by measuring observation-observation closeness [9,13]; these are not considered here, but we cite them for completeness.

The following definitions deal with real-valued quantities and crisp sets, and therefore the symbols $\in$ and $\cup$ have the usual crisp-set-theoretic meaning:

Definition 1 (Fuzzy and possibilistic partitions [4]**).** *Given a set* $X = \{x_1, x_2, \ldots, x_n\}$ *of data items, a set* $\Omega = \{\omega_1, \omega_2, \ldots, \omega_c\}$, *and a membership function* $u(x, \omega)$, $x \in X$, *with* $0 \le u(x, \omega) \le 1 \ \forall x \in X$, $\forall \omega \in \Omega$, *the pair* (Ω, u) *is:*

– *A **possibilistic partition** if*

$$u(x, \omega) \in \mathbb{R} \quad \forall x, \forall \omega \qquad and \qquad 0 < \sum_{i=1}^{c} u(x, \omega_i) < c \quad \forall x \tag{1}$$

– *A **fuzzy partition** if it is a possibilistic partition with*

$$\sum_{i=1}^{c} u(x, \omega_i) = 1 \quad \forall x \tag{2}$$

– *A **crisp partition** if it is a fuzzy partition with*

$$\max_i u(x, \omega_i) = 1 \quad \forall x. \tag{3}$$

$\square$

In the case of central clustering, partitions are represented by centroids.

Definition 2 (Central clustering). *A **central clustering** is a (crisp, fuzzy, possibilistic) partition of a metric data space* Ξ *whose membership functions are monotonically dependent on the similarity of objects to a set of centroids* $\{y_1, \ldots, y_c\} \subset \Xi$. $\square$

Some methods not dealt with in this work, for instance those based on medoids or landmarks, require $\{y_1, \ldots, y_m\} \subset X$.

Central clustering is especially interesting as a concept representation tool because it can be learned from a training set X and applied to the whole data space Ξ. Many other approaches to clustering do not possess this *out-of-sample extension* property and can therefore only be used to partition the given data set.

The most widely used fuzzy clustering method is probably the *Fuzzy c-Means/Fuzzy ISODATA* [6,12,29] (FCM) algorithm, which is a "fuzzy relative" to the simple c-Means technique [5]. FCM defines the ω_j as fuzzy partitions of the data set X.

Well-known limitations of FCM include the need for fixing a fuzziness parameter in addition to the number of centroids, dependency on the initialisation, convergence to possibly bad-quality local solutions, the consequent need for many restarts, and a membership function profile that may not discriminate sharply enough between close and far points.

Variations over this basic scheme try to overcome some of these limitations. All of the following methods have membership functions that involve exponentials rather than powers of distance, which are sharper (for a discussion about this point see for instance [19]).

The *Maximum Entropy* (ME) approach, usually but not necessarily associated to the *Deterministic Annealing* optimisation procedure [24,25], does not minimize a simple cost term, but a compound cost function which is the sum of a distortion term $\hat{E}$ and an entropic term $-H$ (see the next section for the mathematical definitions). The optimization is done by fixing a constant value for one of the two terms and minimizing the other; then this step is iterated for decreasing values of the constant, until a global optimum is reached. This alleviates the false minima problem of standard c-Means and (to a lesser extent) of FCM.

In decision-making and classification applications, algorithms should feature several desirable properties in addition to the basic discrimination or decision function. For instance, it is usually required that in certain configurations a decision is not made (*pattern rejection*). This situation typically occurs in the presence of outliers. This problem is very well-known and well studied (see for instance [7,8,11]), and is tackled in a convenient way within the framework of soft-computing, fuzzy, and neural approaches [10,17,23].

However, the clustering problem as stated above implies that the outlier rejection property cannot be achieved. This is because the membership values are constrained to sum to 1. By giving up the requirement for strict partitioning, and by resorting to a "mode seeking" algorithm, Krishnapuram and Keller proposed the so-called *possibilistic approach* [18,19], where this constraint is relaxed essentially to

$$u_{jl} \in [0,1] \quad \forall l, \forall j \qquad (4)$$

With this model outlier rejection can be achieved, but at the expense of a clear cluster attribution and other computational drawbacks. The same issue of analysing the membership interactions on a local basis, as opposed to the global effects induced by the probabilistic model, is considered in [14].

An additional clustering model that can be thought of as a generalization of all those outlined above can be devised starting from the following observations.

Crisp partitions constrain membership in a very strong way: For a given object, memberships to all clusters must be zero except one. Fuzzy partitions relax this constraint in the sense that all membership can be non-zero, provided that their sum is still one. This means that membership to one cluster directly affects the membership to all other clusters. Finally, possibilistic partitions don't impose any constraint on memberships.

However, it is possible (and in practice it is frequent) that pairs of events are not mutually independent, but are not completely mutually exclusive either. Instead, events can provide *partial information* about other events. To model this idea, we could require the membership to one cluster to have an influence on the other memberships, but not so strong as to determine it directly.

This brings us to the concept of *graded possibility*. An example of such concept is given by a glass and by the fuzzy concepts of "full" and "empty". If the glass is full or almost full, its membership to the concept "empty" should clearly be close to zero, and similarly for the empty or almost empty case. However, if the glass is half filled, it is much more difficult to assess the membership in the concept "empty" with similar confidence. The profile of the membership functions in this case should be decided according to further considerations.

These ideas form the rationale of the Graded possibilistic c Means clustering methods, described in the following.

3   Some Popular Clustering Algorithms: A Unified View

3.1   The c-Means Family

We will now review some clustering algorithms derived from the basic c-Means: ("hard" or "crisp") c-Means (HCM) [5], Minimum-Entropy fuzzy clustering by Deterministic Annealing (ME) [24], Possibilistic c-Means with an entropic cost term (PCM-II) [19], Fuzzy c-Means (FCM) [12], Graded Possibilistic c-Means (GPCM) [21]. All of these techniques are based on minimizing the following cost function:

$$\hat{E} = \sum_{j=1}^{c} \sum_{l=1}^{n} u_{jl} d_{jl}. \tag{5}$$

(this includes also FCM, although in the usual formulation this is not evident; see [22]). We will refer collectively to these algorithms as the c-Means (CM) family.

Here u_{jl} is the degree of membership of pattern x_l to cluster ω_j and $Y = \{y_1, \ldots, y_c\}$. $\hat{E}$ can be termed approximation error in data analysis problems, distortion or quantization error in signal processing contexts, energy in physical analogies, risk in decision-theoretic and statistical learning frameworks.

Miyamoto and Mukaidono [22] show that these algorithms are obtained by adding to the basic cost $\hat{E}$ in (5) either regularization terms or the maximum-entropy term

$$- H = \sum_{j=1}^{c} \sum_{l=1}^{n} u_{jl} \log u_{jl} \tag{6}$$

which represents the (negative) entropy of the clustering defined by Y, U.

Figure 1 shows how the effect of fuzziness parameters on the objective function corresponds to regularization.

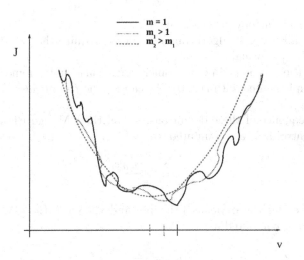

Fig. 1. Regularizing effect of fuzziness parameters on the objective function.

In clustering problems the focus is commonly placed on the analysis of data and clusters themselves, rather than on minimization of a global error criterion. We are often more interested in characterizing (hopefully significant) groups of data than in representing the details of the data with a faithful approximation. As an example, *model-based* clustering approaches focus on cluster modeling rather than performance optimization, and the cluster identification technique called *Alternating Cluster Estimation* [28] does not even assume the existence of a cost function.

Therefore we will introduce a formalism to provide an alternative, unified perspective on these clustering algorithms, focused on the memberships u_{jl} rather than on the cost function.

3.2   A Unifying Formalism

A CM clustering problem is defined by fixing the pair $\{J, \psi\}$, where:

– J is the cost function
– ψ is the constraint on the set of cluster memberships, such that

$$\psi(u_{1l}, \ldots, u_{cl}) = 0 \quad \forall l \in \{1, n\}$$

All the CM algorithms considered here define either:

$$J = \hat{E} \tag{7}$$

or:

$$J = \hat{E} - H \tag{8}$$

where the cluster entropy acts as a regularizer.

Moreover, all the CM algorithms considered require that $u_{jl} \in [0, 1] \; \forall j \in \{1, c\}, \forall l \in \{1, n\}$ (normality condition).

Let v_{jl} be the solution of a CM problem with constraint ψ removed (formally this can be implemented with $\psi \equiv 0$). We call v_{jl} the *free membership* of pattern x_l in cluster ω_j.

As a consequence of these definitions, for all the CM algorithms considered the cluster centroids Y are computed as:

$$y_j = \frac{\sum_{l=1}^{n} u_{jl} x_l}{\sum_{l=1}^{n} u_{jl}} \tag{9}$$

which characterizes the c-Means principle and therefore the CM family. The memberships are computed as:

$$u_{jl} = \frac{v_{jl}}{Z_l}, \tag{10}$$

where Z_l is the (generalized) partition function, which is computed as a function of the conventional partition function $\zeta_l = \sum_{j=1}^{c} v_{lj}$:

$$Z_l = f(\zeta_l) \tag{11}$$

Since the specific form of $f()$ is given by the constraint ψ, a member of the CM family is equivalently defined by the pair (J, f) or (J, Z_l).

With the above set of definitions, the CM algorithms of interest are compactly described as in Table 1.

All algorithms are fuzzy techniques, since they adopt the concept of "partial membership" in a set. HCM itself can be cast without imposing the constraint of binary memberships. The relationships among these algorithms are clear from the table.

A method to allow for non-extreme solutions is the maximum entropy criterion, which is implemented in the ME and PCM-II algorithms. They are related by the use of the entropic term $-H$, implying a parameter β_j. This parameter is different for each cluster and fixed in PCM-II, while it is constant for all clusters and varying with the algorithm progress in ME.

Table 1. The CM family of clustering algorithms

	J	v_{jl}	Z_l	Notes
ME	$\hat{E} - H$	$e^{-d_{jl}\beta}$	$\sum_{j=1}^{c} v_{jl}$	$\beta \in \mathbb{R}$, $\beta > 0$ is the inverse temperature parameter to be increased during the "annealing" process
PCM-II	$\hat{E} - H$	$e^{-d_{jl}\beta_j}$	1	$\beta_j \in \mathbb{R}$, $\beta_j > 0$ are cluster width parameters to be selected a priori before optimization or using heuristic criteria
FCM	$\hat{E}$	$1/d_{jl}$	$\left(\sum_{j=1}^{c} v_{jl}^{1/(m-1)}\right)^{m-1}$	$m \in \mathbb{R}$, $m > 1$ is the fuzzification parameter
HCM	$\hat{E}$	*See note*	*See note*	v_{jl} and Z_l can be written as for FCM, but their values have to be computed in the limit for $m \to 1$
GPCM	$\hat{E}$	$e^{-d_{jl}\beta_j}$	$\left(\sum_{j=1}^{c} v_{jl}\right)^{\alpha}$	$\beta_j \in \mathbb{R}$, $\beta_j > 0$ are cluster width parameters to be selected a priori before optimization or using heuristic criteria. $\alpha \in [0, 1]$ is the degree of probabilistic tendence

4   Membership Function Parametrization

All soft clustering methods require at least one model parameter, which in general terms decides the degree of fuzziness of the solution.

Since Miyamoto and Mukaidono [22] showed that the power membership function of FCM can be transformed into the exponential one of the other methods, the following discussion will only focus on the methods featuring the latter form, i.e., ME, PCM-II, GPCM.

4.1   Possible Parametrizations in the CM Family

The original formulation of free membership in ME features one global parameter β, interpreted as a global temperature, energy, disorder, or resolution.

$$v_{lj} = \exp\left(-\beta\|\boldsymbol{x}_l - \boldsymbol{y}_j\|^2\right) \tag{12}$$

The Deterministic Annealing optimization procedure fixes the temperature at each optimisation step, making it effectively a regularisation coefficient rather than a model parameter.

In contrast, PCM-II features one parameter β_j per centroid.

$$v_{lj} = \exp\left(-\beta_j\|\boldsymbol{x}_l - \boldsymbol{y}_j\|^2\right) \tag{13}$$

In this case the parametrization can be considered that of a system with non-constant energy, i.e., out of thermodynamic equilibrium.

It is also possible to write a free membership function with parameters that differ for each of the vector components of the centroid, although to the best of our knowledge no popular method from the literature features the anisotropic parametrizations described in the following.

Using one *vector* parameter per centroid, with one component β_{ji} per centroid j per component i of the space Ξ, we obtain the following free membership function:

$$v_{lj} = \exp\left(-\sum_{i=1}^{d}(x_{li} - y_{ji})^2 \beta_{ji}\right).\tag{14}$$

In this case, parameters β_{ji} form a $c \times d$ matrix

$$\boldsymbol{B} = \begin{pmatrix} \beta_{11} & \cdots & \beta_{1d} \\ & \vdots & \\ & \beta_{ji} & \\ & \vdots & \\ \beta_{c1} & \cdots & \beta_{cd} \end{pmatrix}\tag{15}$$

and, indicating with $\boldsymbol{b}_j$ the j-th column of $\boldsymbol{B}$, the argument of the exponential can be written in vector-matrix notation:

$$v_{lj} = \exp\left(-(\boldsymbol{x}_l - \boldsymbol{y}_j)^T \operatorname{diag}(\boldsymbol{b}_j)(\boldsymbol{x}_l - \boldsymbol{y}_j)\right)\tag{16}$$

where $\operatorname{diag}(\boldsymbol{v})$ denotes the diagonal matrix that has vector $\boldsymbol{v}$ as its diagonal.

This case is equivalent to a non-equilibrium, anisotropic system with axis-parallel principal directions of anisotropy.

The most general parametrization is obtained when the principal directions of anisotropy are not necessarily the coordinate axes. In this case there is a matrix of coefficients for each centroid, not necessarily diagonal, using a generalised (Mahalanobis) distance [20]:

$$v_{lj} = \exp\left(-\sum_{i=1}^{d}\sum_{k=1}^{d}(x_{li} - y_{ji})(x_{lk} - y_{jk})\boldsymbol{B}_{jik}\right)\tag{17}$$

or

$$v_{lj} = \exp\left(-(\boldsymbol{x}_l - \boldsymbol{y}_j)^T \boldsymbol{B}_j(\boldsymbol{x}_l - \boldsymbol{y}_j)\right)\tag{18}$$

This case implies that the model parameters are contained in a rank-3 tensor of shape (c, d, d). For each j, the corresponding $d \times d$ slice $\boldsymbol{B}_j$ is analogous to an inverse covariance matrix as used in the multidimensional form of the Gaussian density function and consequently in the expression of the Mahalanobis distance.

In addition to these model parameters, GPCM also has an additional parameter α that can be used to set the balance between a possibilistic and a probabilistic behaviour. In the first formulation [21], an interval-valued variable was used. In subsequent works, see for instance [3], a simpler formulation was adopted where $\alpha \in [0, 1] \subset \mathbb{R}$.

4.2 Roles of Parameters

According to the original statistical mechanics analogy, the parameter β in EM can be interpreted as an inverse temperature. From the point of view of information representation, it plays the role of a degree of fuzziness: When β increases (i.e., temperature decreases), the memberships of data observations to clusters become crisper. Finally, from a geometrical interpretation, β is a global resolution parameter that defines the minimum distance between centroids to be considered as distinct; below this distance, centroids collapse into each other.

The limit cases are:

– for $\beta \to 0^+$, we have $u_{lj} = 1/c$ for all l, j, i.e., each instance is equally associated with each cluster;

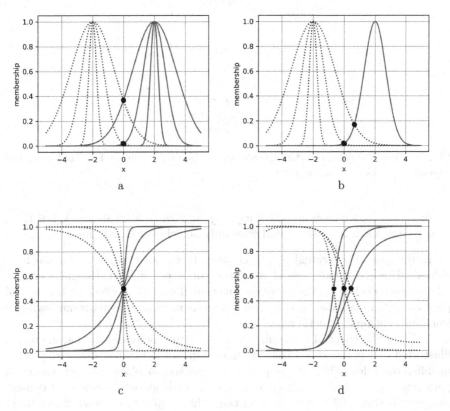

Fig. 2. Effect of varying β or b on the membership of point x to cluster 1 (dotted blue) and 2 (solid red). (Color figure online)

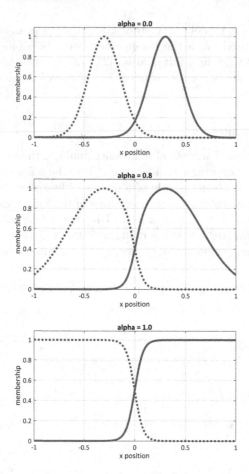

Fig. 3. Effect of varying α on the membership of point x to cluster 1 (dotted blue) and 2 (solid red). (Color figure online)

- for $\beta \to +\infty$, we have $u_{lj} = 1$ if $x_l \in \omega_j$, and $u_{lk} = 0$ for all $k \neq j$, $k \in [1, c]$, i.e., each instance is associated with only one cluster (hard limit).

In the case of individual β_j per cluster, the size of clusters is affected individually. However, in all cases that are not purely possibilistic, the memberships influence each other via the partition function. This has an effect on the critical position for an observation, the point where its maximum membership switches from one centroid to another.

In Fig. 2 the effect of changing the temperature or resolution parameters is illustrated in a two-centroid case. Membership to the two centroids are plotted in different styles. The critical points are marked in black for each choice of parameter values. On the left (graphs a, c) a single global parameter β is used, assigning it three different values; on the right (graphs b, d) individual parameters for each centroid are used, resulting in a vector $\boldsymbol{b} = [\beta_1, \beta_2]$, and only β_1

is changed, again using three values. The top graphs (a, b) are the possibilistic cases; the bottom graphs (c, d) are the probabilistic ones. The effect of having different resolution factors for different centroids on the critical point is clearly visible in graphs b and d.

The global model parameter $\alpha \in [0, 1]$ sets the nature of the clustering model, with $\alpha = 0$ corresponding to a fully possibilistic model (pure mode-seeking), $\alpha = 1$ to a probabilistic model, and intermediate values corresponding to a partly possibilistic behaviour where the generalized partition function does not normalize the sum of memberships to a fixed value of 1 but to a value that depends on the values of all free memberships. An illustration of the effect of varying α in a 2-cluster problem is presented in Fig. 3.

4.3   Factorisation of Parameters

As already noted, the single parameter β of ME is used both as a model parameter, acting on the structure of the final clustering, and as an optimisation parameter, influencing the convergence of the optimisation itself.

It may be useful to express the two concepts in an uncoupled way to allow both actions simultaneously. To this end, we rewrite the most general parametrization (rank-3 tensor) as

$$\beta_{jik} = b\overline{\beta}_{jik} \tag{19}$$

where $\overline{\beta}_{jik}$ expresses the relative magnitude of parameters with respect to each other and b is a global scale factor that can be used for annealing. Disregarding a change of units, all choices for this decomposition are equivalent; we can fix the ideas by setting $\max\{\overline{\beta}_{jik}\} = 1$ which results in $\max\{\beta_{jik}\} = b$, i.e., the global scale parameter is the magnitude of the largest β_{jik}.

In the following we discuss some possible criteria to estimate the model parameters just discussed.

5   Setting the Model Parameters

With respect to the optimization, model parameters can be set beforehand, at each iteration, or at the end. While setting the parameters before the beginning only works in the presence of a good initialisation, the criteria here presented can easily be applied during the iterations or after their end.

By necessity, all criteria ultimately depend on some user-selected parameters. The focus of the methods that are discussed in this section is to reduce the number of these parameters to a minimum and to provide an intuitive interpretation to make it possible for the user to assign meaningful values to these residual degrees of freedom.

In the following we only cover the case of vector scale parameter, $\beta = [\beta_1, \beta_2, \ldots, \beta_c]$. The scalar case is similar but obviously simpler, and the matrix and tensor cases are not as common.

5.1   Setting the Resolution Parameters Using Free Memberships v

Criteria for setting β can be obtained by analysing inter-centroid distance and imposing a bias toward fuzzy solutions, similarly to what was done in the possibilistic approach in [18,19]. The first proposed method uses free memberships. For each centroid $\boldsymbol{y}_j$ we measure the free membership to its cluster ω_j of all other centroids:

$$v(\boldsymbol{y}_h, \boldsymbol{y}_j) = \exp\left(-\|\boldsymbol{y}_h - \boldsymbol{y}_j\|^2 \beta_j\right) \tag{20}$$

Note that this measure is taken using $\boldsymbol{y}_j$ as a reference and is asymmetric, i.e., $v(\boldsymbol{y}_h, \boldsymbol{y}_j) \neq v(\boldsymbol{y}_j, \boldsymbol{y}_h)$.

We define the minimal-overlap condition by setting a threshold $t \in (0,1)$. Membership of centroid h to centroid j should not be larger than this threshold. Enforcing this for the nearest centroid guarantees that this is true also for all other centroids. To guarantee absolutely no overlap, the value should be $t = 1/2$. Other values can be used if some overlap is acceptable ($t > 1/2$) or if narrower boundaries are desired ($t < 1/2$).

The criterion is therefore:

$$\max_{h \neq j} v(\boldsymbol{y}_h, \boldsymbol{y}_j) \leq t$$

$$\Rightarrow \quad \max_{h \neq j} \exp\left(-\|\boldsymbol{y}_h - \boldsymbol{y}_j\|^2 \beta_j\right) \leq t$$

$$\Rightarrow \quad \min_{h \neq j} \|\boldsymbol{y}_h - \boldsymbol{y}_j\|^2 \beta_j \geq -\ln t \tag{21}$$

Let $h^* = \arg\min_{h \neq j} \|\boldsymbol{y}_h - \boldsymbol{y}_j\|$. Note that being the nearest neighbour is not a symmetric relation, so in general β_j and β_{h^*} will be different.

The above inequality yields the final criterion:

$$\Rightarrow \quad \beta_j = -\frac{\ln t}{\|\boldsymbol{y}_{h^*} - \boldsymbol{y}_j\|^2} \tag{22}$$

where the numerator can be used as a global degree of freedom, for instance for regularisation or annealing during the optimization (see Subsect. 4.3).

5.2   Setting the Resolution Parameters Using Memberships u

In this case the function to be used is the fuzzy probabilistic one:

$$u(\boldsymbol{y}_k, \boldsymbol{y}_j) = \frac{\exp\left(-\|\boldsymbol{y}_h - \boldsymbol{y}_j\|^2 \beta_j\right)}{\sum_{k=1, k \neq j}^{c} \exp\left(-\|\boldsymbol{y}_k - \boldsymbol{y}_j\|^2 \beta_j\right)} \tag{23}$$

In this case the minimal-overlap condition:

$$\max_{h \neq j} u(\boldsymbol{y}_h, \boldsymbol{y}_j) \leq t \tag{24}$$

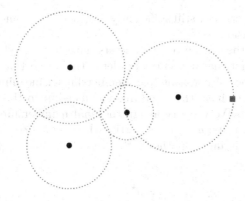

Fig. 4. Setting the value of parameter α by assigning a desired outlier membership. (Color figure online)

is much less simple to solve for β_j. However the value of the partition function (the denominator) can be estimated by using a very rough approximation. We fix an integer number c_{NN} between 1 and c. Among the centroids, we decide to take into account the nearest c_{NN}. The value of $v(\boldsymbol{y}_h, \boldsymbol{y}_j)$ for the neighbours is approximated as:

- For the c_{NN} nearest neighbours, $v(\boldsymbol{y}_h, \boldsymbol{y}_j) \approx 1$
- For the remaining $1 - c_{NN}$ (farthest) neighbours, $v(\boldsymbol{y}_h, \boldsymbol{y}_j) \approx 0$

So we can estimate $\sum_{k=1}^{c} v(\boldsymbol{y}_k, \boldsymbol{y}_j)$ to be approximately equal to the number c_{NN} of neighbours sufficiently close to j. The criterion thus obtained is:

$$\beta_j = -\frac{\ln{(c_{NN}t)}}{\|\boldsymbol{y}_{h^*} - \boldsymbol{y}_j\|^2} \tag{25}$$

where the numerator, a positive real number, can again be used as a global degree of freedom.

5.3 Setting the Possibility Degree α with an Outlier Rejection Criterion

In contrast to the resolution parameters, it is difficult to visualize the effect of α on cluster shape in geometric terms. This is a global parameter that influences the global configuration of clusters and interacts with the other model parameters.

A guideline for the selection of α is to set it in relation to the desired degree of outlier rejection. An outlier is an observation that has low membership to all clusters. We remark that outlier rejection is a crucial property to avoid meaningless generalisation due to extrapolation. However, complete outlier insensitivity makes the clustering model miss potentially meaningful observations. So our goal here is to set a desired worst-case membership u^* sufficiently small so as to clearly indicate outliers, but still sufficiently large to allow some effect of outliers in the centroid equations.

Supposing that the resolution parameters have been fixed, it is possible to calculate v_j for a point that lies on the border of the support of clusters. In Fig. 4 dotted circles are loci of constant free membership v, meaning that all points falling on dotted lines have the same free membership to the cluster to which the circle is referred. We want to assign the final membership u of the outlier (red square) a given value $u^* \leq v_j$ by setting the value of α.

Under the simplifying hypothesis that $v_h = 0 \ \forall h \neq j$, so that $Z = \sum_{h=1}^{c}(v_h)^\alpha = v_j^\alpha$:

$$\frac{v_j}{v_j^\alpha} = u^*$$

$$\Rightarrow \quad v_j^{1-\alpha} = u^*$$

$$\Rightarrow \quad \alpha = 1 - \log u^* / \log v_j \tag{26}$$

5.4 Setting the Possibility Degree α as an Independent Parameter

The value of α can also be assigned independently as a degree of freedom for regularisation or annealing. However, since it acts as an exponent, the effect of changes is much stronger when close to 1 than close to 0. Experimentally, it can be observed that the values between 0.9 and 1 are the most interesting, with values below 0.75 establishing an essentially pure possibilistic behaviour.

It is therefore advisable to set the value of α by means of an auxiliary variable that is related to it logarithmically. A suggested technique is to set $a \in [0, 1]$ so that

$$\alpha = (\log_2(a + 1))^{0.2} \tag{27}$$

where the exponent 0.2 is chosen such that, for $a = 0.5$, $\alpha \approx 0.9$. In this way the interesting range $(0.9, 1.0)$ is mapped onto half the range of variation of the control variable a. See Fig. 5 for a graph illustrating this effect.

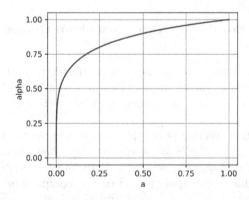

Fig. 5. Setting the value of parameter α via an auxiliary variable a.

6   Conclusion

In this paper we have reviewed a family of central soft clustering methods. Their relevance as feature learning methods for subsequent recognition, approximation, and forecasting tasks has been mentioned.

A key issue of these variations over HCM is the larger number of model parameters. Therefore, several criteria for setting these parameters have been discussed.

Current work on this topic involves the on-line adaptation of model parameters to non-stationary stream learning [1,3].

References

1. Abdullatif, A., Masulli, F., Rovetta, S., Cabri, A.: Graded possibilistic clustering of non-stationary data streams. In: Petrosino, A., Loia, V., Pedrycz, W. (eds.) WILF 2016. LNCS (LNAI), vol. 10147, pp. 139–150. Springer, Cham (2017). https://doi.org/10.1007/978-3-319-52962-2_12
2. Abdullatif, A., Rovetta, S., Masulli, F.: Layered ensemble model for short-term traffic flow forecasting with outlier detection. In: 2016 IEEE 2nd International Forum on Research and Technologies for Society and Industry Leveraging a better tomorrow (RTSI), pp. 1–6, September 2016. https://doi.org/10.1109/RTSI.2016.7740573
3. Abdullatif, A., Masulli, F., Rovetta, S., Cabri, A.: A fuzzy clustering approach to non-stationary data streams learning. In: Lintas, A., Rovetta, S., Verschure, P.F., Villa, A.E. (eds.) ICANN 2017, Part II, pp. 768–769. Springer, Cham (2017)
4. Anderson, D.T., Bezdek, J.C., Popescu, M., Keller, J.M.: Comparing fuzzy, probabilistic, and possibilistic partitions. IEEE Trans. Fuzzy Syst. 18(5), 906–918 (2010). https://doi.org/10.1109/TFUZZ.2010.2052258
5. Ball, G., Hall, D.: ISODATA, an iterative method of multivariate analysis and pattern classification. Behav. Sci. 12, 153–155 (1967)
6. Bezdek, J.C.: Pattern Recognition with Fuzzy Objective Function Algorithms. Kluwer Academic Publishers, Norwell (1981)
7. Chow, C.K.: An optimum character recognition system using decision function. IRE Trans. Electron. Comput. 6, 247–254 (1957)
8. Chow, C.: An optimum recognition error and reject tradeoff. IEEE Trans. Inf. Theory 16, 41–46 (1970)
9. Corsini, P., Lazzerini, B., Marcelloni, F.: A new fuzzy relational clustering algorithm based on the fuzzy C-means algorithm. Soft Comput. 9(6), 439–447 (2005). https://doi.org/10.1007/s00500-004-0359-6
10. Drago, G.P., Ridella, S.: Possibility and necessity pattern classification using an interval arithmetic perceptron. Neural Comput. Appl. 8(1), 40–52 (1999)
11. Duda, R.O., Hart, P.E.: Pattern Classification and Scene Analysis. Wiley, New York (1973)
12. Dunn, J.C.: A fuzzy relative of the ISODATA process and its use in detecting compact well-separated clusters. J. Cybern. 3, 32–57 (1974)
13. Filippone, M., Camastra, F., Masulli, F., Rovetta, S.: A survey of kernel and spectral methods for clustering. Pattern Recogn. 40(1), 176–190 (2008)
14. Flores-Sintas, A., Cadenas, J.M., Martin, F.: Local geometrical properties application to fuzzy clustering. Fuzzy Sets Syst. 100, 245–256 (1998)

15. Goodfellow, I., Bengio, Y., Courville, A.: Deep Learning. MIT Press, Cambridge (2016)
16. Huang, L., Joseph, A.D., Nelson, B., Rubinstein, B.I., Tygar, J.D.: Adversarial machine learning. In: Proceedings of the 4th ACM Workshop on Security and Artificial Intelligence, AISec 2011, pp. 43–58. ACM, New York (2011). https://doi.org/10.1145/2046684.2046692
17. Ishibuchi, H., Nii, M.: Neural networks for soft decision making. Fuzzy Sets Syst. **115**(1), 121–140 (2000)
18. Krishnapuram, R., Keller, J.M.: A possibilistic approach to clustering. IEEE Trans. Fuzzy Syst. **1**(2), 98–110 (1993)
19. Krishnapuram, R., Keller, J.M.: The possibilistic C-means algorithm: insights and recommendations. IEEE Trans. Fuzzy Syst. **4**(3), 385–393 (1996)
20. Mahalanobis, C.P., et al.: On the generalised distance in statistics. Proc. National Inst. Sci. India **2**(1), 49–55 (1936)
21. Masulli, F., Rovetta, S.: Soft transition from probabilistic to possibilistic fuzzy clustering. IEEE Trans. Fuzzy Syst. **14**(4), 516–527 (2006). https://doi.org/10.1109/TFUZZ.2006.876740
22. Miyamoto, S., Mukaidono, M.: Fuzzy C-means as a regularization and maximum entropy approach. In: Proceedings of the Seventh IFSA World Congress, Prague, pp. 86–91 (1997)
23. Ridella, S., Rovetta, S., Zunino, R.: K-winner machines for pattern classification. IEEE Trans. Neural Netw. **12**(2), 371–385 (2001)
24. Rose, K., Gurewitz, E., Fox, G.: A deterministic annealing approach to clustering. Pattern Recogn. Lett. **11**, 589–594 (1990)
25. Rose, K., Gurewitz, E., Fox, G.: Statistical mechanics and phase transitions in clustering. Phys. Rev. Lett. **65**, 945–948 (1990)
26. Rovetta, S., Masulli, F.: Online spectral clustering and the neural mechanisms of concept formation. In: Bassis, S., Esposito, A., Morabito, F.C. (eds.) Advances in Neural Networks: Computational and Theoretical Issues. SIST, vol. 37, pp. 61–72. Springer, Cham (2015). https://doi.org/10.1007/978-3-319-18164-6_7
27. Rovetta, S., Masulli, F., Cabri, A.: Measuring clustering model complexity. In: Lintas, A., Rovetta, S., Verschure, P.F.M.J., Villa, A.E.P. (eds.) ICANN 2017. LNCS, vol. 10614, pp. 434–441. Springer, Cham (2017). https://doi.org/10.1007/978-3-319-68612-7_49
28. Runkler, T.A., Bezdek, J.C.: Alternating cluster estimation: a new tool for clustering and function approximation. IEEE Trans. Fuzzy Syst. **7**(4), 377–393 (1999)
29. Ruspini, E.H.: A new approach to clustering. Inf. Control **15**(1), 22–32 (1969)

Recent Applications of Fuzzy Logic

A Cloud Fuzzy Logic Framework for Oral Disease Risk Assessment

Gloria Gonella[1]([✉]), Elisabetta Binaghi[1], Alberto Vergani[1],
Irene Biotti[2], and Luca Levrini[2]

[1] Department of Theoretical and Applied Science,
University of Insubria Varese, Varese, Italy
ggonella@uninsubria.it
[2] Department of Medicine and Surgery,
University of Insubria Varese, Varese, Italy
luca.levrini@uninsubria.it

Abstract. This paper presents a fuzzy logic framework for dental caries and erosion risk assessment. Two interdependent modules are implemented within a cloud architecture. The first module is a fuzzy expert system designed for physicians and expert users, able to provide an active support in formulating risk judgements. The second module is oriented to generic users for oral health promotion. Conceptual ingredients of the fuzzy logic framework are principally defined by eliciting knowledge from a group of experts. The generation of rules involves both structured interviews and data driven learning procedures based on the use of neuro-fuzzy techniques.

Keywords: Fuzzy expert system · Medical knowledge acquisition · Neuro-fuzzy system · Cloud computing

1 Introduction

Dental disorders such as dental caries and erosion have become recognised as a major cause of public health concern, profoundly affecting quality of life worldwide [1, 2]. These diseases have begun to receive more attention and several decision support systems have been proposed during the last three decades to help dentists for investigation, diagnosis and treatment of them [3–5]. Despite the sizable achievement obtained, the information in the field is not yet well assessed and is mostly fragmented in several sources that deal separately with the diverse conditions affecting the risk of both dental caries and erosion.

Proceeding from these considerations we developed an interdisciplinary study conducted with the aim of implementing a complete framework for the automated assessment of dental caries and erosion risks.

Even if these diseases have begun to receive more attention, the information in the field is not yet well assessed and is mostly fragmented in several sources that deal separately with the diverse conditions affecting the risk of both dental caries and erosion [6, 7].

R. Fullér et al. (Eds.): WILF 2018, LNAI 11291, pp. 85–96, 2019.
https://doi.org/10.1007/978-3-030-12544-8_7

Proceeding from these considerations we developed an interdisciplinary study conducted with the aim of implementing a framework for the automated assessment of dental caries and erosion risks. The objective of the overall framework is to facilitates the following main issues [8–10]:

- *active support and transparency*: rules, modelling multifactorial evaluations, are coded by the system in an explicit form and can be retrieved to explain the inferred risk judgements; the system can be used in educational context or at the same time in clinical activity;
- *information sharing and formalisation*: risk evaluation procedures are formalized allowing a reference point against which to collect agreement and disagreement among physicians and creating the premise for a systematic coherent framework;
- *public awareness enhancement and oral health promotion*: a wide range of users interacts friendly with the system that make accessible and understandable the multifactorial conditions that influence dental caries and erosion.

The conceptual model underlying the design of the overall framework is drawn from Fuzzy Logic (FL) [11, 12]. FL techniques have had an enormous impact on computer-assisted medical diagnosis in which uncertainty plays a key role and we experienced the advantages in several previous works [10, 13]. Uncertainty originates within the process of assigning a risk value based on complex and vague signs and factors. Under these critical conditions, risk assessment has to be properly modelled as a matter of degrees in order to completely represent the expert decisional attitudes and preserve the natural "qualitative" way of their reasoning originating from complexity and interdependency of factors not always completely exploited [14].

The remainder of this paper is organized as follows. Section 2 illustrates the overall architecture of the proposed framework. Section 3 describes the FL expert system devoted to the assessment of the risk of caries and erosion. The knowledge acquisition strategies based on a combined use of structured interviews and neural learning procedures are detailed. Section 4 illustrates the user oriented module for oral health promotion and the relationships with the expert system. Section 5 illustrates program runs and interfaces of the implemented s/w packages. Finally, Sect. 6 concludes the paper.

2 The Architecture of the FL Framework

The proposed framework is implemented within a cloud architecture (see Fig. 1) to ensure portability, scalability and to efficiently manage session data for further analysis.

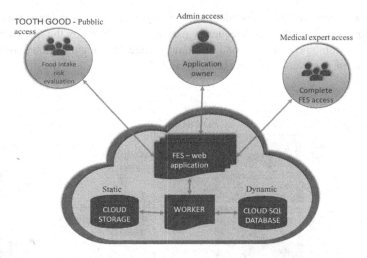

Fig. 1. Architecture of the FL risk assessment framework

The overall framework includes a fuzzy expert system, named FES, designed for physicians and expert users, able to provide an active support in formulating risk judgements basing on a coherent, complete set of factors. A second module, named Tooth-Good, is oriented to generic users that provide in input information about beverages, food habits and oral hygiene and receive judgements about the level of risk. These judgements are produced by mapping the user provided information into the factors of risk modelled within the expert system and activating corresponding rules. The core of the implemented application uses web roles, worker roles, and storage. Three types of users may have access to the application: application owner, public and medical expert users. The application uses SQL Database allowing expert users to dump results of their sessions into a relational database and then to analyse the results in detail. It addresses common multi-tenant challenges such as partitioning, extensibility, provisioning, testability, and customization. An authentication mechanism is provided by the application with a subscriber's own security infrastructure by using a federated identity with multiple partner model [15].

3   Expert System for Oral Disease Risk Assessment

The evaluation of the risk of caries and erosion involves a large number of input factors and complex interactions between them. To manage this complexity and preserve the natural way in which experts formulate their judgements, we designed the expert system with a two-stage hierarchical structure. In the first stage five subsystems compute partial risks for both caries and erosion, basing on different groups of factors.

They are conceived as independent systems; the risk assessment framework can be applied for them separately. Subsequently partial risks inferred by the five expert sub-systems are combined in a second reasoning stage to compute global risks. The architecture of the overall FES is shown in Fig. 2.

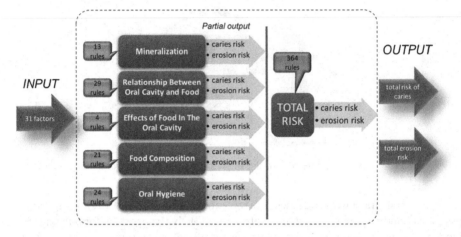

Fig. 2. Structure of the FL expert system

The design process involved the definition of the FL ingredients of the above mentioned sub-systems. Three conceptual steps are involved [11, 12]:

1. Linguistic labelling of risk factors and risk classes concerned
2. Diagnostic rules definition
3. Inference formalisation to deduce risk judgements.

Conceptual ingredients involved in 1. and 2. are principally defined by eliciting knowledge from a group of experts. The generation of rules involves both structured interviews and data driven learning procedures based on the use of neuro-fuzzy techniques.

3.1   Linguistic Labelling of Risk Factors and Risk Classes

The risk assessment is based on the combined evaluation of 31 factors modelled within the FL framework as linguistic variables. Table 1 lists the set of factors of risk considered, subdivided in groups.

Table 1. Factors of risk considered in caries and erosion risk assessment subdivided in groups

Mineralization	Relation oral cavity food	Effects of food	Food composition	Oral hygiene
pH	Consistency	Buffer	Vitamin A	Toothbrush
Calcium	Adhesiveness	capacity	Vitamin B1	Brushing
Phosphate	Intake Frequency	Bacterio-	Vitamin B2	Duration
	Intake Mode Meal	static	Vitamin B9	Brushing
	Frequency Mode		Vitamin B12	Frequency
	Meal Period		Vitamin C	Use of interdent.
			Vitamin D	Aids
			Vitamin E	Fluorine
			Probiotics	
			Magnesium	
			Sugar and	
			sweeten	
			Arginine	
			Anthocyanins	
			Polyphenols	
			Omega 3	

The linguistic description of each factor is given in terms of fuzzy declarative proposition of the form X *is* A [11]. X is the linguistic variable denoting a given factor. A is a term belonging to a given term set and represents a fuzzy set in the corresponding universe of discourse U with membership function $\mu_A : U \rightarrow [0,1]$.

In our context, the experts constitute the principal source from which knowledge may be acquired. Membership functions of fuzzy sets associated with terms in declarative propositions are defined by adopting two elicitation techniques [14, 16, 17]. Initially we assume that values of membership of fuzzy sets corresponding to a given term, such as *low, medium and high*, should be related to the difficulty in attributing this term to numerical values of the universe of discourse. Scaled responses are acquired by a group of experts expressing the degree of certainty with which a given term may be associated with numerical values in the corresponding universe of discourse. In more detail, crisp values (C) in the universe of discourse of a given term (T) are sampled and corresponding grades are generated asking expert if C values are compatible with T. Experts give a YES/NO answer together with a score (S) in the interval (0, 5) expressing the level of certainty in providing the Boolean answer. These answers are processed to compute grades (G) in the initial rough distribution as follows:

```
if (answer is YES)
  G= 0.5 + (S) / 10
Else
  G = 0.5-(S) / 10
```

A rough distribution is obtained from the collected data by averaging G values. The second technique is based on the assumption that experts can refer the distribution of possibility associated with a given term to a standard piecewise function. We adopt *bell functions* having the following form [11, 12]:

$$f(x; a, b, c) = \frac{1}{1 + \left|\frac{x-c}{a}\right|^{2b}} \quad (1)$$

Predefined bell functions are visualized to the experts that provide directly parameters for their specification. The resulting bell functions are compared with the rough distributions and, if case, refined to obtain a best fitting. To exemplify the elicitation methods, Fig. 3 shows the rough distributions obtained for the terms *Low,*

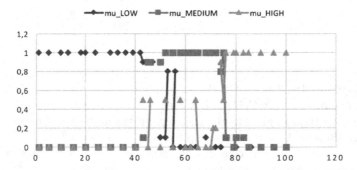

Fig. 3. Rough distributions obtained for the terms Low, Medium and High of the factor Adhesiveness

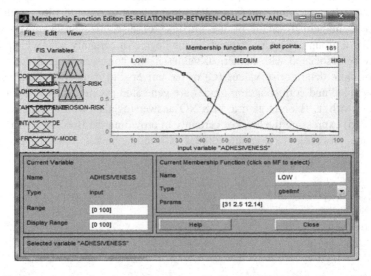

Fig. 4. Final membership functions of terms *Low, Medium* and *High* of the factor *Adhesiveness*

Medium and High of the factor *Adhesiveness* belonging to the group *Relationship Between Oral Cavity and Food* and Fig. 4 shows the corresponding membership functions obtained by using the Matlab Function Editor (MathWorks®) and by filling the derived parameters in predefined bell functions.

Both risk of erosion and risk of caries are modelled as linguistic variables having terms *Low, Medium and High* and modelled as fuzzy sets with bell membership functions.

3.2 Rules Generation and Inference

For each group of factors a subset of rules is generated expressing the relationship between multifactorial evaluations and class of risks. The outcomes of each subsets of rules is provided in input to a final reasoning stage to compute final risk judgements (see Fig. 2). The hierarchical structure of the rules, aggregating multisource information of each group of risk factors separately at the first level, and fusing partial results in a final risk judgments at the second level, reduces the complexity in the knowledge acquisition phase allowing experts to focus on limited chunks of knowledge and derived risk assessment rules. In the proposed framework risk assessment consists in the deduction of a conclusion regarding a specific case. When specific measurements values are set for each factor, the fuzzy logic inference mechanism interprets the set of fuzzy production rules and deduce the risk judgements. The well-known inference mechanism provided by Mamdami is adopted in our framework [18].

An initial set of rules has been generated basing on a direct interview with experts. These rules have been subsequently refined basing on the results obtained by a machine learning strategy. Results inferred by each subset of rules directly elicited from experts have been compared with results generated by a neuro-fuzzy learning procedure based on ANFIS model configured to support Sugeno type 0 order [19]. Five networks corresponding to the five risk groups and one network for final risk computation, have been trained with supervised sets of examples provided by experts for each group of risk factors and for the second level of risk assessment respectively. Examples are supervised case-specific crisp patterns. We collected from the experts a total of 369 supervised examples. A cross-validation procedure has been developed subdividing supervised examples in the proportion of 2/3 and 1/3 for training and test respectively. The accuracy obtained by the neuro fuzzy strategy was equal to 82%.

Attention was focused on test cases that generate disparity between the results obtained by using the initial set of rules and the neuro-fuzzy system. Experts were invited to re-examine and modify the formulation of rules involved in the light of the comparison of results. The refined rules have been validated using a new supervised set of 184 examples. The accuracy obtained was comparable with the accuracy of the neuro-fuzzy procedure and equal to 80%. Figure 5 shows a subset of rules using the linguistic variable "Adhesiveness", generated for the group "Relationship Between Oral Cavity and Food", and rules with multifactorial evaluations, generated for the "Mineralisation" group.

If (ADHESIVENESS is HIGH) then (DENTAL-CARIES-RISK is HIGH)(DENTAL-EROSION-RISK is HIGH) (1)
If (ADHESIVENESS is MEDIUM) then (DENTAL-CARIES-RISK is MEDIUM)(DENTAL-EROSION-RISK is MEDIUM) (1)
If (ADHESIVENESS is LOW) then (DENTAL-CARIES-RISK is LOW)(DENTAL-EROSION-RISK is LOW) (1)
If (pH is VERY-LOW) then (DENTAL-CARIES-RISK is HIGH) (1)
If (pH is LOW) then (DENTAL-EROSION-RISK is HIGH) (1)
If (pH is VERY-MEDIUM) then (DENTAL-CARIES-RISK is MEDIUM)(DENTAL-EROSION-RISK is MEDIUM) (1)
If (pH is HIGH) then (DENTAL-CARIES-RISK is LOW)(DENTAL-EROSION-RISK is LOW) (1)
If (pH is LOW) and (CALCIUM is SI) then (DENTAL-EROSION-RISK is MEDIUM) (1)
If (pH is MEDIUM) and (CALCIUM is SI) then (DENTAL-EROSION-RISK is LOW) (1)
If (pH is LOW) and (PHOSPHATE is SI) then (DENTAL-EROSION-RISK is HIGH) (1)

Fig. 5. Subset of rules using the linguistic variable "Adhesiveness", generated for the group "Relationship Between Oral Cavity and Food", and rules with multifactorial evaluations, generated for the "Mineralisation" group

4   User-Oriented Risk Assessment for Oral Health Promotion

Users interact with Tooth-Good module to know the level of risk associated with the consumption of drinks. They select a drink from a list including drinks with and without sugar, water and alcoholic drinks, and choose their dietetic profile.

These elements are translated into a set of crisp values associated with factors modelled within the expert system, as in the example illustrated below:

1. *Selected Drink*: Wine
   Factors:
   pH:3, Sugar and sweeteners:2, Bacteriostatic:0, Buffer capacity:0, Intake Frequency:4;

2. *Dietetic profile*:
   Fruit (also juices) and Vegetables: 0/1 daily dose
   Dairy Products (milk and yogurt): 0 daily dose
   Cheeses: 0 weekly dose
   Meat, Fish, Eggs: 3 weekly doses
   Carbohydrates and Cereals: 5 daily doses
   Cold Cuts, Candy, Root Vegetables: 6 weekly doses
   Factors:
   a:0, b2:1, b9:0, b12:0, c:0, d:0, e:0, Probiotics:0, Arginine: 0, Anthocyanins: 0, Polyphenols:0, omega3:0, b1:0, Magnesium: 0, Calcium:0, Phosphate:1, Consistency:3, Adhesiveness:90, Intake Mode:1, Meal Frequency Mode:1, Meal Period:0;

3. *Hygiene*: No
   Factors:
   Toothbrush:0, Brushing Duration:0, Brushing Frequency:0, Use of interdental aids:0, Fluorine:1.

The expert system receives the crisp input values and fires the activated rules. The computed risk judgements are presented finally to the users. For the example illustrated the result is *High Risk* for both caries and erosion diseases.

5 Graphical User Interfaces (GUI's) and Sample Runs of the Implemented Framework

The above illustrated FL framework was implemented within a cloud architecture as illustrated in Sect. 2. The software design started with the collection and analysis of requirements in which the expert-user and generic-user models and operation conditions are outlined. The FES procedures are implemented in MATLAB (MathWorks®). Figure 6 shows the interface of FES documenting the initial phase of the session. Commands to run separately sub-expert systems corresponding to the five group of risks and the final set of rules are included.

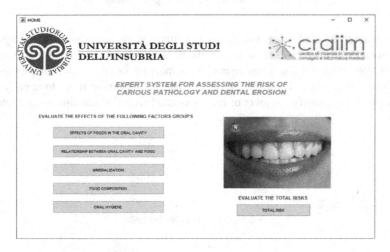

Fig. 6. Interface of FES documenting the initial phase of the session

Options made available by FES for specifying input values related to factors of Oral Hygiene group are illustrated in Fig. 7.

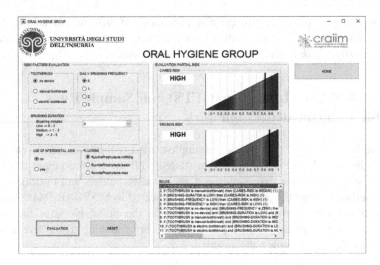

Fig. 7. FES interface for oral hygiene subsystem.

The inferred level of risk is visualized; the interface specifies the deduced linguistic term associated to caries and erosion variable, the corresponding grade of matching and the fired rules. Figure 8 illustrates inputs and outputs of the subsystem computing final risks. Figure 9 shows a Tooth-Good system interface allowing users to specify their dietetic profile and verify the level of risk associated with the assumption of a selected drink.

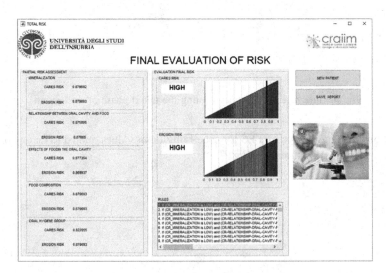

Fig. 8. FES interface for final risk assessment

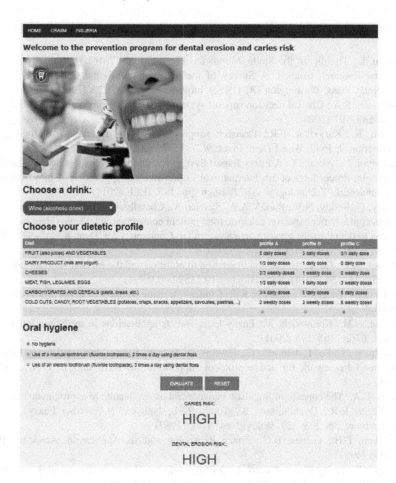

Fig. 9. Tooth-Good system interface

6   Conclusions

The aim of the present work was to design a Fuzzy Logic framework for dental caries and erosion risk assessment. The solutions adopted are implemented within a cloud architecture and are oriented both to medical expert and generic users in an attempt to satisfy both standardisation and prevention requirements.

Fuzzy reasoning and Neuro-fuzzy tools confirm their validity in modelling medical knowledge and decisional attitudes. Cloud architecture exploits the universal connectivity and scalability. Future plans include an extension of the proposed solutions implemented in the Tooth-Good application to allow more flexibility in defining user profiles and the design of an automated refinement strategy aimed to improve usability and accuracy on the base of data collected by the user sessions.

References

1. Levrini, L.: The diet of the Smile. Mondadori Electa, Milan (2016). (in Italian)
2. National Research Council: A Survey of the Literature of Dental Caries. The National Academies Press, Washington DC (1952). https://doi.org/10.17226/21295
3. Mendonça, E.A.: Clinical decision support systems: perspectives in dentistry. J. Dent. Educ. **68**(6), 589–597 (2004)
4. Vikram, K., Karjodkar, F.R.: Decision support systems in dental decision making: an introduction. J. Evid. Based Dent. Pract. **9**(2), 73–76 (2009)
5. Allahverdi, N., Akcan, T.: A Fuzzy Expert System design for diagnosis of periodontal dental disease. In: Proceedings of 5th International Conference on Application of Information and Communication Technologies, AICT, Baku, pp. 1–5. IEEE (2011)
6. Pandey, P., Reddy, N.V., Rao, V.A.P., Saxena, A., Chaudhary, C.P.: Estimation of salivary flow rate, pH, buffer capacity, calcium, total protein content and total antioxidant capacity in relation to dental caries severity, age and gender. Contemp. Clin. Dent. **6**, S65–S71 (2015). https://doi.org/10.4103/0976-237X.152943
7. Barbour, M.E., Lussi, A.: Erosion in relation to nutrition and the environment. Monogr. Oral Sci. **25**, 143–154 (2014). https://doi.org/10.1159/000359941
8. Berner, E.S., Ball, M.J., Hannah, K.J.: Clinical Decision Support Systems. Springer, Berlin (1998)
9. Phoung, N.H., Kreinovich, V.: Fuzzy logic and its applications in medicine. Int. J. Med. Inform. **62**(2), 165–173 (2001)
10. Binaghi, E., Gallo, I., Ghiselli, C., Levrini, L., Biondi, K.: An integrated fuzzy logic and web-based framework for active protocol support. Int. J. Med. Inform. **77**(4), 256–271 (2008)
11. Zadeh, L.A.: The concept of linguistic variable and its application to approximate reasoning. In: Yager, R.R., Ovchinnikov, S., Tong, R.M., Nguyen, H.T. (eds.) Fuzzy Sets and Applications, pp. 293–329. Wiley, New York (1987)
12. Mamdani, E.H., Gaines, B.G.: Fuzzy Reasoning and Its Application. Academic Press, London (1981)
13. Binaghi, E.: Fuzzy logic inference model for a rule-based system in medical diagnosis. Int. J. Expert Syst. **7**, 134–141 (1990)
14. Boegl, K., Adlassnig, K.P., Hayashi, Y., Rothenfluh, T.E., Leitich, H.: Knowledge acquisition in the fuzzy knowledge representation framework of a medical consultation system. Artif. Intell. Med. **30**(1), 1–26 (2004)
15. Homer, A., Betts, D., Jezierski, A., Narumoto, M., Zhang, H.: Developing Multi-tenant Applications for the Cloud, 3rd edn. Microsoft (2012). https://www.microsoft.com/en-us/download/details.aspx?id=29263
16. Chameau, J., Santamarina, J.C.: Membership functions I: comparing method of measurement. Int. J. Approximate Reasoning **1**, 287–301 (1987)
17. Turksen, I.B.: Measurement of membership functions and their acquisition. Fuzzy Sets Syst. **40**(1), 5–38 (1991). https://doi.org/10.1016/0165-0114(91)90045-R
18. Mamdani, E.H., Assilian, S.: Advances in the linguistic synthesis of fuzzy controllers. In: Mamdani, H., Gaines, B.R. (eds.) Fuzzy Reasoning and Its Applications, pp. 311–323. Academic Press, London (1981)
19. Jang, J.R.: ANFIS: adaptive-network-based fuzzy inference system. IEEE Trans. Syst. Man Cybern. **23**, 665–685 (1993)

A Fuzzy Rule-Based Decision Support System for Cardiovascular Risk Assessment

Gabriella Casalino[1,2], Giovanna Castellano[1,2(✉)], Ciro Castiello[1,2], Vincenzo Pasquadibisceglie[1], and Gianluca Zaza[1]

[1] Computer Science Department, Università degli Studi di Bari Aldo Moro, Bari, Italy
{gabriella.casalino,giovanna.castellano,ciro.castiello, vincenzo.pasquadibisceglie,gianluca.zaza}@uniba.it
[2] INdAM Research group GNCS, Rome, Italy

Abstract. In medical problems both the information and the reasoning used by clinicians for drawing conclusions about patients' health are inherently uncertain and vague. Fuzzy logic is a powerful tool for representing and handling this uncertainty, leading to fuzzy systems that can support decisions in medical diagnosis. In this work we propose a fuzzy rule-based system to support the expert in decision making for cardiovascular diseases that are of particular interest due to their obvious medical diagnostic importance. Preliminary experimental results on both healthy and ill people show the effectiveness of the fuzzy system in simulating the decision of the expert.

Keywords: Intelligent Data Analisys (IDA) ·
Decision Support System (DSS) · Fuzzy logic · Cardiovascular disease

1 Introduction

Medical Informatics is a recent multidisciplinary field dealing with the use of the information technology for the healthcare industry.

The amount of patient health data is increasing exponentially. The volume of healthcare data in 2013 has been estimated at 153 Exabytes and it will reach 2314 Exabytes by 2020[1]. Traditional manual data analysis techniques have became unsuitable to extract useful information from this big amount of data, thus automatic mechanisms are necessary [1,2]. However, expert knowledge cannot be completely replaced by machines. Intelligent data analysis (IDA) aims at combining human expertise and computational models for advanced data analysis [3–5], in order to narrow the gap between data gathering and their comprehension [6]. In the medical field, more than in others, this interaction is mandatory:

---

[1] https://www.cio.com/article/2860072/healthcare/how-cios-can-prepare-for-healthcare-data-tsunami.html.

© Springer Nature Switzerland AG 2019
R. Fullér et al. (Eds.): WILF 2018, LNAI 11291, pp. 97–108, 2019.
https://doi.org/10.1007/978-3-030-12544-8_8

on the one hand the experts need automatic tools to transform raw and complex data into easily interpretable information, on the other hand algorithm outputs alone are not sufficient for medical diagnosis, since expert knowledge is needed to understand them. Several IDA methods have been applied for supporting decision making in medicine [6–9].

The representation of medical knowledge and the decision making in the presence of uncertainty and imprecision are of fundamental importance to derive a suitable model for medical decision making. Indeed, in medical problems, both patient information and the reasoning used by clinicians for drawing conclusions about patients' health, are inherently uncertain and vague [10]. Among the different IDA methods, fuzzy logic is the most suitable mean for representing and handling this uncertainty. In particular, fuzzy logic proved to be a powerful tool for decision support systems (DSSs), such as medical rule-based systems [11]. Several medical Decision Support Systems (DSSs) have been developed using fuzzy rule-based systems [10–20]. These fuzzy systems use linguistic terms to represent the patients' symptoms, and a fuzzy inference mechanism to derive a suggestion. The domain knowledge is embedded into the knowledge base in form of fuzzy rules.

In this paper we propose a fuzzy rule-based system to support the medical expert in decision making for cardiovascular risk assessment. Starting from the patients' vital signs such as heart rate (HR), breath rate (BR), peripheral oxygen saturation (SpO2) and lips color, we designed a fuzzy rule-based system that can suggest a level of cardiovascular risk. The fuzzy rules are defined according to the expert knowledge with the help of the FISDeT tool [21].

The rest of the paper is organized as follows. In Sect. 2 the vital signs related to cardiovascular diseases are introduced. The fuzzy rule-based decision support system is described in Sect. 3. Section 4 reports preliminary results of experiments aimed to prove the accuracy of the fuzzy system in simulating the expert reasoning. In Sect. 5 we draw conclusions and outline future works.

2   Vital Signs of Cardiovascular Disease

Heart rate (HR), breath rate (BR), and peripheral oxygen saturation (SpO2) are parameters typically considered by physicians to formulate a diagnosis of cardiovascular disease. All of them are descriptive enough of the human health condition providing also the additional benefit of being easily detectable.

HR is defined as the speed of the heartbeat, i.e., the number of heart contractions per minute (BPM). Such a value is varying according to a number of conditions affecting the human organism, ranging from the physical exercise to the stress, the illness, and the drug consumption. Even age, sex, and physical fitness provoke change in the HR values. However, the average HR of a resting male adult falls in the range of 60 to 90 BPM.

BR is defined as the speed of the breath sequence, i.e., the number of breaths occurring per minute. The common factors influencing the BR evaluation are age and physical exercise. However, the average BR of a resting male adult falls in the range of 12 to 18 breaths per minute. A modified value of BR (which can be a reduced rate, bradypnea, or an augmented rate, tachypnea) is commonly associated to various illness conditions.

SpO2 is evaluated as the percentage of oxygen-saturated hemoglobin with respect to the total hemoglobin (unsaturated and saturated) present in the blood. SpO2 values are considered normal when falling in the range of 95 to 100%. Values below 90% indicate pathological conditions (hypoxemia), inducing organ impairment when falling below 80%.

Different methods can be adopted to measure the vital signs previously described. Among them, photoplethysmography (PPG) is commonly employed in several medical settings and is implemented in simple devices that are commercially available at the present days. By means of photoplethysmograph techniques it is possible to perform optical measurements to detect volumetric change of organs and to assess skin perfusion [22]. PPG is easy to use, noninvasive and is founded on the idea that plethysmoograph signals, acquired through the enlightenment of the skin, provide information concerning changes in blood flow, thus contributing to design a picture of the cardiovascular state [23]. Some PPG systems are applied directly on specific anatomical parts (which can be fingers, forearms, etc.). Some other systems are contactless, thus constituting a kind of remote-PPG (rPGG) systems which typically rely on facial examination. The simple employment of computer webcams proved to be effective in detecting the vital signs of interest for subsequent analysis [24–28].

The human face provides also several clues about the health condition. Some kinds of pathologies can be identified through the analysis of some face features. In particular, a specific element useful to assess human wellness is the color of lips. Normal people show a pinkish nuance in their lips, while altered states or illness may provoke a modification of this color. Pale lips are a symptom of different problems, ranging from vitamin deficiency to anemia. Lips appearing purplish or bluish can refer to cardiovascular or respiratory disorders which may require a punctual medical consulting. Automatic analysis of the lips color can be suitably performed by means of image processing techniques applied to a specific ROI (region of interest) extracted from the image of the patient's face.

In the following section we discuss how the described vital signs have been involved in the design of a fuzzy inference system capable to provide a risk level of cardiovascular disease.

3   The Fuzzy Rule-Based Decision Support System

The aim of this work is to set up a fuzzy rule-based system which can support the diagnosis of cardiovascular diseases by assessing a risk level for each patient according to her measured vital signs.

To design the rule base of the fuzzy inference system (FIS) we exploited FISDeT (Fuzzy Inference System Development Tool) [21], a software conceived to facilitate the creation and the management of fuzzy rule-based systems. Keypoints of FISDeT are the adoption of the FCL standard for the description of a

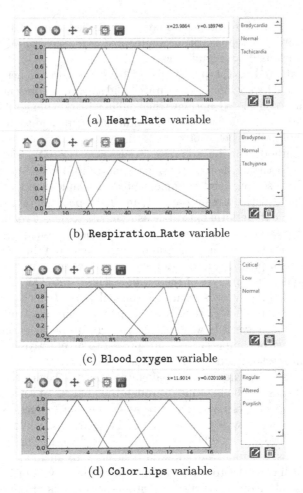

(a) `Heart_Rate` variable

(b) `Respiration_Rate` variable

(c) `Blood_oxygen` variable

(d) `Color_lips` variable

Fig. 1. Fuzzy sets partitioning the domain of the linguistic variables related to the vital signs.

FIS, the freely availability through the open-source development methodology, and a general-purpose approach which allows both the creation of a knowledge base and the inference of results from the analysis of input data. Developed in Python, FISDeT is endowed with a GUI supporting the user through all the steps required to define a FIS. FISDeT has been successfully applied to create FIS for classification problems [29].

The input-output configuration we considered to design the FIS draws a relationship between the four vital signs (HR, BR, SpO2, lips color) and a risk level referred to cardiovascular diseases. The parameters involved in the FIS design have been investigated with the support coming from a physician. Specifically, the fuzzy variables and their fuzzy sets have been arranged as follows.

HR This parameter is associated with the linguistic input variable Heart_rate, whose domain is the numerical range [10–180]. Such a linguistic variable may assume the values corresponding to three linguistic terms: *Bradycardia, Normal,* and *Tachycardia*. Triangular fuzzy sets are associated to the linguistic terms, partitioning the domain of the Heart_rate variable as follows (triangle vertices are reported in parenthesis as coordinates):

    –*Bradycardia*: (30, 0) (35, 1) (52, 0);
    –*Normal*: (48, 0) (75, 1) (100, 0);
    –*Tachycardia*: (95, 0) (110, 1) (180, 0).

Figure 1(a) shows the FISDeT GUI illustrating the fuzzy sets involved in the definition of the Heart_rate variable.

BR This parameter is associated to the linguistic input variable Respiration_rate, whose domain is the numerical range [0–80]. Such a linguistic variable may assume the values corresponding to three linguistic terms: *Bradypnea, Normal,* and *Tachypnea*. Triangular fuzzy sets are associated to the linguistic terms, partitioning the domain of the Respiration_rate variable as follows:

    –*Bradypnea*: (0, 0) (6, 1) (8, 0);
    –*Normal*: (7, 0) (15, 1) (23, 0);
    –*Tachypnea*: (20, 0) (35, 1) (80, 0).

Figure 1(b) shows the FISDeT GUI illustrating the fuzzy sets involved in the definition of the Respiration_rate variable.

SpO2 This parameter is associated to the linguistic input variable Blood_oxygen, whose domain is the numerical range [75–100]. Such a linguistic variable may assume the values corresponding to three linguistic terms: *Critical, Low,* and *Normal*. Triangular fuzzy sets are associated to the linguistic terms, partitioning the domain of the Blood_oxygen variable as follows:

    –*Critical*: (75, 0) (83, 1) (90, 0);
    –*Low*: (87, 0) (93, 1) (95, 0);
    –*Normal*: (94, 0) (97, 1) (100, 0).

Figure 1(c) shows the FISDeT GUI illustrating the fuzzy sets involved in the definition of the Blood_oxygen variable.

Lips color This parameter is associated to the linguistic input variable Color_lips, whose domain is identified in the numerical range [0–14]. Such a domain derives from the identification of 15 hues in the color scale which can be properly labeled through linguistic expressions. They are altogether reported in Fig. 2, where the hues are grouped into three reference categories, corresponding to the linguistic terms related to the Color_lips variable. Triangular fuzzy sets are associated with the linguistic terms, partitioning the domain of the Color_lips variable as follows:

    –*Regular*: (0, 0) (3, 1) (6, 0);
    –*Altered*: (5, 0) (7.5, 1) (10, 0);
    –*Purplish*: (8, 0) (12, 1) (16, 0).

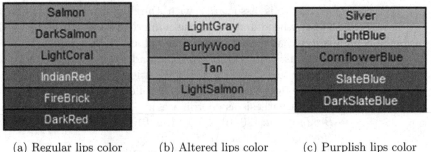

(a) Regular lips color   (b) Altered lips color   (c) Purplish lips color

Fig. 2. The set of 15 hues describing the domain of the `Color_lips` variable. They are grouped into three categories: Regular, Altered, and Purplish.

Figure 1(d) shows the FISDeT GUI illustrating the fuzzy sets involved in the definition of the `Color_lips` variable.

Risk level This parameter is associated to a linguistic output variable named `Risk_level` that assumes the values of four linguistic terms to be intended as class labels: *Risk_low*, *Risk_medium*, *Risk_high*, and *Risk_very_high*.

As concerning the structural organization of the FIS designed by FISDeT, we adopted the common choices regarding the t-norm and the t-conorm operators. The inference of the fuzzy system is carried on through the employment of the min and the max functions, determining the rule activation strength and the aggregation of rules respectively.

Once the input-output configuration has been properly set up, we defined the knowledge base to be embedded in the FIS. We considered all the possible combinations of input values, so that a number of 81 rules has been compiled. The rules have been crafted following some general guidelines collected during an interview with the physician. Such guidelines can be sketched as follows:

- when all the vital signs exhibit standard values, the risk level is low;
- when one vital sign exhibits a nonstandard value, the risk is medium;
- when two vital signs exhibit some nonstandard values, the risk is high;
- when three vital signs exhibit some nonstandard values, the risk is very high.

Following such guidelines, we compiled the fuzzy rule base of the decision-support FIS. The derived fuzzy rules embed the expert knowledge in a very interpretable linguistic form. This can be appreciated by the illustrative excerpt shown in Table 1.

4   Experimental Results

To test the effectiveness of the fuzzy inference system, we performed an evaluation based on real data coming from the examination of 116 persons. The vital

Table 1. Excerpt of the fuzzy rule base

Premise (IF)	Consequent (THEN)
Heart_rate is *Normal* AND Respiration_rate is *Normal* AND Blood_oxygen is *Normal* AND Color_lips is *Regular*	Risk_level is *Risk_low*
Heart_rate is *Normal* AND Respiration_rate is *Bradypnea* AND Blood_oxygen is *Normal* AND Color_lips is *Regular*	Risk_level is *Risk_medium*
Heart_rate is *Normal* AND Respiration_rate is *Tachypnea* AND Blood_oxygen is *Critical* AND Color_lips is *Regular*	Risk_level is *Risk_high*
Heart_rate is *Tachycardia* AND Respiration_rate is *Tachypnea* AND Blood_oxygen is *Critical* AND Color_lips is *Regular*	Risk_level is *Risk_very_high*
Heart_rate is *Tachycardia* AND Respiration_rate is *Tachypnea* AND Blood_oxygen is *Low* AND Color_lips is *Purplish*	Risk_level is *Risk_very_high*

signs related to the HR, BR, and SpO2 parameters have been obtained through the collection of PPG signals. To acquire the information concerning the lips color, we processed the face image of each person so as to identify the ROI related to the lips. Subsequently the ROI was processed to derive the dominant color information. To do this, the *K-means* clustering algorithm was applied to perform a quantization of the color into $K = 3$ levels (see Fig. 3). Finally, the K colors were averaged to derive a unique dominant color.

Once collected the data related to vital signs, we asked the physician to associate a risk level to each sample. Table 2 reports an illustrative excerpt from the dataset. Then, we applied the FIS to each sample in order to compare the inferred result with the human decision. In practice, we intended the physician's hints as the actual classes to be considered against the risk levels provided by the fuzzy system. The results of comparison were examined at different levels.

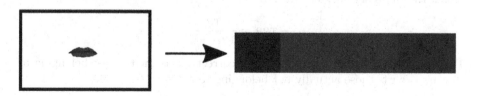

Fig. 3. Example of lips color quantization using *K-means*.

Table 2. Some samples from the dataset.

Subject	HR	BR	SpO2	Lips color	Risk level
S1	73	10.7	98.9	12	Risk_low
S2	98.3	9.4	98.4	12	Risk_medium
S3	136.6	9	94	12	Risk_very_high
S4	79.1	10.8	93.6	1	Risk_medium
S5	88.4	21.2	98	1	Risk_low
S6	70.8	31.4	92	1	Risk_high

As a first note, we observe that the overall value of classification accuracy is 68.97%. However, accuracy alone may be a misleading index, especially when it is considered during the analysis of unbalanced datasets (which is the case at hand, since the individuals who underwent the screening were mostly healthy persons). Therefore, we performed a further analysis evaluating the accuracy related to each of the four output classes, together with additional measures that are commonly considered in classification tasks. In particular, while analyzing a single class c, we consider true positive (TP), true negative (TN), false positive (FP), and false negative (FN) classification results, and we take into account the following measures:

Accuracy: ratio of correct discriminations w.r.t. class c

$$\text{ACC} = \frac{\text{TP} + \text{TN}}{\text{TP} + \text{FP} + \text{FN} + \text{TN}}$$

Positive Predictive Value: ratio of correctly classified samples w.r.t. those identified as pertaining to class c

$$\text{PPV} = \frac{\text{TP}}{\text{TP} + \text{FP}}$$

Negative Predictive Value: ratio of correctly classified samples w.r.t. those identified as not pertaining to class c

$$\text{NPV} = \frac{\text{TN}}{\text{TN} + \text{FN}}$$

True Positive Rate: ratio of samples correctly classified as belonging to class c w.r.t. those actually belonging to class c

$$\text{TPR} = \frac{\text{TP}}{\text{TP} + \text{FN}}$$

True Negative Rate: ratio of samples correctly classified as not belonging to class c w.r.t. those actually not belonging to class c

$$\text{TNR} = \frac{\text{TN}}{\text{FP} + \text{TN}}$$

Table 3. Evaluation measures derived for each output class.

	ACC	TNR	TPR	PPV	NPV
Risk_low	0.83	1	0.77	1	0.6
Risk_medium	0.75	0.76	0.57	0.13	0.96
Risk_high	0.91	0.94	0.50	0.4	0.96
Risk_very_high	0.88	0.96	0.40	0.6	0.91

Table 3 reports the values of these measures evaluated for each class. It can be observed how the TNR and NPV values are generally greater than those of TPR and PPV. This means that the knowledge embedded into the FIS is more effective in determining the non-membership to each class than the sensitivity to each specific risk level. This could be related to the fact that an unbalanced dataset is tackled by a set of rules crafted while keeping in mind a more general setting.

Table 4. Confusion matrix.

		Fuzzy decision system			
		Risk_low	Risk_medium	Risk_high	Risk_very_high
	Risk_low	66	17	2	1
Expert	Risk_medium	0	4	2	1
	Risk_high	0	2	4	2
	Risk_very_high	0	7	2	6

The obtained results can be further analyzed by considering the information conveyed by the overall confusion matrix depicted in Table 4. Such an overview allows to better focus a specific feature of the classification problem at hand: the involved classes are ranked in a range going from a *low* to a *very high* risk level. In this sense, a misclassification involving classes that are distant in this rank is more troublesome than others involving one class next to another. From the analysis of Table 4 we can argue that only 66 out of 86 low risk samples have been correctly identified. However, almost every misclassified low risk sample has been associated with the most similar class (Risk_medium). The same argument goes with the misclassification of medium risk samples (only one case has been shifted toward a very high risk) and high risk samples (misclassified samples are related to adjacent classes). On the other hand, management of the Risk_very_high class is somewhat troublesome since 7 out of 15 cases have been incorrectly related to a medium risk level.

As a conclusive remark, we point out that the misclassifications produced by the fuzzy system in most cases represent an overestimation of the risk level. In medical contexts this can be read as a problem with reduced harm, the opposite occurrence being regarded as the cause of much more serious consequences.

5   Conclusions

In this work we have presented a fuzzy rule-based system for decision support in the medical realm of cardiovascular diseases. Preliminary experimental results on both healthy and ill people show the effectiveness of the fuzzy system in simulating the decision of the expert. The fuzzy rules developed so far rely only on four main vital signs of a person, namely heart rate, breath rate, blood oxygen saturation and lips color. The choice of these parameters lies in the simplicity of their measurement together with the reliability of their associated information. For these reasons they represent the ideal parameters to be involved in a wearable device or in a domotic system endowed with the inferring capabilities provided by our fuzzy system. As a further improvement, we intend to enrich the knowledge base of the fuzzy decision support system by including other information about the patient, such as demographic features (age and sex) and information coming from the patient's history and the family history.

Acknowledgement. The authors are thankful to Dr. Ilaria Engaddi from "Istituti Milanesi Martinitt e Stelline e Pio Albergo Trivulzio" (Milan, Italy) for providing her knowledge and expertise useful to define the fuzzy rule base.

References

1. Casalino, G., Castiello, C., Del Buono, N., Mencar, C.: Intelligent Twitter data analysis based on nonnegative matrix factorizations. In: Gervasi, O., et al. (eds.) Computational Science and Its Applications – ICCSA 2017. LNCS, vol. 10404, pp. 188–202. Springer, Cham (2017). https://doi.org/10.1007/978-3-319-62392-4_14
2. Del Buono, N., Mencar, C., Casalino, G., Castiello, C.: A framework for intelligent Twitter data analysis with non-negative matrix factorization. Int. J. Web Inf. Syst. **14**(3), 334–356 (2018)
3. Berthold, M., Hand, D.J. (eds.): Intelligent Data Analysis: An Introduction, 1st edn. Springer, New York (1999). https://doi.org/10.1007/978-3-662-03969-4
4. Berthold, M.R., Borgelt, C., Höppner, F., Klawonn, F.: Guide to Intelligent Data Analysis: How to Intelligently Make Sense of Real Data. TCS, 1st edn. Springer, London (2010). https://doi.org/10.1007/978-1-84882-260-3
5. Casalino, G., Del Buono, N., Mencar, C.: Nonnegative matrix factorizations for intelligent data analysis. In: Naik, G.R. (ed.) Non-negative Matrix Factorization Techniques. SCT, pp. 49–74. Springer, Heidelberg (2016). https://doi.org/10.1007/978-3-662-48331-2_2
6. Bellazzi, R., Zupan, B.: Intelligent data analysis in medicine and pharmacology: a position statement. In: IDAMAP Workshop Notes at the 13th European Conference on Artificial Intelligence, ECAI, vol. 98 (1998)
7. Lavrač, N., Kononenko, I., Keravnou, E., Kukar, M., Zupan, B.: Intelligent data analysis for medical diagnosis: using machine learning and temporal abstraction. AI Commun. **11**(3,4), 191–218 (1998)

8. Lavrač, N., Keravnou-Papailiou, E., Zupan, B.: Intelligent Data Analysis in Medicine and Pharmacology, vol. 414. Springer, New York (2012). https://doi.org/10.1007/978-1-4615-6059-3

9. Magdalena-Benedito, R.: Medical Applications of Intelligent Data Analysis: Research Advancements: Research Advancements. IGI Global, Hershey (2012)

10. Adlassnig, K.P.: Fuzzy set theory in medical diagnosis. IEEE Trans. Syst. Man Cybern. **16**(2), 260–265 (1986)

11. Phuong, N.H., Kreinovich, V.: Fuzzy logic and its applications in medicine. Int. J. Med. Inform. **62**(2–3), 165–173 (2001)

12. Begum, S.A., Devi, O.M.: Fuzzy algorithms for pattern recognition in medical diagnosis. Assam Univ. J. Sci. Technol. **7**(2), 1–12 (2011)

13. Sanz, J.A., Galar, M., Jurio, A., Brugos, A., Pagola, M., Bustince, H.: Medical diagnosis of cardiovascular diseases using an interval-valued fuzzy rule-based classification system. Appl. Soft Comput. **20**, 103–111 (2014)

14. Tsipouras, M.G., et al.: Automated diagnosis of coronary artery disease based on data mining and fuzzy modeling. IEEE Trans. Inf. Technol. Biomed. **12**(4), 447–458 (2008)

15. Dagar, P., Jatain, A., Gaur, D.: Medical diagnosis system using fuzzy logic toolbox. In: 2015 International Conference on Computing, Communication & Automation (ICCCA), pp. 193–197. IEEE (2015)

16. Rana, M., Sedamkar, R.R.: Design of expert system for medical diagnosis using fuzzy logic. Int. J. Sci. Eng. Res. **4**(6), 2914–2921 (2013)

17. Awotunde, J.B., Matiluko, O.E., Fatai, O.W.: Medical diagnosis system using fuzzy logic. Afr. J. Comput. ICT **7**(2), 99–106 (2014)

18. Gorgulu, O., Akilli, A.: Use of fuzzy logic based decision support systems in medicine. Stud. Ethno-Med. **10**(4), 393–403 (2016)

19. Alonso, J.M., Castiello, C., Lucarelli, M., Mencar, C.: Modeling interpretable fuzzy rule-based classifiers for medical decision support. In: Medical Applications of Intelligent Data Analysis: Research Advancements, pp. 255–272. IGI Global (2012)

20. Cannone, R., Castiello, C., Fanelli, A.M., Mencar, C.: Assessment of semantic cointension of fuzzy rule-based classifiers in a medical context. In: 2011 11th International Conference on Intelligent Systems Design and Applications, pp. 1353–1358, November 2011

21. Castellano, G., Castiello, C., Pasquadibisceglie, V., Zaza, G.: FISDeT: Fuzzy inference system development tool. Int. J. Comput. Intell. Syst. **10**(1), 13–22 (2017). https://doi.org/10.2991/ijcis.2017.10.1.2

22. Challoner, A.V.J.: Photoelectric plethysmography for estimating cutaneous blood flow. Non-invasive Physiol. Meas. **1**, 125–151 (1979)

23. Kamal, A.-A.M., Gomaa, A., El Kafif, M., Hammad, A.S.: Plasma lipid peroxides among workers exposed to silica or asbestos dusts. Environ. Res. **49**(2), 173–180 (1989)

24. Hu, S., Peris, V.A., Echiadis, A., Zheng, J., Shi, P.: Development of effective photoplethysmographic measurement techniques: from contact to non-contact and from point to imaging. In: 2009 Annual International Conference of the IEEE Engineering in Medicine and Biology Society, EMBC 2009, pp. 6550–6553. IEEE (2009)

25. Wieringa, F.P., Mastik, F., van der Steen, A.F.W.: Contactless multiple wavelength photoplethysmographic imaging: a first step toward "SpO2 camera" technology. Ann. Biomed. Eng. **33**(8), 1034–1041 (2005)

26. Rouast, P.V., Adam, M.T.P., Chiong, R., Cornforth, D., Lux, E.: Remote heart rate measurement using low-cost RGB face video: a technical literature review.

Front. Comput. Sci. **12**(5), 858–872 (2018). https://doi.org/10.1007/s11704-016-6243-6

27. Hassan, M.A., et al.: Heart rate estimation using facial video: a review. Biomed. Signal Process. Control **38**, 346–360 (2017)
28. Poh, M.-Z., McDuff, D.J., Picard, R.W.: Non-contact, automated cardiac pulse measurements using video imaging and blind source separation. Opt. Express **18**(10), 10762–10774 (2010)
29. Castellano, G., Castiello, C., Fanelli, A.M.: The FISDeT software: application to beer style classification. In: Proceedings of the IEEE International Conference on Fuzzy Systems (FUZZ-IEEE 2017), Naples, Italy, 9–12 July 2017. https://doi.org/10.1109/FUZZ-IEEE.2017.8015503

Enhancing the DISSFCM Algorithm
for Data Stream Classification

Gabriella Casalino[1,2], Giovanna Castellano[1,2(✉)], Anna Maria Fanelli[1],
and Corrado Mencar[1,2]

[1] Computer Science Department, University of Bari "Aldo Moro", Bari, Italy
{gabriella.casalino,giovanna.castellano,annamaria.fanelli,
corrado.mencar}@uniba.it
[2] INdAM Research Group GNCS, Rome, Italy

Abstract. Analyzing data streams has become a new challenge to meet
the demands of real time analytics. Conventional mining techniques are
proving inefficient to cope with challenges associated with data streams,
including resources constraints like memory and running time along with
single scan of the data. Most existing data stream classification methods require labeled samples that are more difficult and expensive to
obtain than unlabeled ones. Semi-supervised learning algorithms can
solve this problem by using unlabeled samples together with a few labeled
ones to build classification models. Recently we proposed DISSFCM,
an algorithm for data stream classification based on incremental semi-
supervised fuzzy clustering. To cope with the evolution of data, DISS-
FCM adapts dynamically the number of clusters by splitting large-scale
clusters. While splitting is effective in improving the quality of clusters,
a repeated application without counter-balance may induce many small-
scale clusters. To solve this problem, in this paper we enhance DISSFCM
by introducing a procedure that merges small-scale clusters. Preliminary
experimental results on a real-world benchmark dataset show the effectiveness of the method.

Keywords: Data stream classification ·
Semi-supervised fuzzy clustering · Incremental adaptive clustering

1 Introduction

Data stream mining is a recent methodology that deals with the analysis of
large volumes of ordered sequences of data records. Data streams are a manifes-
tation of Big Data, which are characterized by the four 'V' dimensions, namely
Volume, Velocity, Variety and Veracity [1]. In particular, data stream mining
assumes that the volume of the sequence of data is so large that records can
be used few times (or just once) for the analysis. Data streams are produced
by sensor networks, e-mails, online transactions, network traffic, weather forecasting, health monitoring, social networks, etc., just to cite the most common
applications made available by current technology [2,3].

© Springer Nature Switzerland AG 2019
R. Fullér et al. (Eds.): WILF 2018, LNAI 11291, pp. 109–122, 2019.
https://doi.org/10.1007/978-3-030-12544-8_9

The requirement of using data records few times for extracting useful information involves the development of special-purpose data analysis methods, which should not require to store the whole stream of data in memory [4–6]. An approach to analyze data streams exploits an incremental generation of informational patterns, which represent a synthesized view of all data records analyzed in past and progressively evolve as new data records are available. Incremental and on-line algorithms are potentially useful to deal with continuous arrival of data in rapid, time-varying, and potentially unbounded streams since they continuously incorporate information into their model [7,8].

Data stream mining is applied for different tasks, such as classification, clustering and frequent pattern mining. In this paper, we focus on classification of data records in a stream, which is deeply studied in literature [4,9–14]. Differently from most works in literature, which focus on supervised methods [15,16], we specialize into semi-supervised methods as we do not assume that all data records are completely labeled; on the other hand, we recognize that, in many contexts, labeled samples are difficult or expensive to obtain, meanwhile unlabeled data are relatively easy to collect. For example it is quite easy to collect new sensor data coming from continuous streams but it may be difficult or even impossible to manually label all such data. Semi-supervised learning in the context of data streams is relatively new when compared to supervised and unsupervised learning [17–20]. Despite several semi-supervised learning methods have been developed in the literature [21], only few of them have been applied to classify data streams [22,23]. Moreover, there are few attempts of using fuzzy clustering for data stream mining, despite fuzzy clustering could be particularly useful to capture the continuous changes in the clustering structure [24–28].

Based on the idea of combining the benefits of semi-supervised learning and fuzzy clustering, recently we developed an incremental semi-supervised clustering method for data stream classification [29], which applies the Semi-Supervised Fuzzy C-Means algorithm (SSFCM) [30] to data chunks. The method has been further refined by enabling the dynamic determination of the number of clusters through an appropriate splitting procedure, leading to the DISSFCM (Dynamic Incremental Semi-Supervised FCM) algorithm [31]. In essence, DISSFCM applies SSFCM to data chunks that correspond to a fixed-size collection of contiguous data records coming from a stream. Furthermore, SSFCM is modified in order to allow the incremental evolution of clusters; cluster quality is evaluated by reconstruction error so that, when the quality goes below a threshold, a splitting procedure is applied in order to divide a low-quality cluster into two higher-quality clusters. While splitting is effective in improving the quality of clusters, a repeated application without counter-balance may induce many small-scale clusters that do not represent meaningful patterns.

In this paper we enhance DISSFCM by introducing a merging procedure that merges clusters when there are too many clusters or there are clusters with too few data records. Clusters are merged when they are sufficiently close so as to not hamper the overall quality of the cluster structure.

The organization of the rest of the paper is as follows. Section 2 presents our method for data stream classification and its extension proposed in this work. In Sect. 3 the effectiveness of the extended method is evaluated on a benchmark dataset. The last section draws the conclusion and outlines future work.

2   Dynamic Incremental Semi-Supervised FCM

In this section we describe the complete DISSFCM (Dynamic Incremental Semi-Supervised FCM) algorithm [31], including a merging mechanism to avoid small-scale clusters and improve the structure of clusters.

DISSFCM assumes that data belonging to C different classes are continuously available during time and processed as chunks. Namely, a chunk of N_1 data is available at time t_1, a chunk of N_2 data is available at t_2 and so on[1]. We denote by X_t the data chunk available at time t. No assumption is made on the dimension of chunks that may vary from one chunk to another. One key feature of DISSFCM is the possibility to exploit partial supervision when available. Namely, when some pre-labeled data are available in a chunk, their labels can be used to drive the clustering process. The presence of pre-labeled data is not mandatory but it should be assured in the first chunk in order to initialize properly the cluster prototypes.

The core of DISSFCM is the SSFCM (Semi-Supervised FCM) algorithm [30] that is applied incrementally so as to enable continuous update of clusters based on new data chunks. At each time step SSFCM granulates data in the current chunk by producing a set of K clusters represented by K labeled prototypes $P = \{\mathbf{p}_1, \mathbf{p}_2, \ldots, \mathbf{p}_K\}$ representatives for the local data chunk they model. Each prototype $\mathbf{p}_k$ is a medoid, i.e. it is the datapoint closest to the center $\mathbf{c}_k$. Before starting the clustering process, K labeled data are randomly chosen to initialize the prototypes, so that each cluster prototype is associated to a class label $(K = C)$. To take into account the evolution of the data during the incremental clustering process, the cluster prototypes discovered from the previous chunk are used as pre-labeled prototypes for the current chunk.

To better take into account the data evolution, DISSFCM is equipped with a splitting mechanism [31] that is applied to the current clusters in order to divide a low-quality cluster into two higher-quality clusters. The cluster quality is evaluated in terms of the *reconstruction error* [30]:

$$V_k = \sum_{\mathbf{x}_j \in C_k} \|\mathbf{x}_j - \hat{\mathbf{x}}_j\|^2 \tag{1}$$

that measures the difference between the original data $\mathbf{x}_j$ and their "reconstructed" counterpart $\hat{\mathbf{x}}_j$ that is derived using the clustering outcome (prototypes and membership degrees) as follows:

---

[1] Any stream can be turned into a chunked stream by simply waiting for enough data points to arrive.

$$\hat{\mathbf{x}}_j = \frac{\sum_{k=1}^{K} u_{jk}^m \mathbf{p}_k}{\sum_{k=1}^{K} u_{jk}^m} \qquad (2)$$

The splitting mechanism is activated when the reconstruction error on the current chunk exceeds a tolerance value ϵ the reconstruction error computed on the previous chunk. This means that the current number of clusters is not enough to effectively represent the data, hence the number of clusters should be augmented.

The cluster having the highest value of the reconstruction error, i.e. the cluster with lowest reconstruction ability, is selected as candidate for splitting. The splitting is performed by means of the *conditional fuzzy clustering* [32] applied to the collection of data belonging to the cluster so as to create two novel prototypes. If we denote by S^* the set of data belonging to the cluster k^* selected for splitting and by $\mathbf{z}_1$ and $\mathbf{z}_2$ the two novel prototypes, the conditional clustering minimizes the following objective function:

$$J = \sum_{k=1}^{2} \sum_{j \in S^*} f_{jk}^m \|\mathbf{x}_j - \mathbf{z}_k\|^2 \qquad (3)$$

under the constraint $f_{j1} + f_{j2} = u_{jk^*}$ where f_{jk} is the membership degree of $\mathbf{x}_j$ to the new cluster k. The objective function (3) is minimized by iteratively computing the membership values f_{jk} and the prototypes $\mathbf{z}_k$ according to:

$$f_{jk} = \frac{u_{jk^*}}{\sum_{c=1}^{2} \left(\frac{\|\mathbf{x}_j - \mathbf{z}_k\|}{\|\mathbf{x}_j - \mathbf{z}_c\|} \right)^{1/(m-1)}} \qquad (4)$$

and

$$\mathbf{z}_k = \frac{\sum_{j \in S^*} f_{jk}^m \mathbf{x}_j}{\sum_{j \in S^*} f_{jk}^m}, \qquad k = 1, 2; \qquad (5)$$

After conditional clustering, the prototype $\mathbf{p}_{k^*}$ is replaced by the two novel prototypes $\mathbf{z}_1$ and $\mathbf{z}_2$ that inherit the class label from $\mathbf{p}_{k^*}$. Then membership values u_{ik} are recomputed as in SSFCM. The splitting can be repeated until the reconstruction error drops below the previous value. A maximum pre-fixed number N_s of splittings is allowed for each chunk.

Since a repeated application of the splitting without counter-balance may induce many small-scale clusters that do not represent meaningful patterns, in this work we enhance DISSFCM by introducing a merging procedure that merges clusters when there are too many clusters or there are clusters with too few data records in a chunk. Clusters are merged when their prototypes are close so as to not hamper the overall quality of the cluster structure. The merging mechanism is activated when one of the following conditions is met:

1. the number of clusters exceeds a predefined threshold θ;
2. the number of data belonging to a cluster is below a predefined threshold λ.

In case 1. we select the nearest prototypes having the same class label as candidates for merging. We denote by $\mathbf{p}_s$ and $\mathbf{p}_l$ the nearest prototypes among all the current cluster prototypes sharing the same label. The new prototype $\mathbf{p}$ obtained by merging $\mathbf{p}_s$ and $\mathbf{p}_t$ is given by the following formula:

$$\mathbf{p} = \frac{\sum_{i=1}^{N} (u_{is} + u_{it})^m \mathbf{x}_i}{\sum_{i=1}^{N} (u_{is} + u_{it})^m} \tag{6}$$

where u_{is} and u_{it} are the membership values of $\mathbf{x}_i$ to cluster s and cluster t. In case 2. the prototype of the cluster with low number of data is merged with the closest cluster prototype, using Eq. (6). In each case, the merging reduces the number of clusters by one. The merging is repeated until there are no small clusters nor too many clusters. However, a maximum pre-fixed number N_m of merges is allowed for each chunk.

Algorithm 1. DISSFCM

Require: Data stream of chunks $X_1, X_2, \ldots$ containing few labeled data belonging to C classes
Require: Initial set P_0 of K labeled prototypes containing at least one prototype per class;
Ensure: P: labeled prototypes; K: number of prototypes
1: $t \leftarrow 1$
2: $K \leftarrow |P_0|$
3: $P \leftarrow P_0$
4: **while** $\exists$ nonempty chunk X_t **do**
5:     $X_t \leftarrow X_t \cup P$ /* Add previous prototypes to the current chunk */
6:     $P, U \leftarrow SSFCM(X_t, K, P)$
7:     $n_s \leftarrow 0$ /* Number of splits */
8:     $V_{max}^{(t)} \leftarrow$ reconstruction_error(X_t, P, U)
9:     **while** $(V_{max}^{(t)} - V_{max}^{(t-1)} > \epsilon)$ **and** $(n_s < MAX_s)$ **do**
10:         $P, U \leftarrow split(X_t, P, U)$
11:         $V_{max}^{(t)} \leftarrow$ reconstruction_error(X_t, P, U)
12:         $n_s \leftarrow n_s + 1$
13:     **end while**
14:     $n_m \leftarrow 0$ /* Number of merges */
15:     **while** $(|P| > \theta$ or $\exists k : \sum_{j=1}^{N_t} u_{jk} < \lambda)$ **and** $(n_m < MAX_m)$ **do**
16:         $P, U \leftarrow merge(X_t, P, U)$
17:         $n_m \leftarrow n_m + 1$
18:     **end while**
19:     $K \leftarrow |P|$
20:     Classify data in X_t using labeled prototypes in P
21:     $t \leftarrow t + 1$
22: **end while**
23: **return** P

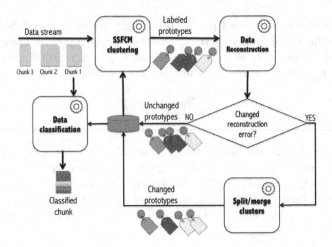

Fig. 1. Outline of DISSFCM. (Color figure online)

The overall scheme of DISSFCM enhanced with merging is shown in Fig. 1 and described in Algorithm 1. The algorithm requires the data stream as a sequence of chunks and an initial collection of labeled prototypes such that each class label is represented by at least one prototype. After application of SSFCM clustering (Step 6) the resulting prototypes are labeled automatically due to the semi-supervised nature of SSFCM. The derived prototypes are the basis for the classification process (Step 20). Indeed, the derived labeled prototypes are used to classify all the data in the current chunk via a matching mechanism. Namely, each data sample is matched against all prototypes and assigned to the class label of the best-matching prototype. The matching mechanism is based on the standard Euclidean distance. At the end, the algorithm returns the most recent collection of the prototypes, reflecting the data structure of the last data chunk. Notice that the returned collection can be used as input for a new run of the algorithm as long as new data are available from the data stream.

3   Experimental Results

Numerical experiments were conducted to evaluate the effectiveness of the proposed algorithm in data stream classification. The Optical Recognition of Handwritten Digits dataset[2] has been considered. It contains $5,620$ images of handwritten digits belonging to 10 classes (namely, $0, 1, 2, \ldots, 9$). We used 10% of the samples as test set, and we partitioned the remaining 90% in a fixed number of chunks in order to simulate a data stream. The class distribution was preserved both in the chunks and in the test set.

---

[2] https://archive.ics.uci.edu/ml/datasets/optical+recognition+of+handwritten+digits.

Table 1. Parameters of the enhanced DISSFCM algorithm.

Parameter	values
MaxSplits	10
MaxMerge	2
%Labeling	75%
#Chunk	5 10 15 20
ϵ	25 50 100

The accuracy measure has been used to evaluate the classification results:

$$Acc = \frac{|\{\mathbf{x}_j | y_j = a_j\}|}{N_t}$$

where $\mathbf{x}_j$ is the j-th data point, y_j is the true class label and a_j is the predicted class label, N_t is the number of data points. After the t-th chunk has been processed, accuracy is computed not only on the test set, but also on the t-th chunk and on the previous processed chunks.

The purity external clustering measure has been used to evaluate the extent to which clusters contain a single class, after each chunk arrival. To compute purity, each cluster C_k is assigned to the class of a_k of its prototype, and then the accuracy of this assignment is measured by counting the number of correctly assigned data points and dividing by the cardinality of the cluster:

$$Pur(k) = \frac{|\{\mathbf{x}_j | \mathbf{x}_j \in C_k \cap y_j = a_k\}|}{|C_k|}$$

Then an average purity is computed on all the clusters.

We carried out some preliminary experiments by varying the parameters of the DISSFCM algorithm. Table 1 summarizes the experimental settings. A first evaluation was done by observing the reconstruction error. As an example, Fig. 2 shows the trend of the reconstruction error with $\#Chunk = 15$ and $\epsilon = 50$. Green dots correspond to the error after processing the current chunk, the blue dots indicate the error after a split and the yellow ones the error after a merge. Numbers on the dots indicate the number of prototypes (clusters). It can be seen that every time the reconstruction error exceeds the previous value plus the threshold ϵ, a split is activated and a new cluster is created (the number of clusters upon the blue dot is increased by one). When a cluster with a small number of samples occurs, a merge is activated and the number of clusters is reduced. It can be seen that most peaks occur when a new chunk arrives. This means that DISSFCM is still learning the correct model to fit the data and it improves the model as soon as a new chunk arrives (i.e. more training data). We observe that the split and merge steps help the model to fit the data. This could be better observed from Fig. 3, where the average purity values obtained

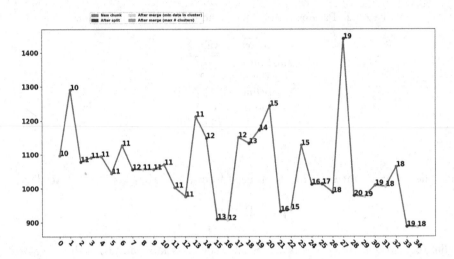

Fig. 2. Trend of the reconstruction error V_{max} with #Chunk = 15 and ϵ = 50. (Color figure online)

Table 2. Number of cluster prototypes for each class at the end of the incremental process with #Chunk = 15, ϵ = 50.

		Tot
Class	0 1 2 3 4 5 6 7 8 9	10
#Cluster	1 1 6 1 2 2 1 1 2 1	18

on single chunks during the learning process are reported. It can be seen that after processing the fifth chunk, the average value of purity decreases. When the sixth chunk arrives one split and one merge are applied (Fig. 2) rising the purity value. The same behavior could be observed after chunks 14-th and 15-th are processed. The processing of all the chunks ends with 18 cluster prototypes that are used to represent the 10 original classes. The number of cluster prototypes for each class is reported in Table 2.

Table 3 reports the accuracy computed on the chunks at each step t_i, during the incremental process with #Chunk = 15 and ϵ = 50. Bold terms represent accuracy values on the current chunk. We observe that the model is properly adapted to the new arrived chunk. At each time step we also evaluated the classification accuracy of the current model on the previously seen chunks to verify if the model still fits the old data.

To assess the effectiveness of DISSFCM, we evaluated the classification accuracy of the final models for each configuration of parameters (#Chunk, ϵ). Results are summarized in Table 4. Both on the test and the training sets we can observe that the impact of the tolerance ϵ is higher when the number of chunks

Table 3. Accuracy obtained on single chunks during the incremental process, with #Chunk = 15, $\epsilon = 50$.

	t_1	t_2	t_3	t_4	t_5	t_6	t_7	t_8	t_9	t_{10}	t_{11}	t_{12}	t_{13}	t_{14}	t_{15}
K	10	10	11	11	11	11	11	11	11	11	11	12	15	19	18
X_1	**0.84**	0.88	0.87	0.85	0.83	0.84	0.84	0.84	0.84	0.84	0.83	0.85	0.86	0.85	0.88
X_2	-	**0.86**	0.88	0.88	0.85	0.82	0.82	0.81	0.82	0.79	0.80	0.79	0.82	0.82	0.82
X_3	-	-	**0.82**	0.82	0.79	0.81	0.81	0.81	0.81	0.79	0.79	0.76	0.79	0.79	0.81
X_4	-	-	-	**0.82**	0.81	0.82	0.82	0.79	0.80	0.80	0.83	0.82	0.83	0.83	0.89
X_5	-	-	-	-	**0.81**	0.81	0.78	0.77	0.79	0.79	0.81	0.76	0.80	0.81	0.80
X_6	-	-	-	-	-	**0.84**	0.83	0.82	0.83	0.84	0.85	0.81	0.82	0.82	0.82
X_7	-	-	-	-	-	-	**0.86**	0.86	0.86	0.86	0.84	0.83	0.87	0.86	0.88
X_8	-	-	-	-	-	-	-	**0.85**	0.85	0.85	0.86	0.83	0.86	0.83	0.88
X_9	-	-	-	-	-	-	-	-	**0.87**	0.86	0.87	0.85	0.89	0.86	0.89
X_{10}	-	-	-	-	-	-	-	-	-	**0.82**	0.83	0.77	0.79	0.80	0.85
X_{11}	-	-	-	-	-	-	-	-	-	-	**0.84**	0.81	0.82	0.80	0.87
X_{12}	-	-	-	-	-	-	-	-	-	-	-	**0.81**	0.84	0.84	0.87
X_{13}	-	-	-	-	-	-	-	-	-	-	-	-	**0.83**	0.84	0.86
X_{14}	-	-	-	-	-	-	-	-	-	-	-	-	-	**0.87**	0.87
X_{15}	-	-	-	-	-	-	-	-	-	-	-	-	-	-	**0.91**

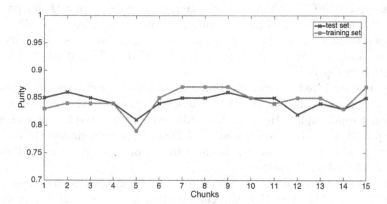

Fig. 3. Average purity obtained on single chunks during the incremental process, with #Chunk = 15, $\epsilon = 50$ on training and test sets. (Color figure online)

grows (i.e. the data samples in each sample decreases). Indeed the accuracy values with 5 and 10 chunks are stable when varying the values of ϵ. With 15 and 20 chunks the accuracy is more sensitive to the value of ϵ. This behavior can be better observed in the plots of Fig. 4 that show the trend of the accuracy on the test set during the processing of the chunks, varying the ϵ tolerance.

This is explained by observing that the higher the number of chunks, the less the number of samples in each chunk; therefore the algorithm has fewer samples to learn from. Thus the number of the samples in each chunk affects the stability of the algorithm. With 5 and 10 chunks (high number of data) the algorithm keeps the same behavior as new chunks arrive (Fig. 4(a) and (b)). As the number of chunks increases (and hence the number of data in each chunk decreases), the algorithm is more unstable and needs more time to converge to an accurate model (Fig. 4(c) and (d)).

Table 4. Classification accuracy on the whole training set (a) and the test set (b), varying the number of chunks and the tolerance ϵ.

(a) Training set					(b) Test set				
	# chunks					# chunks			
ϵ	5	10	15	20	ϵ	5	10	15	20
25	0.84	0.85	0.88	0.93	25	0.86	0.84	0.85	0.89
50	0.84	0.85	0.91	0.83	50	0.86	0.85	0.87	0.78
100	0.84	0.85	0.85	0.85	100	0.86	0.84	0.81	0.79

Finally, DISSFCM enhanced with merge was compared with its previous version [31] and with ILFM (Incremental Learning Fuzzy Measures) [33], which is a supervised incremental method based on Choquet integrals to classify data streams. Comparative results with #chunks = 15, ϵ = 50 and labeling = 75% are plotted in Fig. 5.

It can be seen that the introduction of the merging mechanism in DISSFCM slightly deteriorates the classification results with respect to the previous version which only applies splits. However, it should be noted that the final classification model provided by the novel version of DISSFCM is very simple (18 clusters) in comparison to the final model obtained by the previous version of DISSFCM which was based on 70 clusters.

The models obtained by DISSFCM were also compared to the model built by ILFM. It can be seen that the classification accuracy of ILFM is slightly better. However it should be noted that ILFM is a supervised method, thus it requires completely labeled data, that are difficult to find in real applications. Conversely, DISSFCM works with partially labeled data. Moreover the model produced by ILFM is an ensemble of classifiers, hence it is far more complex than our model. On the overall, DISSFCM achieves a good balance between accuracy and complexity of the classification model, while taking into account the evolution of data.

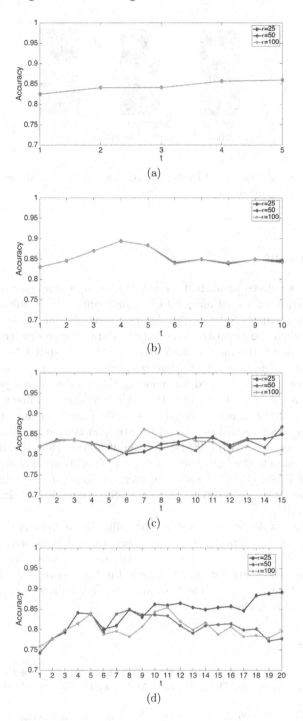

Fig. 4. Accuracy on the test set varying ϵ for #Chunk equal to 5 (a), 10 (b), 15 (c) and 20 (d). (Color figure online)

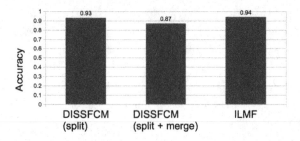

Fig. 5. Comparing the enhanced DISSFCM against its previous version (no merge), and ILMF. (Color figure online)

4   Conclusions

In this work we have described DISSFCM, a dynamic incremental semi-supervised version of the standard FCM clustering that is suitable for data stream classification. DISSFCM enables the structure of clusters to change dynamically: when the reconstruction error of data given a clustering structure becomes inadequate, the most troublesome clusters are split into finer grained clusters that better represent data. Moreover, when few samples are grouped in a cluster, a merge step is activated for reducing the number of groups. Numerical preliminary analysis has shown that the split tolerance ϵ influences the accuracy results when the chunks dimension is small. Finally, it has been observed that the merge mechanism has a small negative impact on the accuracy of the model, when compared with DISSFCM without merge. However, in the face of such accuracy reduction we observe a significant simplification of the final model (18 cluster for DISSFCM with split and merge, against 70 for DISSFCM with split only). Similar considerations can be derived by comparing DISSFCM (with merge) and ILMF.

Further work is devoted to analyze the influence of the chunk composition on DISSFCM, so as to better take into account real data stream scenarios, where the incoming chunks may have different sizes and may contain data with inhomogeneous class distributions. Moreover further research is going on along the direction of introducing a mechanism to detect outliers, concept drift and the emergence of new classes.

References

1. Eaton, C., Zikopoulos, P.: Understanding Big Data: Analytics for Enterprise Class Hadoop and Streaming Data, 1st edn. McGraw-Hill Osborne Media, New York (2011)
2. Casalino, G., Castiello, C., Del Buono, N., Mencar, C.: Intelligent Twitter data analysis based on nonnegative matrix factorizations. In: Gervasi, O., et al. (eds.) ICCSA 2017. LNCS, vol. 10404, pp. 188–202. Springer, Cham (2017). https://doi.org/10.1007/978-3-319-62392-4_14

3. Casalino, G., Castiello, C., Del Buono, N., Mencar, C.: A framework for intelligent Twitter data analysis with nonnegative matrix factorization. Int. J. Web Inf. Syst. **14**(3), 334–356 (2018)
4. Babcock, B., Babu, S., Datar, M., Motwani, R., Widom, J.: Models and issues in data stream systems. In: Proceedings of the Twenty-First ACM SIGMOD-SIGACT-SIGART Symposium on Principles of Database Systems, PODS 2002, pp. 1–16. ACM, New York (2002)
5. Gaber, M.M., Zaslavsky, A., Krishnaswamy, S.: Mining data streams: a review. SIGMOD Rec. **34**(2), 18–26 (2005)
6. Gama, J.: Knowledge Discovery from Data Streams, 1st edn. Chapman & Hall/CRC, Boca Raton (2010)
7. Chandak, M.B.: Role of big-data in classification and novel class detection in data streams. J. Big Data **3**(1), 5 (2016)
8. Chen, M., Mao, S., Liu, Y.: Big data: a survey. Mob. Netw. Appl. **19**(2), 171–209 (2014)
9. Ferranti, A., Marcelloni, F., Segatori, A., Antonelli, M., Ducange, P.: A distributed approach to multi-objective evolutionary generation of fuzzy rule-based classifiers from big data. Inf. Sci. **415**, 319–340 (2017)
10. Ducange, P., Pecori, R., Mezzina, P.: A glimpse on big data analytics in the framework of marketing strategies. Soft Comput. **22**(1), 325–342 (2018)
11. Lughofer, E., Pratama, M.: Online active learning in data stream regression using uncertainty sampling based on evolving generalized fuzzy models. IEEE Trans. Fuzzy Syst. **26**(1), 292–309 (2018)
12. Hyde, R., Angelov, P., MacKenzie, A.R.: Fully online clustering of evolving data streams into arbitrarily shaped clusters. Inf. Sci. **382–383**, 96–114 (2017)
13. Lughofer, E.: A dynamic split-and-merge approach for evolving cluster models. Evol. Syst. **3**(3), 135–151 (2012)
14. Olorunnimbe, M.K., Viktor, H.L., Paquet, E.: Dynamic adaptation of online ensembles for drifting data streams. J. Intell. Inf. Syst. **50**(2), 291–313 (2018)
15. Domingos, P., Hulten, G.: Mining high-speed data streams. In: Proceedings of the Sixth ACM SIGKDD International Conference on Knowledge Discovery and Data Mining, KDD 2000, pp. 71–80. ACM (2000)
16. Hulten, G., Spencer, L., Domingos, P.: Mining time-changing data streams. In: Proceedings of the Seventh ACM SIGKDD International Conference on Knowledge Discovery and Data Mining, KDD 2001, pp. 97–106. ACM (2001)
17. Nguyen, H.-L., Woon, Y.-K., Ng, W.-K.: A survey on data stream clustering and classification. Knowl. Inf. Syst. **45**(3), 535–569 (2015)
18. Mousavi, M., Bakar, A.A., Vakilian, M.: Data stream clustering algorithms: a review. Int. J. Adv. Soft Comput. Appl. **7**(3), 13 (2015)
19. Toshniwal, D.: Clustering techniques for streaming data - a survey. In: 2013 3rd IEEE International Advance Computing Conference, IACC, pp. 951–956, February 2013
20. Ghesmoune, M., Lebbah, M., Azzag, H.: Micro-batching growing neural gas for clustering data streams using spark streaming. Proc. Comput. Sci. **53**, 158–166 (2015). INNS Conference on Big Data 2015 Program, San Francisco, CA, USA, 8–10 August 2015
21. Zhu, X.: Semi-supervised learning literature survey. Technical report 1530, Computer Sciences. University of Wisconsin-Madison (2005)
22. Blum, A., Mitchell, T.M.: Combining labeled and unlabeled data with co-training. In: Bartlett, P.L., Mansour, Y. (eds.) COLT, pp. 92–100. ACM (1998)

23. Zhou, Z.-H., Li, M.: Tri-training: exploiting unlabeled data using three classifiers. IEEE Trans. Knowl. Data Eng. **17**(11), 1529–1541 (2005)
24. Beringer, J., Hüllermeier, E.: Fuzzy clustering of parallel data streams. In: Advances in Fuzzy Clustering and Its Application, pp. 333–352 (2007)
25. Abdullatif, A., Masulli, F., Rovetta, S.: Clustering of nonstationary data streams: a survey of fuzzy partitional methods. Wiley Interdisc. Rev.: Data Min. Knowl. Discov. **8**(4), e1258 (2018)
26. Mostafavi, S., Amiri, A.: Extending fuzzy C-means to clustering data streams. In: 20th Iranian Conference on Electrical Engineering, ICEE 2012, pp. 726–729, May 2012
27. Upadhyay, D., Jain, S., Jain, A.: A fuzzy clustering algorithm for high dimensional streaming data. J. Inf. Eng. Appl. **3**(10), 1–9 (2013)
28. Geweniger, T., Fischer, L., Kaden, M., Lange, M., Villmann, T.: Clustering by fuzzy neural gas and evaluation of fuzzy clusters. Comput. Intell. Neurosci. **2013**, 9 (2013)
29. Castellano, G., Fanelli, A.M.: Classification of data streams by incremental semi-supervised fuzzy clustering. In: Petrosino, A., Loia, V., Pedrycz, W. (eds.) WILF 2016. LNCS, vol. 10147, pp. 185–194. Springer, Cham (2017). https://doi.org/10.1007/978-3-319-52962-2_16
30. Pedrycz, W.: Algorithms of fuzzy clustering with partial supervision. Pattern Recogn. Lett. **3**(1), 13–20 (1985)
31. Casalino, C., Castellano, G., Mencar, C.: Incremental adaptive semi-supervised fuzzy clustering for data stream classification. In: Proceedings of the 2018 IEEE Conference on Evolving and Adaptive Intelligent Systems, EAIS 2018, Rhodes, Greece, 25–27 May 2018, pp. 1–7 (2018)
32. Li, P., Wu, X., Hu, X., Wang, H.: Learning concept-drifting data streams with random ensemble decision trees. Neurocomputing **166**(C), 68–83 (2015)
33. Xuefei, L., Huimin, F., Hongbo, S.: Incremental learning fuzzy measures with Choquet integrals in fusion system. J. Chem. Pharm. Res. **6**, 102–112 (2014)

Fuzzy Similarity-Based Hierarchical Clustering for Atmospheric Pollutants Prediction

F. Camastra[1], A. Ciaramella[1(✉)], L. H. Son[2], A. Riccio[1], and A. Staiano[1]

[1] Department of Science and Technology, University of Naples "Parthenope",
Isola C4, Centro Direzionale, 80143 Naples (NA), Italy
{francesco.camastra,angelo.ciaramella,angelo.riccio,
antonino.staiano}@uniparthenope.it
[2] Vietnam National University, 334 Nguyen Trai, Thanh Xuan, Hanoi, Vietnam
sonlh@vnu.edu.vn

Abstract. This work focuses on models selection in a multi-model air quality ensemble system. The models are operational long-range transport and dispersion models used for the real-time simulation of pollutant dispersion or the accidental release of radioactive nuclides in the atmosphere. In this context, a methodology based on temporal hierarchical agglomeration is introduced. It uses fuzzy similarity relations combined by a transitive consensus matrix. The methodology is adopted for individuating a subset of models that best characterize the predicted atmospheric pollutants from the ETEX-1 experiment and discard redundant information.

Keywords: Fuzzy similarity · Hierarchical agglomeration ·
Ensemble models · Air pollutant dispersion

1 Introduction

The real-time simulation of pollutant dispersion or the accidental release of radioactive substances in the atmosphere is a challenging aspect of many national services and agencies. In particular, releases of harmful radionuclides (e.g. Fukushima, Chernobyl) could be simulated and monitored [1, 10, 13, 20]. In this work we consider atmospheric compounds from the *ENSEMBLE* system [6–8]. ENSEMBLE is a web-based system aiming at assisting the analysis of multi-model data provided by many national meteorological services and environmental protection agencies worldwide. It is worth noting that in the case of multi-model ensemble for atmospheric dispersions, models are certainly more or less dependent from several intrinsic mechanisms (e.g., they often share features, initial/boundary data, numerical methods, parameterizations and emissions). For this reason, results obtained by ensemble analysis may lead to erroneous interpretations and in a multimodel approach the effective number of models may

© Springer Nature Switzerland AG 2019
R. Fullér et al. (Eds.): WILF 2018, LNAI 11291, pp. 123–133, 2019.
https://doi.org/10.1007/978-3-030-12544-8_10

be lower than the total number, since models could be linearly (or nonlinearly) dependent on each other.

To solve this problem, a number of techniques has been proposed in literature. In [15,17,18] the authors present a statistical analysis (i.e., *Bayesian Model Averaging*) for combining predictive distributions from different sources of a multi-model ensemble, and in [16] some basic properties of multi-model ensemble systems are investigated. Moreover, cluster-based approaches have also been proposed [2–4]. In this paper, we introduce a methodology that improves the forecasting by considering observations that may become available during the course of the event. The methodology is based on fuzzy similarity relations that allow to combine multiple hierarchical agglomerations, each for a different forecasting leading time. From the overall temporal agglomeration obtained by a consensus matrix it is possible to select a subset of models and discard redundant information.

The remainder of the paper is organized as follows. In Sect. 2 the proposed methodology is detailed. In particular, some fundamental concepts on t-norms and fuzzy similarity relations (Sect. 2.2) are given and the agglomerative based approach is described in Sect. 2.3. Finally, in Sect. 3 some experimental results, obtained by applying this methodology on an ensemble of prediction models, are described. Conclusions and future remarks are given in Sect. 4.

2   Fuzzy Similarity and Agglomerative Clustering

In general, when one deals with clustering tasks, *fuzzy logic* permits to obtain soft clustering, instead of hard (crisp or non-fuzzy) clustering of data. Hierarchical clustering is a methodology for cluster analysis which seeks to build a hierarchy of clusters and it can be agglomerative or divisive. In this work we consider an agglomerative clustering approach. One of the main aspects of this methodology is the use of a measure of dissimilarity between sets of observations, by using an appropriate metric. On the other hand, a dendrogram is a tree diagram used to illustrate the results produced by hierarchical clustering. In the following, we show that a dendrogram can be associated with a fuzzy equivalence relation based on Łukasiewicz valued fuzzy similarities. Successively, a consensus matrix, that is the representative information of all dendrograms, is obtained by combining multiple temporal hierarchical agglomerations of dispersion models. The main steps of the proposed approach are

1. Membership functions characterization;
2. Fuzzy Similarity Matrix calculation (or dendrogram) for all the models at a fixed time;
3. Consensus matrix construction for temporal hierarchical agglomerations.

2.1   Membership Functions

The effective of *fuzzy logic* is the transformation of linguistic variables in fuzzy sets. Fuzzification is the process of changing a real scalar value into a fuzzy value and it is achieved by using different types of membership functions. The membership function represents the degree of truth to which a given input belongs to a fuzzy set. In the proposed approach, *fuzzy sets* are described by the following *membership functions* [21]

$$\mu(\mathbf{x}_i) = \frac{\mathbf{x}_i - \min(\mathbf{x}_i)}{\max(\mathbf{x}_i) - \min(\mathbf{x}_i)}, \tag{1}$$

where $\mathbf{x}_i = [x_1^i, x_2^i, \ldots, x_L^i]$ is the i-th observation vector of the L considered models.

2.2   Fuzzy Similarity

We observe that fuzzy sets can be combined via the conjunction and disjunction operations and continuous triangle norms or co-norms are adopted, respectively. A *triangular norm* (*t-norm* for short), is a binary operation t on the unit interval $[0, 1]$. In particular, it is a function $t : [0, 1]^2 \rightarrow [0, 1]$, such that it satisfies the following four axioms for all $x, y, z \in [0, 1]$ [11]

$$
\begin{aligned}
t(x, y) &= t(y, x) & (commutativity) \\
t(x, t(y, z)) &= t(t(x, y), z) & (associativity) \\
t(x, y) &\leq t(x, z) \quad \text{whenever } y \leq z & (monotonicity) \\
t(x, 1) &= x & (boundary\ condition)
\end{aligned}
\tag{2}
$$

In practical situations the following four basic t-norms are considered

$$
\begin{aligned}
t_{\mathbf{M}}(x, y) &= \min(x, y) & (minimum) \\
t_{\mathbf{P}}(x, y) &= x \cdot y & (product) \\
t_{\mathbf{L}}(x, y) &= \max(x + y - 1, 0) & (Lukasiewicz\ t\text{-}norm) \\
t_{\mathbf{D}}(x, y) &= \begin{cases} 0 & \text{if } (x, y) \in [0, 1]^2 \\ \min(x, y) & \text{otherwise} \end{cases} & (drastic\ product)
\end{aligned}
\tag{3}
$$

However, in these years, several parametric and non-parametric t-norms have been introduced [11] and generalized versions have also been studied [5]. In the following, we focus on the properties of the Lukasiewicz t-norm ($t_{\mathbf{L}}$). One main operator adopted in fuzzy-based systems (e.g., fuzzy inference systems) is the *residuum* $\rightarrow_t$

$$x \rightarrow_t y = \bigvee \{z | t(z, x) \leq y\} \tag{4}$$

where $\bigvee$ is the *union* operator and, for the left-continuous basic t-norm $t_{\mathbf{L}}$, is given by

$$x \rightarrow_{\mathbf{L}} y = \min(1 - x + y, 1) \, (Lukasiewicz\ implication) \tag{5}$$

Moreover, we also note that letting p be a fixed natural number in a *generalized Lukasiewicz structure*, we obtain

$$t_{\mathbf{L}}(x,y) = \sqrt[p]{\max(x^p + y^p - 1, 0)}$$
$$x \to_{\mathbf{L}} y = \min(\sqrt[p]{1 - x^p + y^p}, 1) \tag{6}$$

Another fundamental operation on a residuated lattice is the *bi-residuum* that will be used for our construction of the fuzzy similarities. It is defined as

$$x \leftrightarrow_t y = (x \to_t y) \wedge (y \to_t x), \tag{7}$$

where $\wedge$ is the *meet*. In the case of the left-continuous basic t-norm $t_{\mathbf{L}}$, we obtain the following *bi-residuum*

$$x \leftrightarrow_{\mathbf{L}} y = 1 - \max(x, y) + \min(x, y) \tag{8}$$

On the other hand, a binary *fuzzy relation* R is defined on $U \times V$ as a fuzzy set on $U \times V$ ($R \subseteq U \times V$). A *similarity matrix* is a fuzzy relation $S \subseteq U \times U$ such that, for each $u, v, w \in U$, the following properties are satisfied

$$S\langle u, u \rangle \quad\quad = \quad 1 \quad\quad (everthing\ is\ similar\ to\ itself)$$

$$S\langle u, v \rangle \quad\quad = S\langle v, u \rangle \quad\quad\quad (symmetric) \tag{9}$$

$$t(S\langle u, v \rangle, S\langle v, w \rangle) \leq S\langle u, w \rangle \quad\quad (weakly\ transitive)$$

It is essential to observe that from fuzzy sets with membership functions $\mu :$ $X \to [0, 1]$, a fuzzy similarity matrix S can be generated as

$$S\langle a, b \rangle = \mu(a) \leftrightarrow_t \mu(b) \tag{10}$$

for all $a, b \in X$.

Moreover, to build the fuzzy similarity matrix a main result is considered [19,21]

Proposition 1. *Consider n Lukasiewicz valued fuzzy similarities S_i, $i = 1, \ldots, n$ on a set X. Then*

$$S\langle x, y \rangle = \frac{1}{n} \sum_{i=1}^{n} S_i \langle x, y \rangle \tag{11}$$

is a Lukasiewicz valued fuzzy similarity on X.

In this work, we consider for Eq. 11

$$S_i \langle x, y \rangle = x \leftrightarrow_{\mathbf{L}} y. \tag{12}$$

Now, let $t_{\mathbf{L}}$ be the Lukasiewicz product, it is worth noting that S is a fuzzy equivalence relation on X with respect to $t_{\mathbf{L}}$ iif $1 - S$ is a *pseudo-metric* on X.

Algorithm 1. Min-transitive closure

1: **Input** R the input relation
2: **Output** R^T the output transitive relation
3: **Elaborate**
   1. Calculate $R^* = R \cup (R \circ R)$
   2. if $R^* \neq R$ replace R with R^* and go to step 1
   else $R^T = R^*$ and the algorithm terminates.

2.3   Dendrogram and Consensus Matrix

We also have to observe that if a similarity relation is *min-transitive* ($t = \min$ in (9)) then it is a *fuzzy-equivalence relation* that can be graphically described by a *dendrogram* [12]. In other words, transitivity implies the existence of the dendrogram.

The min-transitive closure R^T of R can be obtained as follows [14]

$$R^T = \bigcup_{i=1}^{n-1} R^i \tag{13}$$

where R^{i+1} is defined as

$$R^{i+1} = R^i \circ R, \tag{14}$$

and n is the dimension of a relation matrix.

Considering two fuzzy relations R and S, we observe that the composition $R \circ S$ is a fuzzy relation defined by

$$R \circ S\langle x, y \rangle = \mathrm{Sup}_{z \in X}\{R\langle x, z \rangle \odot S\langle z, y \rangle\} \tag{15}$$

$\forall x, y \in X$, where $\odot$ stands for a t-norm (e.g., min operator) [14]. Then we can conclude that the min-transitive closure R^T of a matrix R can be easily computed and the overall process is described in Algorithm 1.

We also observe that to accomplish an agglomerative clustering a dissimilarity relation is needed. Here we considered the following result [14].

Lemma 1. *Letting R be a similarity relation with the elements $R\langle x, y \rangle \in [0, 1]$ and letting D be a dissimilarity relation, which is obtained from R by*

$$D(x, y) = 1 - R\langle x, y \rangle \tag{16}$$

then D is ultrametric iif R is min-transitive.

In other words, we have a one-to-one correspondence between min-transitive similarity matrices and dendrogram and between ultrametric dissimilarity matrices and dendrograms.

Finally, after the dendrograms have been obtained at each time, a consensus matrix, that is the representative information of all temporal dendrograms, is obtained by combining the transitive closures by using Eq. 15 (i.e., max-min) [14]. The overall approach is described in Algorithm 2.

Algorithm 2. Combination of dendrograms

1: **Input** $S^{(i)}$, $1 \leq i \leq L$ L input similarity matrices (dendrograms)
2: **Output** S the resulted similarity matrix (dendrogram)
    1. Aggregate the similarity matrices to a final similarity matrix $S = Aggregate(S^{(1)}, S^{(2)}, \ldots, S^{(L)})$
    a. Let S^* be the identity matrix
    b. For each $S^{(i)}$ calculate e $S^* = S^* \cup (S^* \circ S^{(i)})$
    c. If S^* is not changed $S = S^*$ and goto step 3 else goto step 1.b
3: Create the final dendrogram from S

---

3 Experimental Results

This Section aims to illustrate some results obtained by the proposed approach. In particular, we consider the multi-model ensemble simulated distributions of the ETEX-1 experiment [9]. The ETEX-1 experiment concerned the release of pseudo-radioactive material on 23 October 1994 at 16:00 UTC from Monterfil, southeast of Rennes (France). Briefly, a steady westerly flow of unstable air masses was present over central Europe. Such conditions persisted for the 90 h that followed the release with frequent precipitation events over the advection area and a slow movement toward the North Sea region. Just for an example, in Fig. 1 we show the integrated concentration after 78 h from release. In the experiment, the main objective of the several independent groups worldwide (25 members) was to forecast the observations with different atmospheric dispersion models. Moreover, each simulation was based on weather fields generated by (most of the time) different *Global Circulation Models* (GCM) and all the simulations relate to the same release conditions. For further information on the involved groups and the adopted models the reader can refer to [8] and [9].

Now we apply the proposed approach to analyze data of the ETEX-1 experiment. The preliminary step is the *fuzzification*. In particular, Eq. 1 is applied on the concentrations estimated by models at each time level. Successively, for each concentration at different times a dendrogram (similarity matrix) is produced (Eq. 11 with Łukasiwicz norm and $p = 1$). Finally, the consensus matrix that described the representative dendrogram is estimated by using the approach described in Algorithm 2. In Fig. 2 a particular of the representative dendrogram obtained after 78 h is visualized. We observe that different clusters of similar models are obtained.

To highlight the clustering outcomes, in Fig. 3, we show some representative distributions of the clustered models. For example, as confirmed by dendrogram, the distributions of the models 22 and 24 are very close. See Figs. 3a and b for a comparison. Instead, the model 21 has a very diffusive distribution, as highlighted by the dendrogram. This distribution is visualized in Fig. 3c. At this point, we can identify models that have similar behavior by analyzing the different clusters. In order to identify the group of models that more appropriately describe observations, we compare the distributions of the models by using a Kullback Leibler divergence.

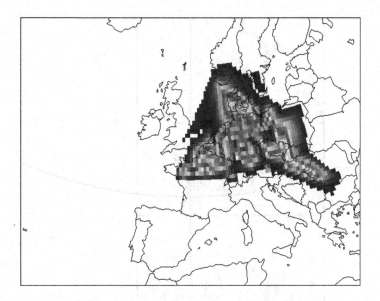

Fig. 1. ETEX-1 temporal integrated observations after 78 h.

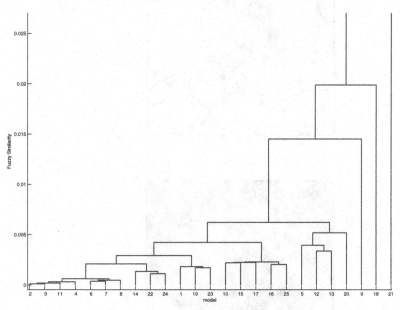

Fig. 2. Representative dendrogram obtained by consensus matrix: x-axis are related to the models and those on the y-axis are related to the model data similarities.

The Kullback Leibler (KL) divergence between two discrete n-dimensional probability density functions $\mathbf{p} = [p_i \ldots p_n]$ and $\mathbf{q} = [q, \ldots q_n]$ is defined as

$$KL(\mathbf{p}||\mathbf{q}) = \sum_{i=1}^{n} p_i \log\left(\frac{p_i}{q_i}\right). \tag{17}$$

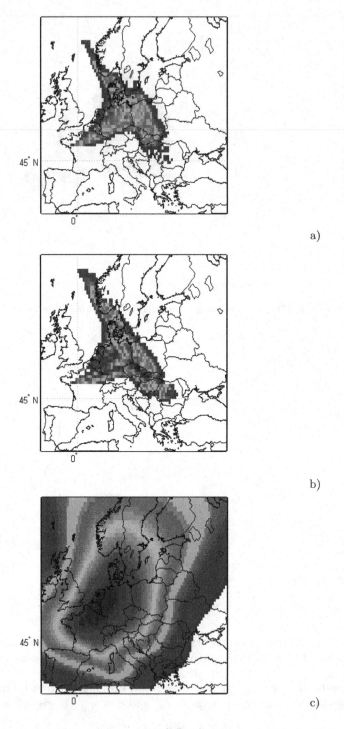

a)

b)

c)

Fig. 3. Model distributions: (a) model 22; (b) model 24; model 21.

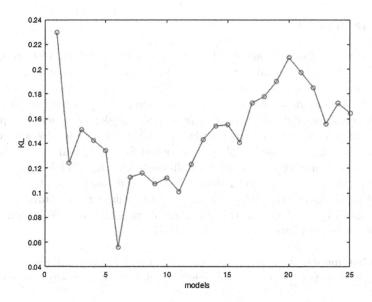

Fig. 4. KL divergence varying the clustering number.

This is known as the relative entropy. It satisfies the Gibbs' inequality

$$KL(\mathbf{p}||\mathbf{q}) \geq 0 \tag{18}$$

where equality holds only if $\mathbf{p} \equiv \mathbf{q}$. In general $KL(\mathbf{p}||\mathbf{q}) \neq KL(\mathbf{q}||\mathbf{p})$. In our experiments we use the symmetric version [2] that can be defined as

$$KL = \frac{KL(\mathbf{p}||\mathbf{q}) + KL(\mathbf{q}||\mathbf{p})}{2}. \tag{19}$$

First of all, we compute the KL divergence between each model and the median value of the overall cluster. Successively, for each cluster, the model with the minimum KL is selected. The *median model* of these considered models is compared with the real observations by KL. In Fig. 4 we show the KL obtained by varying the number of clusters.

We observe that varying the number of clusters this procedure permits to select the models that have the best approximation of the real observation (see [17] and [4] for more details). After our analysis, we conclude that the best approximation is obtained by using 6 clusters. Moreover, we stress that a lower KL does not necessarily correspond to the use of a large number of models. This suggest an approach for systematic reduction of ensemble data complexity and the use of the consensus matrix permits to obtain a more robust and realistic temporal analysis.

4   Conclusions

In this work we focused on models comparison in a multi-model air quality ensemble system. A methodology based on temporal hierarchical agglomeration is introduced for real-time simulation of pollutant dispersion or the accidental release of radioactive nuclides in the atmosphere. The proposed methodology is able to combine multiple temporal hierarchical agglomerations of dispersion models and it is based on fuzzy similarity relations combined by a transitive consensus matrix. The methodology is adopted for individuating models that characterize the predicted atmospheric pollutants from the ETEX-1 experiment. The results show that this methodology is able to discard redundant temporal information, reducing the data complexity. In the next future, further experimentations will be devoted to real pollutant dispersions (e.g., Fukushima) and different similarity relations also using ordinal sums.

Acknowledgments. This work was partially funded by the University of Naples Parthenope (*Sostegno alla ricerca individuale per il triennio 2016–2018* project).

References

1. Ascione, I., Giunta, G., Mariani, P., Montella, R., Riccio, A.: A grid computing based virtual laboratory for environmental simulations. In: Nagel, W.E., Walter, W.V., Lehner, W. (eds.) Euro-Par 2006. LNCS, vol. 4128, pp. 1085–1094. Springer, Heidelberg (2006). https://doi.org/10.1007/11823285_114
2. Ciaramella, A., et al.: Interactive data analysis and clustering of genomic data. Neural Netw. **21**(2–3), 368–378 (2008)
3. Napolitano, F., Raiconi, G., Tagliaferri, R., Ciaramella, A., Staiano, A., Miele, G.: Clustering and visualization approaches for human cell cycle gene expression data analysis. Int. J. Approximate Reasoning **47**(1), 70–84 (2008)
4. Ciaramella, A., Giunta, G., Riccio, A., Galmarini, S.: Independent data model selection for ensemble dispersion forecasting. In: Okun, O., Valentini, G. (eds.) Applications of Supervised and Unsupervised Ensemble Methods. SCI, vol. 245, pp. 213–231. Springer, Heidelberg (2009). https://doi.org/10.1007/978-3-642-03999-7_12
5. Ciaramella, A., Pedrycz, W., Tagliaferri, R.: The genetic development of ordinal sums. Fuzzy Sets Syst. **151**, 303–325 (2005)
6. Galmarini, S., Bianconi, R., Bellasio, R., Graziani, G.: Forecasting consequences of accidental releases from ensemble dispersion modelling. J. Environ. Radioactiv. **57**, 203–219 (2001)
7. Galmarini, S., et al.: Ensemble dispersion forecasting, part I: concept, approach and indicators. Atmos. Environ. **38**, 4607–4617 (2004)
8. Galmarini, S., et al.: Ensemble dispersion forecasting? Part II: application and evaluation. Atmos. Environ. **38**, 4619–4632 (2004)
9. Girardi, F., et al.: The ETEX project. EUR Report 181–43 EN, 108 pp. Office for official publications of the European Communities, Luxembourg (1998)
10. Giunta, G., Montella, R., Mariani, P., Riccio, A.: Modeling and computational issues for air/water quality problems: a grid computing approach. Nuovo Cimento C Geophys. Space Phys. **28**, 215–224 (2005)

11. Klement, E.P., Mesiar, R., Pap, E.: Triangular Norms. Kluwer Academic Publishers, Dordrecht (2001)
12. Meyer, H.D., Naessens, H., Baets, B.D.: Algorithms for computing the min-transitive closure and associated partition tree of a symmetric fuzzy relation. Eur. J. Oper. Res. **155**(1), 226–238 (2004)
13. Montella, R., Giunta, G., Riccio, A.: Using grid computing based components in on demand environmental data delivery. In: Proceedings of the Second Workshop on Use of P2P, GRID and Agents for the Development of Content Networks, UPGRADE-CN 2007, pp. 81–86 (2007)
14. Mirzaei, A., Rahmati, M.: A novel hierarchical-clustering-combination scheme based on fuzzy-similarity relations. IEEE Trans. Fuzzy Syst. **18**(1), 27–39 (2010)
15. Potempski, S., Galmarini, S., Riccio, A., Giunta, G.: Bayesian model averaging for emergency response atmospheric dispersion multimodel ensembles: is it really better? How many data are needed? Are the weights portable? J. Geophys. Res. **115** (2010). https://doi.org/10.1029/2010JD014210
16. Potempski, S., Galmarini, S.: Est modus in rebus: analytical properties of multimodel ensembles. Atmos. Chem. Phys. **9**(24), 9471–9489 (2009)
17. Riccio, A., Giunta, G., Galmarini, S.: Seeking for the rational basis of the median model: the optimal combination of multi-model ensemble results. Atmos. Chem. Phys. **7**, 6085–6098 (2007)
18. Riccio, A., Ciaramella, A., Giunta, G., Galmarini, S., Solazzo, E., Potempski, S.: On the systematic reduction of data complexity in multimodel atmospheric dispersion ensemble modeling. J. Geophys. Res. **117**(D5), D05314 (2012)
19. Sessa, S., Tagliaferri, R., Longo, G., Ciaramella, A., Staiano, A.: Fuzzy similarities in stars/galaxies classification. In: Proceedings of IEEE International Conference on Systems, Man and Cybernetics, pp. 494–4962 (2003)
20. Solazzo, E., Riccio, A., Van Dingenen, R., Valentini, L., Galmarini, S.: Evaluation and uncertainty estimation of the impact of air quality modelling on crop yields and premature deaths using a multi-model ensemble. Sci. Total Environ. **633**, 1437–1452 (2018)
21. Turunen, E.: Mathematics Behind Fuzzy Logic. Advances in Soft Computing. Springer, Heidelberg (1999)

A Neuro-Fuzzy Approach to Assess the Soft Skills Profile of a PhD

Antonia Azzini, Stefania Marrara[✉], and Amir Topalovic

Consortium for the Technology Transfer (C2T),
via Nuova Valassina, Carate Brianza, MB, Italy
{antonia.azzini,stefania.marrara,amir.topalovic}@consorzioc2t.it

Abstract. In this paper a framework aimed at representing the soft skills profile of a job seeker by means of a 2-tuple fuzzy linguistic approach and a Neuro fuzzy controller is presented.

The framework can be used in many contexts, in this work it is employed for designing a recommender system of candidates for recruiting agencies. The recommender system's Neuro fuzzy controller simulates the decision of a Human Resource (HR) manager in evaluating the soft skills profile of a candidate and proposes only the best profiles w.r.t. a set of preferences. The framework has been developed in the context of the Find Your Doctor (FYD) start up and applied to the PhD recruiting task, but it is easily applicable to any recruiting activity.

Keywords: 2-tuple fuzzy linguistic approach ·
Neuro-fuzzy controller · User profile

1 Introduction

In the last few years, several European countries have provided many support programs to help the transition of PhD graduates outside the academic research, but other states, especially in the Mediterranean area, are still far less accustomed to exploit this professional background. In Italy, in particular, the majority of job-placement agencies hardly even handle PhD profiles and Doctors are alone in the task of gaining visibility towards Human Resources (HR) offices and employers, who have little idea on how PhDs' experiences and curriculum vitae may be employed and enhanced in companies.

In fact, the majority of job-matching portals available online used by large companies HR offices and recruitment agencies are systems where PhDs are mostly in disadvantage compared to people with previous experience in business. In this context, most job-matching portals are usually based on searching keywords in a candidates CV, but the taxonomy used in job advertisings (also called *job vacancies*) is set on the vocabulary of the employers and usually does not match the words that a PhD would use to describe his/her experience. Many candidates and HR managers report that PhDs often score well in job-interviews,

© Springer Nature Switzerland AG 2019
R. Fullér et al. (Eds.): WILF 2018, LNAI 11291, pp. 134–147, 2019.
https://doi.org/10.1007/978-3-030-12544-8_11

but are mostly cut out from the selection at an earlier stage, due to low-level keywords filtering.

As a consequence, the need to define a system able to support a HR team in the recruitment activity of PhDs candidates is compelling. The idea has been born in the context of the Find Your Doctor (FYD) startup, which aims at becoming an important instrument dedicated to PhDs who are undergoing the transition outside Academia, with the mission of outlining the value of the research background as an asset for the development of companies and society as a whole. Within FYD a novel job-matching semi-automatic tool called SOON "Skills Out Of Narrative" is developed; it is based on a *narrative* approach for the soft skills: starting with a questionnaire of open questions, a semi-structured interview leads the candidate to reason on a given number of macro-skills usually considered important by employers, such as communication, relation, rigor, ability to face uncertainty and more. The focus is on the so-called soft skills [1], since the words used to express comparable content may vary more across contexts than for technical expertise. The questionnaire is designed to promote the candidate to first describe the meaning that he/she attributes to a given skill and only then to self-evaluate with respect to it, possibly grounding this evaluation in an actual experience. By the analysis of the text is then possible to infer a-posteriori a taxonomy that covers the possible meanings attributed to a certain skill by the respondent as described in [3].

Aim of this paper is to present an approach aimed at building a decision-support, pre-filtering tool able to guide the choices of a HR manager of a company in the PhD's profiles evaluations. In particular the representation of the soft skills of a PhD' profile is carried out by applying the fuzzy linguistic model presented by Porcel and Herrera-Viedma in [18]. Then, the obtained profile representation is used to design a novel recommendation system for PhD employment in the context of the FYD activity. The designed recommender system is able to periodically select and highlight in a pool of new profiles, those that better fulfill a set of preferences the HR Manager stated beforehand. The approach proposed in this paper has been studied in the PhD recruiting use case, but it can be easily extended for any kind of recruitment activity just by changing the reference skills taxonomy.

The novelty of this work is based on the way in which the profiles used by the recommender system are obtained, through the implementation of a Neuro fuzzy controller. Such a controller is indeed capable of learning the soft skills and of calculating a set of inference rules that are shown to be very similar to those that an HR human expert should otherwise calculate each time for each selected profile and for each individual skill. The claim of this work is emphasized by the definition of a trained model, whose behavior will be fully personalized since computed on a training dataset of pairs (profile, set of soft skills) that represents the decision behavior of a certain HR Manager rather than another. The remaining of the paper is organized as follow. After a brief summary of the related works already presented in the literature in Sect. 2, and a brief overview of the fuzzy linguistic model in Sect. 3, the soft skills profiles are defined in

Sect. 4. The overall architecture of the implemented approach is then reported in Sect. 5, while Sect. 6 describes the Neuro Fuzzy controller. Some preliminary experiments are presented and discussed, together with the obtained results, in Sect. 7, while Sect. 8 concludes with some final remarks.

2 Related Work

Several literature's studies, commissioned by the European Union, show that a good percentage of PhDs, who graduate across Europe, are not going to find long-term occupation in the Academia, but will eventually migrate towards both private and public companies and organizations [5]. In this area the skills identification is becoming one of the most important aspects for the HR team that have to spend the most part of the time in the profile analyses. Some works implement, for this reason, approaches based on machine learning and fuzzy systems to handle, for example the employability [12], that, together with skills, takes into account personal attributes for the teaching strategies development. Other works are based on cloud profile matching systems [6], while others again examine human resource (HR) practitioners subjective evaluations of job applicants as a function of specific traits, together with the assessment methods used to measure those traits [22].

On a different perspective, the automatic extraction of meaningful information from unstructured texts has been mainly devoted to support the e-recruitment process [14], e.g., to help human resource departments to identify the most suitable candidate for an open position from a set of applicants or to help a job seeker in identifying the most suitable open positions. For example, the work described in [20] proposes a system which aims to analyze candidate profiles for jobs, by extracting information from unstructured resumes through the use of probabilistic information extraction techniques as Conditional Random Fields [13]. Differently, in [23] the authors define Structured Relevance Models (SRM) and describe their use to identify job descriptions and resumes vocabulary, while in [10] a job recommender system is developed to dynamically update the job applicant profiles by analyzing their historical information and behaviors. Finally, the work described in [17] illustrates the use of supervised and unsupervised classifiers to match candidates to job vacancies suggesting a ranked list of vacancies to job seekers.

In a previous paper [3], a methodology based on machine learning aimed at extracting the soft skills of a PhD from a textual, self-written, description of her competencies was described: in that work, the soft skills were classified with respect to a proprietary taxonomy that includes around 60 different soft skills gathered into 6 skills areas. In this paper, the fuzzy representation of the PhD' soft skills profile based on the 2-tuple FML presented in Sect. 3 has been adopted. This proposal allows the HR operators to deal with the vagueness and uncertainty that are common when they assess the soft skills of a candidate during an interview, allowing a very flexible representation. Moreover, this proposal facilitates the creation of an enhanced PhD's profile by suggesting skills recognized by the ML classifier to be included into the evaluation performed by the HR operator.

3   A Brief Overview on the 2-Tuple Fuzzy Linguistic Approach

The Fuzzy Linguistic Model (FML) is based on the concept of linguistic variable [24], which has been successfully applied in many contexts as, for instance, Information Retrieval (IR) [9] and decision making [7]. The 2-tuple Fuzzy Linguistic Model [8] is built on FML with the aim to create a continuous representation model of information. This work considers the 2-tuple FLM [8], in order to create a continuous representation model of information.

Let $S = \{s_0, ..., s_n\}$ be the set of linguistic terms with odd cardinality where $s_{n/2}$ represents an indifference value, the other terms are symmetrically distributed around it. Each label is assumed to be represented by means of a triangular membership function, and all terms are distributed on an ordered scale. In this context, if a linguistic aggregation operator [7] computes a value $\alpha \in [0, n]$, and $\alpha \notin \{0, ..., n\}$, then an approximate function is used to represent the result in S. In this framework α is represented by means of 2-tuples (s_i, β_i), $s_i \in S$ and $\beta_i \in [-0.5, 0.5)$, where s_i is the linguistic label of the information, and β_i is a numerical value expressing the translation of the original result α to the closest index label i, within the linguistic term set S. This 2-tuple representation model defines a set of transformation functions between numeric values and 2-tuples: $\Delta(\alpha) = (s_i, \beta_i)$ and $\Delta^{-1}(s_i, \beta_i) = \alpha \in [0, n]$. The model also includes a negation operator, and a comparison of 2-tuple and aggregation operators [8]. Another important parameter is the "granularity of uncertainty", i.e, the cardinality of the set S of linguistic terms. This granularity can be different concept by concept, therefore the full approach also proposes a linguistic hierarchy based model to deal with it.

4   Dealing with the PhD' Soft Skills Profiles by Using the 2-Tuple Fuzzy Linguistic Approach

At the core of its activity, this system provides companies with PhDs' profiles that include both structured information, as for instance names, date of graduation, and non-structured or partially structured information, as curriculum vitae, textual descriptions of their competencies, reports of talks with the HR team.

4.1   Creating the PhDs Profiles

There are several techniques that can be applied to create the PhD profile, at present the vector based representation developed by the Information Retrieval researchers is investigated for the documents representation. In the vector based model a document D is represented as an m-*dimensional* vector, where each dimension corresponds to a distinct term and m is the total number of terms used in the collection of documents.

From this basis, in this approach a profile RP is composed by two vectors, $RP = (H, S)$ where H is the vector representing the hard skills of the PhD, while S represents her soft skills. The hard skills vector H is written as $(w_1, ..., w_m)$, where w_i is the weight of skill h_i that indicates its importance, while m is the number of skills defined in the European Skills/Competences, Qualifications and Occupations (ESCO) taxonomy[1]. The soft skills vector S is written as $(x_1, ..., x_n)$, where x_j is the weight of skill s_j and n is the number of skills defined in the soft skills FYD taxonomy described in [2]. If the profile RP does not contain a skill s_j or h_i then the corresponding weight is zero. The vector H is extracted from the PhD cv text, while the vector S is created extracting the skills information from the pills questionnaire. After a preprocessing phase in which the raw text is divided into sentences, each sentence is analyzed to extract the skills. At the moment two different solutions are developed and tested for this task: the first proposal is based on machine learning techniques and it has been presented in [3]. The second solution is based on Language Models and is an on-going research. At the end of this phase the PhD's profile RP is stored in the *Profile DB* and sent to the Evaluation Module for the recommendation phase.

PhD Soft Skills Representation. As already described, a PhD Profile RP is represented as a couple of skills vectors (H, S). H represents the PhD's hard skills while S is a vector of soft skills. In this paper the focus is on the soft skills vector. Every item of the vector is a linguistic 2-tuple value representing the degree the PhD possesses that soft skill. Note that a positional notation is used: $S = (s_1, s_2, .., s_k)$, where $s_j \in S$, with $j = \{1, ..., 60\}$, describes the linguistic degree assigned to the $j - th$ skill of the PhD. In order to allow a high flexibility we adopt a representation with 11 labels (L^{11}) to assess each skill (s_j).

- $L^{11} = \{L_0 = Null = N, l_1 = VeryVeryLow = VVL, l_2 = VeryLow = VL, l_3 = Low = L, l_4 = AlmostMedium = AM, l_5 = Medium = M, l_6 = MoreThanMedium = MM, l_7 = AlmostHigh = AH, l_8 = High = H, l_9 = VeryHigh = VH, l_{10} = Full = F\}$.

PhD Profile Computation. The vector of soft skills S is computed by taking into account two contributions. The first contribution to S is a vector HR of 60 skills, which represents the assessment the HR operator performs during an interview with the candidate. To allow a flexible assessment, but avoiding at the same time an excessive overhead for the HR operator, this vector adopts a representation with 5 labels (L^5) plus the NC value (NC = not classified) to describe each skill. Note that during an interview the HR operators explicitly assess only a few skills (usually 6 or 7), all other skills are set to NC by default.

---

[1] http://ec.europa.eu/social/main.jsp?catId=1326&langId=en.

- $L^5 = \{l_0 = VeryLow = VL, l_1 = Low = L, l_2 = Medium = M, l_3 = High = H, l_4 = Full = F\}$.

The second contribution to S is the vector ML of 60 skills that represents the automatic assessment of the candidate performed by the machine learning based classifier, presented in [3]. The ML classifier analyses the textual self description each PhD is required to provide when she enrolls to the database.

Each skill s_j of the PhD profile RP is computed in this way: (1) if HR($s_j = NC$) and ML($s_j = value$) then RP($s_j = value$); (2) otherwise the output value is computed by a fuzzy controller employing a Sugeno approach [21] with a *center of area* defuzzification. The set of rules describing the controller behavior has been manually evaluated on a set of 580 profiles with the contribution of the HR team in order to simulate the decision process of the HR operator. One of the core activity was to compare the set of rules automatically computed on a training dataset given by a HR manager with the rules manually created for a Mamdani [15] controller by the same person. Aim of this comparison is to show that the training of a Sugeno controller produces a set of rules so similar to those manually created by the HR manager to be considered an "automatic substitute" of the HR manager herself.

4.2   Recommending the Best PhD Profile for a Given Company

In this phase the HR manager receives, twice a week, a notification containing the best PhDs profiles w.r.t. the vision of the company represented as a vector of soft skills preferences (the *Soft-Skills Manager Interests HRMI*). The *Evaluation module* in Fig. 1 computes the distance between $HRMI$ and the soft skills component of the profiles RPs (S). At present the distance between S and $HRMI$ is simply computed by using the cosine similarity as follows:

$$sim(HRMI, SSRP) = \frac{HRMI * S}{\|HRMI\|\|S\|} \tag{1}$$

The Profiles are ranked w.r.t. this soft skills similarity value, while at this stage the hard skills vector allows the HR manager to automatically filter out profiles not in line with her actual interests on technical competencies.

5   Profiles Recommendation to Support the HR Manager Activity

In this section, the architecture of a personalized decision support tool able to recommend the best PhDs' profiles to a given HR manager is presented. As shown in Fig. 1, the HR manager receives, twice a week, a report containing the best profiles analized by the tool w.r.t. the company vision in terms of employed soft skills, in figure the *Interesting R.Profiles*. As previously described, the PhD registering to the job agency portal is asked to provide two textual descriptions of her competencies: a *curriculum vitae (CV)* and the questionnaire called "pills".

The technical competencies (or *hard skills*) of the PhD are extracted from the CV and saved in a vector as presented in [4]. However, no further discussion is given to those skills: the focus of this paper is indeed on the reasoning that computes the soft skills vector and on the selection of the PhDs profiles that better fulfill a set of preferences given by a HR manager. This set of preferences is manually assessed by the HR manager by using the HR Manager interface. The output produced is a vector, the *Soft-Skills Manager Interests*, containing a preference value for all the 60 skills in the FYD taxonomy [2].

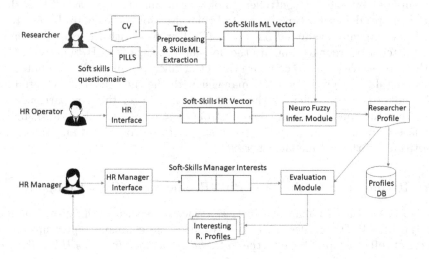

Fig. 1. The architecture of the recommender tool SOON.

Besides the hard skills, the *Text Preprocessing & Skills ML Extraction* module is in charge to extract a soft skills vector (in figure is the *Soft-Skills ML Vector*) from the textual content of the "pills". In most organization, recruiting agencies in particular, the HR manager does not meet personally all candidates but she is helped by a team of collaborators (in figure *HR Operators*), who are in charge to interview the PhDs. In this approach the HR Operator, during the interview, compiles a report regarding his evaluation of the soft skills of the candidate via a simple interface (in figure *HR Interface*) that guides the compilation of the *Soft-Skills HR Vector*. Please note that even if a complete evaluation of the candidate would require an assessment of each of the 60 soft skills available in the FYD taxonomy, this is not necessary here, and the operator is required to give an explicit assessment only to the few skills (usually 6/7) he really saw during the interview. The others skills in the *Soft-Skills HR Vector* are automatically set to NC (*not classified*). The two soft skills vectors are used by the *Neuro Fuzzy Inference Module* to compute the final soft skill vector that composes the PhD Profile RP. At this point, all profiles are saved in the *Profiles DB*, while the *Evaluation Module* compares the PhD Profiles w.r.t. the *Soft-Skills Manager*

Interests vector, and the most interesting profiles emerged in this comparison are then proposed to the HR Manager.

6 The Neuro-Fuzzy Approach

As previously reported, the approach implements a Neuro-fuzzy system aimed at supporting the evaluation of PhDs candidates by the HR manager. Given a certain dataset, the Neuro-fuzzy controller "learns" the rules that, from a certain input, produce a given output simulating the decision pattern the HR manager used when creating the dataset. In other words, the Neuro-fuzzy controller learns the reasoning used by the manager when evaluating profiles: in particular, the approach automates the definition of all the inference rules referring to a certain set of soft-skills inputs that otherwise have to be manually defined each time by the HR operator, and checked by the HR manager.

Generally speaking, Neuro-Fuzzy computing is a well defined methodology that integrates neural and fuzzy metrics. As also reported in the literature [25], one of the most important aspects regards the capability to incorporate the generic advantages of artificial neural networks, like massive parallelism, robustness, and learning in data-rich environments into a fuzzy system, where the modeling of imprecise and qualitative knowledge as well as the transmission of uncertainty is possible through the use of fuzzy logic.

The Adaptive Neuro Fuzzy Inference System (ANFIS) provides a systematic and directed approach for model building and gives the best possible design settings. Inspired by the idea of basing the fuzzy inference procedure on a feed-forward network structure, Jang proposed a fuzzy neural network model [11], employed to model nonlinear functions and identify nonlinear components online in a complex control system. In fact, such a system is able to bring the learning capabilities of neural networks for the fuzzy inference system definition.

In this work the defined Neural Network is responsible of the soft skills learning. The implemented Sugeno Fuzzy Inference System calculates a set of inference rules and selects the most performing skill. The final soft skill vector S that composes the overall PhD Profile RP is then produced as output.

6.1 The Neural Network Architecture

In order to incorporate the capability of learning from input/output data sets into a fuzzy inference system, a corresponding adaptive neural network is generated. An adaptive network is a multilayer neural network consisting of nodes and directional links through which nodes are connected.

As shown in Figure layer 1 is the input layer, layer 2 describes the membership functions of each fuzzy input. Layer 3 is the inference layer and normalization is performed in layer 4. Layer 5 gives the output and layer 6 is the defuzzification layer. The layers consist of fixed and adaptive nodes. Each adaptive node performs a particular function (the node function) on incoming signals, as explained in [25].

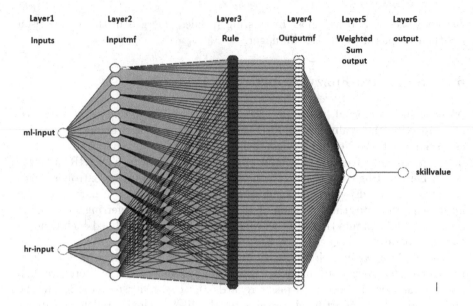

Fig. 2. Backpropagation neural network architecture

The learning module tunes the membership functions of a Sugeno-type fuzzy inference system by using the training input output data. In particular, it may consist of either BackPropagation (BP) [19], based on gradient descent optimization algorithm, or hybrid learning algorithm, based on the combination of Least Square Error (LSE) and gradient used into the BP [16] (Fig. 2).

The learning rule specifies how the parameters of adaptive nodes should be changed to minimize a prescribed error measure [11]. The change in values of the parameters results in change in shape of membership functions associated with fuzzy inference system. After a learning phase, the controller is able to generate the appropriate actions for the desired task.

The set of produced inference rules are shown to be very similar to those that an HR human expert should otherwise calculate each time for each selected profile and for each individual skill. We can claim the Neuro-fuzzy controller learns the reasoning used by the manager when evaluating profiles, therefore we can create a "personalized automatic manager" for any Company, just having the Company old candidates profiles dataset to train the controller.

7 A Preliminary Set of Experiments and Discussion

A first set of experiments has been carried out in order to test and validate the defined architecture. The details of the dataset and the parameter's setting are reported in the following.

The input variables of the designed Sugeno ANFIS correspond, respectively, to the skill evaluation vector ML produced by the machine learning based classifier previously implemented [3] (mlinput) and to the human resource expert evaluation vector HR (hrinput). As previously reported, the output corresponds to the skill value produced by the Neuro-Fuzzy evaluation that will contribute to define the resource profile soft skills vector.

The inputs are preprocessed and defined in a range from -0.5 to 0.5 (not included) values. The 75% of the dataset is used for the Neural Network training, while the remaining 25% is used for checking to validate the model. More precisely 435 PhDs define the training set and 145 define the test set.

Then, after a first initial tuning, as explained in Sect. 4 the inputs adopt "triangular" membership functions, with, respectively, 11 and 5 nodes for "machine learning based classifier vector" ML and "human resource expert evaluation vector" HR input variables.

The membership functions are aggregated by using a "T-Norm Product" operator to construct the Fuzzy $IF - THEN$ rules with a fuzzy antecedent part and "linear" consequent. The total number of rules is equal to 55, corresponding to the vector product of the nodes defined at level 2 of the neural network. All the rules detailed into $Layer3$ are calculated by the Sugeno Fuzzy Inference System.

The Neuro Fuzzy controller has been trained for 150 epochs, by using the most simple and widely used for neural network training, BackPropagation algorithm, particularly suitable for the learning of supervised multi layer neural networks [19]. After the training phase, the overall Neuro Fuzzy performances have been evaluated by testing the ANFIS model.

This Neuro Fuzzy approach has been validated by comparing the results obtained by applying the Neuro Fuzzy "trained" rules with the results obtained by applying a Fuzzy Mamdani controller where the rules are defined by an HR expert team. The surfaces in Fig. 3a and b show the results obtained. In particular, Fig. 3a shows the results coming from the Fuzzy Mamdani controller according to the rules manually defined by the HR manager on the test dataset, while Fig. 3b shows the results coming from the Neuro Fuzzy controller simply trained on the same dataset, with the rules defined by the ANFIS evaluation.

The surface values are displayed on a color scale from -0.5 (blue color) to 0.5 (not included) (green color) showing that the Neuro Fuzzy approach associates more importance to the HR evaluation with respect to the machine learning based one. The machine learning based evaluation is aimed at integrating and supporting the HR evaluation but it never overcomes it. Note that, according to the figure, when $hrinput$ is high, $skilloutput$ is high even with low $mlinput$ values, while with low $hrinput$ values $skilloutput$ is still low even if $mlinput$ values are high.

Even if the results are very similar (and this shows that the Neuro fuzzy controller "simulates" the HR expert reasoning), the graphic representation of the surfaces highlights how the Neuro Fuzzy approach presents a more harmonious surface, in which maximum and minimum values are reached with more precision (represented by the greater intensity of colors displayed by the surface). The results obtained from the two surfaces allow to say that the Neuro Fuzzy

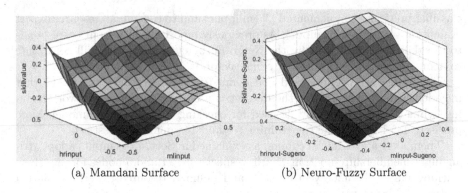

(a) Mamdani Surface   (b) Neuro-Fuzzy Surface

Fig. 3. Mamdani and Neuro fuzzy surface's comparison (Color figure online)

approach can be put in place of Mamdani, whenever an HR human expert is not available, thus representing a good automation of his way of reasoning when assessing profiles.

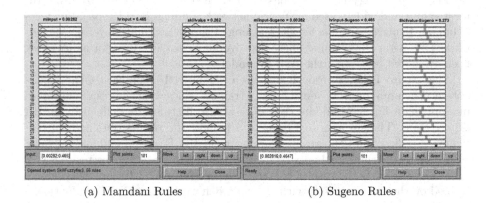

(a) Mamdani Rules   (b) Sugeno Rules

Fig. 4. Comparative example between Mamdani and Sugeno rules.

As also shown by the triangular rules, in Fig. 4a and b it is possible to see how the Sugeno model follows the trend of the human expert supported by the evaluation of the machine learning approach. Such a trend assumes, however, an attitude that takes into account both the assessments made by the human expert and those obtained from the machine learning.

Anyway, as reported in Table 1, the evaluations carried out on the skills leaving the Neuro Fuzzy system show a cautious and prudent attitude, especially in cases that report opposite assessments between the human expert and the machine learning approach.

Table 1. Example of the skill values obtained from the rules application.

ML input	HR input	Output skill value	
		Mamdani	Sugeno
0.0028	0.45	0.2500	0.2620
0.0028	−0.45	−0.1960	−0.1920
0.0028	0.0028	0.0042	−0.0007
0.45	0.45	0.3870	0.4230
0.45	−0.45	7.17e−4	1.84e−05
0.45	0.0028	0.1100	0.1000
−0.45	0.45	0.2240	0.2620
−0.45	−0.45	−0.3240	−0.3460
−0.45	0.0028	−0.1930	−0.230

8   Conclusions

In this paper a framework to represent the soft skills profile of a PhD by means of a 2-tuple fuzzy linguistic approach is presented. The framework can be used in many contexts: in this work it has been applied for designing a recommender system of candidates for recruiting purposes. The recommender system is defined through the implementation of a Neuro fuzzy controller that aims to simulate the decision of a HR team in evaluating the soft skills profile of a candidate. Among all the profiles assessed by the HR team with the support of the ML soft skills classifier, the system proposes to the HR manager only the profiles evaluated as the best w.r.t. a set of preferences representing the Company vision in terms of employees soft skills. The approach is capable of calculating a set of inference rules that are shown to be very similar to those that an HR human expert should otherwise calculate each time for each selected profile and for each individual skill.

Since the outcome of the framework is a vector representation of a profile, any matching function derived from the vector space model can be easily applied. The overall model behavior will be fully personalized since computed on a training dataset of pairs (profile, set of soft skills) that represents the decisional behavior of a certain HR Manager rather than another. Future work will investigate the application of the implemented Neuro Fuzzy recommender system in other contexts, like those regarding the human resources or recruitment areas of big companies, thus collecting more datasets to better test the hypothesis of creating a good "HR manager bot".

References

1. Allen, J., van der Velden, R.: The Flexible Professional in the Knowledge Society: New Challenges for Higher Education. Springer, Dordrecht (2011). https://doi.org/10.1007/978-94-007-1353-6

2. Azzini, A., Galimberti, A., Marrara, S., Ratti, E.: A taxonomy of researchers soft skills. Internal Report, Consortium for the Technology Transfer, C2T (2017)

3. Azzini, A., Galimberti, A., Marrara, S., Ratti, E.: A classifier to identify soft skills in a researcher textual description. In: Sim, K., Kaufmann, P. (eds.) EvoApplications 2018. LNCS, vol. 10784, pp. 538–546. Springer, Cham (2018). https://doi.org/10.1007/978-3-319-77538-8_37

4. Azzini, A., Galimberti, A., Marrara, S., Ratti, E.: SOON: supporting the evaluation of researchers' profiles. In: Uden, L., Hadzima, B., Ting, I.-H. (eds.) KMO 2018. CCIS, vol. 877, pp. 3–14. Springer, Cham (2018). https://doi.org/10.1007/978-3-319-95204-8_1

5. Box, S.: Transferable skills training for researchers supporting career development and research: supporting career development and research. OECD Publishing (2012)

6. Buga, A., Freudenthaler, B., Martinez-Gil, J., Nemes, S.T., Paoletti, L.: Management of accurate profile matching using multi-cloud service interaction. In: Proceedings of the 19th International Conference on Information Integration and Web-based Applications & Services, pp. 161–165. ACM, New York (2017)

7. Herrera, F., Herrera-Viedma, E., Verdegay, J.: Direct approach processes in group decision making using linguistic OWA operators. Fuzzy Sets Syst. 79(2), 175–190 (1996)

8. Herrera, F., Martínez, L.: A 2-tuple fuzzy linguistic representation model for computing with words. IEEE Trans. Fuzzy Syst. 8(6), 746–752 (2000)

9. Herrera-Viedma, E.: Modeling the retrieval process for an information retrieval system using an ordinal fuzzy linguistic approach. J. Assoc. Inf. Sci. Technol. 52(6), 460–475 (2001)

10. Hong, W., Zheng, S., Wang, H.: Dynamic user profile-based job recommender system. In: 2013 8th International Conference on Computer Science & Education (ICCSE), pp. 1499–1503. IEEE (2013)

11. Jang, J.: ANFIS adaptive-network-based fuzzy inference system 23, 665–685 (1993)

12. Kumar, S., Kumari, R., Sharma, V.: Adaptive neural fuzzy inference system for employability assessment. In: Int. J. Comput. Appl. Technol. Res

13. Lafferty, J., McCallum, A., Pereira, F.: Conditional random fields: probabilistic models for segmenting and labeling sequence data. In: Proceedings of the Eighteenth International Conference on Machine Learning, ICML, vol. 1, pp. 282–289 (2001)

14. Lee, I.: Modeling the benefit of e-recruiting process integration. Decis. Support Syst. 51(1), 230–239 (2011)

15. Mamdani, E., Assilian, S.: An experiment in linguistic synthesis with a fuzzy logic controller. Int. J. Man Mach. Stud. 7(1), 1–13 (1975)

16. Pfister, M., Rojas, R.: Hybrid learning algorithms for feed-forward neural networks. In: Reusch, B. (ed.) Fuzzy Logik, pp. 61–68. Springer, Berlin, Heidelberg (1994). https://doi.org/10.1007/978-3-642-79386-8_8

17. Poch, M., Bel, N., Espeja, S., Navıo, F.: Ranking job offers for candidates: learning hidden knowledge from big data. In: Language Resources and Evaluation Conference (2014)

18. Porcel, C., Herrera-Viedma, E.: Dealing with incomplete information in a fuzzy linguistic recommender system to disseminate information in university digital libraries. Knowl.-Based Syst. **23**(1), 32–39 (2010)
19. Rumelhart, D.E., Hinton, G.E., Williams, R.J.: Parallel distributed processing: explorations in the microstructure of cognition. In: Learning Internal Representations by Error Propagation, vol. 1, pp. 318–362. MIT Press, Cambridge, MA, USA (1986)
20. Singh, A., Rose, C., Visweswariah, K., Chenthamarakshan, V., Kambhatla, N.: PROSPECT: a system for screening candidates for recruitment. In: Proceedings of the 19th ACM International Conference on Information and Knowledge Management, pp. 659–668. ACM (2010)
21. Takagi, T., Sugeno, M.: Fuzzy identification of systems and its applications to modeling and control. IEEE Trans. Syst. Man Cybern. **SMC–15**(1), 116–132 (1985)
22. Topor, D.J., Colarelli, S.M., Han, K.: Influences of traits and assessment methods on human resource practitioners' evaluations of job applicants. J. Bus. Psychol. **21**(3), 361–376 (2007)
23. Yi, X., Allan, J., Croft, W.B.: Matching resumes and jobs based on relevance models. In: Proceedings of the 30th Annual International ACM SIGIR Conference on Research and Development in Information Retrieval, pp. 809–810. ACM (2007)
24. Zadeh, L.A.: The concept of a linguistic variable and its application to approximate reasoningi. Inf. Sci. **8**(3), 199–249 (1975)
25. Zaheeruddin, Garima: A neuro-fuzzy approach for prediction of human work efficiency in noisy environment. Appl. Soft Comput. **6**, 283–294 (2006)

Fuzzy Decision Making

Two SMART Fuzzy Aggregation Operators

Andrea Capotorti[1]([⊠]) [iD] and Gianna Figà-Talamanca[2] [iD]

[1] Dip. Matematica e Informatica, Università degli Studi di Perugia, Perugia, Italy
andrea.capotorti@unipg.it
[2] Dip. Economia, Università degli Studi di Perugia, Perugia, Italy
gianna.figatalamanca@unipg.it

Abstract. In this paper we introduce new disjunction and conjunction (named SMART) for merging, without any exogenous components, any number of fuzzy memberships. The present proposals are n-ary operators, based on a specific adaptation of Marzullo's algorithm, that depart from the usual fuzzy mean on the base of the agreement/disagreement among the different memberships. These different operators are suitable to be applied in any model where the same quantity (usually a parameter) can be measured (estimated) through different fuzzy memberships stemming by different sources of information. In our previous contributions we have considered the special case of two fuzzy memberships that were elicited for the volatility parameter in an hybrid fuzzy-stochastic model for option pricing. Here we adopt the same example to have an application at hand and to compare our new proposed operators with the ordinary fuzzy mean; nevertheless, the operators can be applied to merge any n memberships which are candidates to represent the same fuzzy number.

Keywords: Smart average operators · Fuzzy mean · Merging · Fuzzy option pricing

1 Introduction and Motivation

The need to define a suitable merging operator arises in our contribution [1]; in the quoted paper we introduced a methodology for membership elicitation on the hidden volatility of a risky asset through both the historical volatility estimator $\hat{\sigma}$ and the estimator $\nu = \mathrm{VIX}/100$, based on VIX [19]. Our elicitation proposal was based on the Coletti and Scozzafava [3] interpretation of membership functions as coherent conditional probability assessments, integrated with observed data, expert evaluations and simulation results. The peculiarity of our procedure was to deal with alternative sources of information, though leaving as an open problem the search for proper *fusion operators* to merge the different memberships stemmed by the different sources.

The choice of a fusion operator, given the variety of information items, is not unique and heavily context-dependent. Authors in [14] affirm that there are

© Springer Nature Switzerland AG 2019
R. Fullér et al. (Eds.): WILF 2018, LNAI 11291, pp. 151–163, 2019.
https://doi.org/10.1007/978-3-030-12544-8_12

more than 90 different fuzzy operators proposed in the literature for fuzzy set operations and there is a wide family of aggregation functions with predictable and tailored properties (e.g. those analyzed in [9,10]) related to different areas and disciplines.

Classes of aggregation functions include triangular norms and conorms, copulas, means and averages, and those based on nonadditive integrals [8].

But the main point is that the role of fuzzy sets in merging information can be understood in two ways: either as a tool for extending estimation techniques to fuzzy data (this is done applying the extension principle to classical estimators, and methods of fuzzy arithmetics - see [5] for a survey); or as a tool for combining possibility distributions that represent imprecise pieces of information (then fuzzy set-theoretic operations are instrumental for this purpose - see [7] for a survey).

In view of this dichotomy, a bridge between the "estimation" and the "fusion" is represented by fuzzy arithmetic mean (named simply "fuzzy mean" in the sequel). In fact it is a basic operation for estimation and also a fuzzy set-theoretic connective. Anyhow it is well known that fuzzy mean, even more than the crisp one, suffers from the drawback of being insensitive with respect the agreement or disagreement among original values (we can have the same mean between two very close values ad well as between two very distant). Hence, fuzzy mean remains as a reference operator from which we want to move away.

A first attempt to solve such problem has been the proposal of *constrained merging* in [6, Sect. 6.6.2]. We borrow from it the motivation of including a "smart" component in the averaging process to address conflicts in the data to be fused. Since the *constrained merging* was based on an original Yager's "intelligent" component [20], we have named our operators as "SMART". SMART is in fact both a synonym of "intelligent" and a, commonly used, acronym for "**S**pecific, **M**easurable, **A**chievable, **R**ealistic and **T**ime-related", most of which are also goals of our approach. The main difference with respect Yager's proposal is that we do not make use of any exogenous "combinability function" that was instead used in [20].

Note that we do not look to t-norms and t-conorms, as e.g. done in [3], because the different memberships we want to merge are not different "claims" (estimations) given by the same "subject" (estimator) over an unknown parameter, but are the same "claim" (estimation) obtained by different "subjects" (estimators). Hence we need to "average" the memberships we have at hand; to this aim we resort to something similar to the fuzzy mean, but with different behaviors if the estimators give alternative ("disjunctive") or concomitant ("conjunctive") information.

In [2] we have given a first formulation of merging operators characterized by two distinct deformations with respect to the usual fuzzy mean, anyhow such proposal suffered from being tailored just for merging two fuzzy numbers, with two ad hoc binary deformations. Here, we consistently overcome such drawbacks with a new proposal so that we can fuse any n fuzzy numbers with a *disjunctive* and with a *conjunctive* mean operators. In fact, in the new operators we avoid any extra, and hence arbitrary, component.

For the sake of clarity, we set up our framework by giving in Sect. 2 the preliminary concepts at the base of our proposal; in Sect. 3 we give details for our n-ary operators for disjunction and conjunction; while in Sect. 4 we furnish some preliminary result about their entailment on the pricing for options based on the $S\&P500$ Market Index. Section 5 gives some concluding remarks.

2   Preliminaries

Membership functions $\mu : \mathbb{R} \to [0,1]$ of the fuzzy set of possible values of a random variable X are usually viewed as imprecise values.

From a practical point of view, we will profit from membership characterization through α-cuts, that for a generic j-th membership result as

$$\mu j^{\alpha} = \{x \in \mathbb{R} : \mu(x) \geq \alpha\}, \quad \alpha \in [0,1]. \tag{1}$$

In particular, since we will deal with fuzzy numbers, i.e. memberships with nested, compact and close α-cuts, they reduce to closed intervals characterized by a left and a right extreme:

$$\mu j^{\alpha} = [\mu j_l^{\alpha}, \mu j_r^{\alpha}]. \tag{2}$$

Our operators work on these extremes, α-cut by α-cut, hence, in line with [11], "horizontally" with respect the membership function definition. Figure 1 shows such quantities for two memberships, together with other specific quantities that will be involved in our averaging operators as described below.

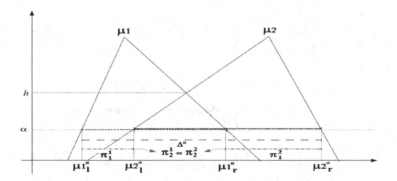

Fig. 1. α-cuts of two memberships $\mu1$ and $\mu2$ and their characteristic values for their SMART averages

The agreement/disagreement among n different α-cuts can be expressed by exploiting the q-relaxed intersection computation applied in [12] and based on the Marzullo's algorithm [15]. This choice avoids any external, hence arbitrary, imposition.

Marzullo's algorithm efficiently computes the shortest interval shared by the maximum number of intervals. But the original Marzullo's algorithm just returns the optimal, i.e. shortest, interval which is consistent with the maximum number of inputs. For example, among $\mu1^\alpha = [8,9]$, $\mu2^\alpha = [8,12]$ and $\mu3^\alpha = [10,12]$ it will produce $[8,9]$ as the shortest intersection between two of the original intervals.

We actually want to control for intersections among all subsets of the n α cuts; to this aim we can modify the algorithm by taking trace of the different numbers of intersecting intervals, so that the results are now specific weights π_f^j representing the overlap lengths among f α-cuts, $f = 1, \ldots, n$, inside the j-th α-cut, $j = 1, \ldots, n$ (the detailed procedure will appear in a forthcoming publication). Just to give an idea, with the three α-cuts mentioned before we would obtain:

$$\pi_1^1 = 0 \qquad \pi_1^2 = 10 - 9 = 1 \qquad \pi_1^3 = 0 \tag{3}$$
$$\pi_2^1 = (9-8) = 1 \; \pi_2^2 = (9-8) + (12-10) = 3 \; \pi_2^3 = (12-10) = 2 \tag{4}$$
$$\pi_3^1 = 0 \qquad \qquad \pi_3^2 = 0 \qquad \qquad \pi_3^3 = 0. \tag{5}$$

Others π_f^j, with $f, j = 1, \ldots, 3$, stemming from α-cuts of three membership functions are depicted in Fig. 2, where we also visually anticipate the aforementioned deformation of our disjunction towards the canonical operator, i.e. the max, with respect the fuzzy arithmetic mean.

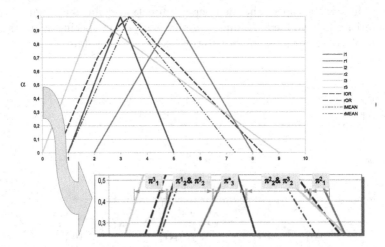

Fig. 2. SMART disjunction (dashed line) among 3 fuzzy numbers (solid lines) compared to the fuzzy arithmetic mean (dashed-dotted line). The zoom shows the relaxed intersections computed through adapted Marzullo's algorithm.

In few words, we can say that each π_f^j measures the part of the j-th α-cut involved into the intersection among f α-cuts. It comes straightforwardly that the sum of the various weights associated to a specific α-cut gives its length:

$$\sum_{f=1}^{n} \pi_f^j = \delta_j = \mu j_r^\alpha - \mu j_l^\alpha. \tag{6}$$

3 The General SMART Disjunction and Conjunction

As already outlined, the SMART fuzzy disjunction $\underline{\vee}$ and conjunction $\overline{\wedge}$ operators, that we are going to define, are specific modifications of fuzzy arithmetic mean. They are based on full/partial overlap among the α-cuts $[\mu j_l^\alpha, \mu j_r^\alpha]$, $j = 1, 2, \ldots, n$, of the fuzzy memberships to be merged.

Since the same procedure is applied for each level α, we omit the superscript α whenever not strictly necessary.

Arithmetic, crisp as well as fuzzy, mean is characterized by a convex combination of the α-cuts' extremes with uniform coefficients $\frac{1}{n}$. Our operators realize in its modifications by tuning such coefficients inducing a specific aimed behavior of the merging: deforming the mean towards the canonical maximum for the disjunction and towards the canonical minimum for the conjunction. In the former case, since the α-cuts of the maximum are simply the intervals with the more external extremes, our operator gives more weight to the outer bounds and less weight to the inner ones. On the contrary, convex combination for the latter give more weight to the inner bounds with respect the outer ones. Let us propose in what follows a possible choice of proper weights to achieve such behaviors.

As already anticipated in the previous section, the new coefficients will be based on the weights π_f^j obtained by the adapted Marzullo's algorithm to emphasize the partial agreement or disagreement among the different memberships.

3.1 SMART Disjunction

For the disjunctive operator the extremes of the α-cuts

$$[(\mu 1 \underline{\vee} \ldots \underline{\vee} \mu n)_l^\alpha, (\mu 1 \underline{\vee} \ldots \underline{\vee} \mu n)_r^\alpha] \tag{7}$$

are computed as convex combinations of the original ones with coefficients

$$\frac{1}{n}(1 + \epsilon_j) \qquad j = 1, \ldots, n-1. \tag{8}$$

with

$$\epsilon_j = \begin{cases} \frac{\sum_{f=1}^{n} \frac{1}{f} \pi_f^j}{\Delta} & \text{if } \Delta \neq 0 \\ 0 & \text{otherwise} \end{cases}, \quad j = 1, \ldots, n-1, \tag{9}$$

and $\Delta = \max\{\mu i_r^\alpha\}_{i=1}^n - \min\{\mu i_l^\alpha\}_{i=1}^n$. Equations (8, 9) display the weighted contributions of the $n-1$ more relevant extremes, i.e. the first $n-1$ outer ones. The n-th coefficient, associated to the inner extreme, is simply given by

$$\frac{1}{n}(1 - \sum_{j=1}^{n-1} \epsilon_j). \tag{10}$$

Note that the division by Δ in (9) makes the contribution of each j-th α-cut relative with respect to the length of the α-cut of the maximum. Moreover, whenever the original memberships have at least two distinct cores (i.e. distinct values with maximum membership 1) the various ϵ_j tend to zero whenever α tends to one so that the upper side of the merged membership converge to the core of the fuzzy arithmetic mean.

Let us also stress that, since each weight π_f^j is shared exactly by f α-cuts, the fractions $\frac{1}{f}$ let $\sum_{j=1}^{n-1} \sum_{f=1}^{n} \frac{1}{f} \pi_f^j \leq \Delta$ so that $0 \leq \sum_{j=1}^{n-1} \epsilon_j \leq 1$ and hence coefficients (8, 10) are well defined to be used for a convex combination of α-cuts' extremes.

The result of such operator has been already shown in Fig. 2 where the deformation with respect the usual fuzzy mean is evident. In fact, for each α-cut, except for the core associated to $\alpha = 1$, we obtain wider intervals with extremes stretched toward those of the canonical disjunction, i.e. the max operator. It is clear from Fig. 2 that taking into account the strength of the disagreements among the different α-cuts leads to a wider membership, and hence a more vague estimation, with respect to the fuzzy mean.

3.2   SMART Conjunction

The construction of the conjunctive operator $\bar{\wedge}$ follows the same basic rule, but needs a more articulated formulation.

First of all, we need to take as a reference value the level h of not empty intersection among all the n α-cuts of the original fuzzy numbers (as visualized in Fig. 1 for the case $n = 2$). In fact, the extremes of the α-cuts of the merging operator

$$[(\mu 1 \bar{\wedge} \ldots \bar{\wedge} \mu n)_l^\alpha, (\mu 1 \bar{\wedge} \ldots \bar{\wedge} \mu n)_r^\alpha], \tag{11}$$

are computed differently whether α is below or above such level h. In the former case the extremes are again obtained as direct convex combinations of the original ones with coefficients

$$\frac{1}{n}(1 + \gamma_j) \qquad j = 1, \ldots, n-1, \tag{12}$$

where now the quantities

$$\gamma_j = \frac{\sum_{f=1}^{n} \frac{1}{n+1-f} \pi_f^j}{\sum_{k=1}^{n} \sum_{f=1}^{n} \frac{1}{n+1-f} \pi_f^k}, \tag{13}$$

reflect the weighted normalized contribution of the $n-1$ more relevant extremes, here the first $n-1$ inner ones.

Whenever all the n α-cuts coincide the γ_j can be consistently set to zero.

Note that coefficients $\frac{1}{n+1-f}$ let each part in the numerator of the γ_j to be proportional to the number of overlaps (representing the agreement).

Again, the n-th coefficient, i.e. that associated to the outer extreme, is simply given by

$$\frac{1}{n}(1 - \sum_{j=1}^{n-1} \gamma_j). \tag{14}$$

Since the denominator in (13) is simply a normalizing constant, it comes to the fore that $\sum_{j=1}^{n-1} \gamma_j \leq 1$ so that (12, 14) define a proper convex combinations.

For the other cases, that is when α is above the level h, the definition is more subtle since many subgroups of intersections involving two or more memberships can be identified. In fact, assume we have k subgroups of indexes $J_l \subset \{1, \ldots, n\}$, each with cardinality n_l, $l = 1, \ldots, k$. Note that, in general, subgroups share some elements, hence we have $\sum_{l=1}^{k} n_l \geq n$. The logic underlying our approach is to compute first the conjunctive operator $\overline{\wedge}$ within each subgroup J_l, $l = 1, \ldots, k$, obtaining k intermediate α-levels, and secondly to merge them by applying the disjunctive operator $\underline{\vee}$ above defined.

In formulas, for each $j \in J_l$, we compute

$$\overline{\gamma}_j = \frac{\sum_{f=1}^{n_l} \frac{1}{n_l + 1 - f} \pi_f^j}{\sum_{k \in J_l} \sum_{f=1}^{n} \frac{1}{n+1-f} \pi_f^k} \tag{15}$$

and we make the convex combination of the extremes of the α-cuts in J_l with coefficients

$$\frac{1}{n_l} \left((1 + \overline{\gamma}_1) \ldots, (1 + \overline{\gamma}_{n_l - 1}), (1 - \sum_{j=1}^{n_l - 1} \overline{\gamma}_j) \right) \tag{16}$$

where the order is from the inner to the outer. Once we obtain these k conjunctions of subgroups, characterized by the extremes

$$\overline{\mu}i_l \quad \text{and} \quad \overline{\mu}i_r \ , \quad i = 1, \ldots, k, \tag{17}$$

we compute the new relaxed intersection coefficients $\overline{\pi}_f^l$, $l, f = 1, \ldots, k$, and consequently obtain new weights

$$\overline{\epsilon}_i = \frac{\sum_{f=1}^{k} \frac{1}{f} \overline{\pi}_f^i}{\overline{\Delta}}, \quad \text{with } \overline{\Delta} = \max\{\overline{\mu}i_r\}_{i=1}^{k} - \min\{\overline{\mu}i_l\}_{i=1}^{k}, \tag{18}$$

that can be plugged in new coefficients for the convex combination to take into account the cardinality of each contribution:

$$\frac{1}{\sum_{i=1}^{k} n_i} \left(n_1(1 + \overline{\epsilon}_1), \ldots, n_{k-1}(1 + \overline{\epsilon}_{k-1}), (n_k - \sum_{l=1}^{k-1} n_l \overline{\epsilon}_l) \right). \tag{19}$$

Note that in this case the order is from the outer to the inner.

These steps must be iterated for increasing values of α for which the partial intersections may change, and this happens whenever one of indexes $\overline{\pi}_f^j$ vanish.

It remains the problem that the merged α-cuts obtained through this two-steps aggregation procedure must be "glued" to those of the levels below.

This can be obtained by a proper translation and deformation of the extremes. Due to the necessary discretization of the α levels to be actually considered (see [1]), we can formulate the transformation by referring to two consecutive values α_1 and α_2 so that the transformed new α_2-cut will have extremes computed recursively by:

$$\widehat{\mu}_l^{\alpha_2} = \widehat{\mu}_l^{\alpha_1} + \varrho^{\alpha_1} |\overline{\mu}_l^{\alpha_2} - \overline{\mu}_l^{\alpha_1}| \tag{20}$$

$$\widehat{\mu}_r^{\alpha_2} = \widehat{\mu}_r^{\alpha_1} - \varrho^{\alpha_1} |\overline{\mu}_r^{\alpha_2} - \overline{\mu}_r^{\alpha_1}| \tag{21}$$

with

$$\varrho^{\alpha_1} = \frac{\widehat{\mu}_r^{\alpha_1} - \widehat{\mu}_l^{\alpha_1}}{\overline{\mu}_r^{\alpha_1} - \overline{\mu}_l^{\alpha_1}} \tag{22}$$

and where the "overlined" extremes are those obtained by the inter/intra sub-group merging and the "hatted" ones are those obtained by the "gluing" transformation at the specified levels.

Result of such procedure can be seen in Fig. 3 where it is possible to appreciate the aforementioned deformation of our conjunction towards the canonical operator, i.e. the min, with respect the fuzzy arithmetic mean; the convex combinations vary for each level α according to changes of partial overlaps. Note that in the picture this last aspect is emphasized by the rough discretization adopted for the α levels; for a finer mesh, as the one used in the empirical application, the "gluing" process produces smoother memberships.

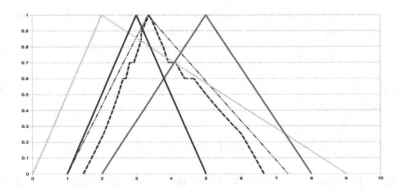

Fig. 3. SMART conjunction (dashed line) among 3 fuzzy numbers (solid lines) compared to the fuzzy arithmetic mean (dashed-dotted line).

It is clear from Fig. 3 that taking into account the strength of the agreements among the different α-cuts leads to a narrower membership, and hence a less vague estimation, with respect to the fuzzy mean.

4   A Practical Application of the SMART Operators to Option Pricing

As an illustrative example we apply the generalized SMART operators described in previous sections in the option pricing framework of CRR [4]. In particular, we describe the dynamics of the risky asset by a binomial tree with $N = 6$ periods. The crucial point is to obtain a fuzzy number for the volatility parameter σ; as also well described in [17], we can infer about the volatility through different estimators. Since we aim at eliciting the volatility parameter of the $S\&P500$ index, we build the membership elicitation for σ on both the historical volatility estimator $\hat{\sigma}$ and the estimator $\nu = \text{VIX}/100$ (based on the VIX Index that represents the one-month ahead integrated volatility implied by option prices on the $S\&P500$ index). The elicitation procedure we adopt here follows the idea we developed in [1]; however aggregations of different memberships are now performed with our new generalized SMART operators.

We can summarize the main steps in the following items (for further details refer to the cited paper):

1. The available time series are considered in order to elicit a pseudo-membership for each estimator;
2. on the test date the most probable scenarios are selected according to the current value of the estimator and depending on the pseudo-memberships obtained in step 1;
3. a probability-possibility transformations is applied to the simulating distributions (Uniform, LogNormal, Gamma) corresponding to the selected scenario(s);
3a. whenever there are more than one membership associated to a simulating model, they are merged via our **disjunction** operator;
4. the memberships, stemming from the different simulation models, are merged into a single membership through our **disjunction**.
5. steps 1–4 are performed for both $\hat{\sigma}$ and ν leading to two different fuzzy numbers;
6. the memberships associated to the two different estimators are merged via our **conjunction** to obtain a single fuzzy number for the volatility σ.

Once we have obtained a single aggregated membership function for σ we can pass to compute fuzzy prices for options traded on the test date.

Market bid-ask prices for the quoted options can be compared to the corresponding fuzzy option prices, obtained either via our approach or via fuzzy mean. Comparisons of the two can be based on the computation of a proper similarity index (see, e.g., [18]).

In the following we can briefly describe the main crucial points of the previous items (a detailed analysis is demanded to a forthcoming publication).

For the elicitation step n.1 the $S\&P500$ daily returns from January 1960 to September 2016 and daily observations of the VIX index from January 1990 to September 2016 are considered respectively for $\hat{\sigma}$ and $\hat{\nu}$. Based on the values of the two estimators on the test date the current scenario is selected; on October 5, 2016 all simulating models agree on the "low volatility" scenario for both estimators.

After obtaining the three memberships associated to the simulating distributions (Uniform, LogNormal, Gamma) corresponding to the "low volatility" scenario, we can merge them with the $\vee$ operator; this is done for both estimators. Finally the two outcomes are fused via the $\overline{\wedge}$, obtaining the fuzzy representation $\tilde{\sigma} = \mu_{\widehat{\sigma}_{obs}} \overline{\wedge} \mu_{\nu_{obs}}$ for the volatility parameter σ, which is shown in Fig. 4.

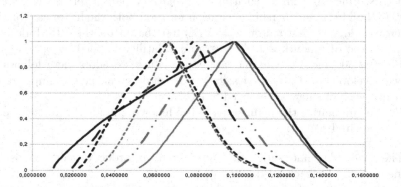

Fig. 4. The merging results: disjunction of the fuzzy numbers stemming from different models for $\widehat{\sigma}$ (dashed lines on the left) and for ν (solid lines on the right) and their final conjunction (dashed-dotted lines on the center), by applying our $\vee$ and $\overline{\wedge}$ (black) or the fuzzy mean (gray).

Once we have such fuzzy number for $\tilde{\sigma}$, it is possible to price options by a straightforward extension of standard CRR model in [4] to a fuzzy multi-period binomial model. Our explicit numerical evaluation of each α-cut of the fuzzy number for $\tilde{\sigma}$ allows us to take advantage of others contributions available in literature for each step of the pricing procedure. In particular:

- from $\tilde{\sigma}$ to the fuzzy "UP" and "DOWN" jump factors *(Zadeh's extension principle [21])*

$$[\underline{u}^\alpha, \overline{u}^\alpha] = [e^{\underline{\sigma}^\alpha \sqrt{\Delta t}}, e^{\overline{\sigma}^\alpha \sqrt{\Delta t}}] \qquad [\underline{d}^\alpha, \overline{d}^\alpha] = [e^{-\overline{\sigma}^\alpha \sqrt{\Delta t}}, e^{-\underline{\sigma}^\alpha \sqrt{\Delta t}}] \; ; \qquad (23)$$

- from $\tilde{u}$ and $\tilde{d}$ to the fuzzy risk neutral probabilities *(Muzzioli and Torricelli [16])*

$$[\underline{p}_u^\alpha, \overline{p}_u^\alpha] = \left[\frac{e^{r \Delta t} - \overline{d}^\alpha}{\overline{u}^\alpha - \overline{d}^\alpha}, \frac{e^{r \Delta t} - \underline{d}^\alpha}{\underline{u}^\alpha - \underline{d}^\alpha} \right] \; [\underline{p}_d^\alpha, \overline{p}_d^\alpha] = \left[\frac{\underline{u}^\alpha - e^{r \Delta t}}{\underline{u}^\alpha - \underline{d}^\alpha}, \frac{\overline{u}^\alpha - e^{r \Delta t}}{\overline{u}^\alpha - \overline{d}^\alpha} \right] \; ;$$
$$(24)$$

- from $\tilde{p}_u$ and $\tilde{p}_d$ to option price (e.g. call) *(Li and Han [13])*

$$[\underline{C}_0^\alpha, \overline{C}_0^\alpha] = e^{-r N \Delta t} \left[\sum_{i=0}^{N} (\underline{p}_u^\alpha)^i (\underline{p}_d^\alpha)^{N-i} \underline{C}_{N,i}^\alpha, \sum_{i=0}^{N} (\overline{p}_u^\alpha)^i (\overline{p}_d^\alpha)^{N-i} \overline{C}_{N,i}^\alpha \right] \qquad (25)$$

with

$$[\underline{C}_{N,i}^{\alpha}, \overline{C}_{N,i}^{\alpha}] = \left[\max(S_0(\underline{u}^{\alpha})^i(\underline{d}^{\alpha})^{N-i}- K, 0), \max(S_0(\overline{u}^{\alpha})^i(\overline{d}^{\alpha})^{N-i}- K, 0)\right].$$

(26)

The overall performance can be evaluated by measuring a proper fuzzy similarity of our option prices with respect to the market bid-ask spread, considered as a "crisp interval", for quoted options on a test date. This is done for options quoted on October 5, 2016 by using he similarity index proposed in [18]. A comparison with an analogous procedure where the fuzzy arithmetic mean is applied in place of our disjunction and conjunction operators has been made.

In Fig. 5 we plot the fuzzy price obtained by applying either our SMART merging operators (solid) or the fuzzy mean (dashed) for two examples of options traded on October, 5, 2016; the bid-ask interval is also included in the picture. In both examples our procedure gives narrower memberships and in the plot on the right the core value is closer to the mid-point of the bid-ask. The similarity values are also better for our merging operators rather than for the fuzzy mean option prices; in particular for the above examples the similarity values are 39% and 28% of our merging operators against 27% and 20% obtained through the fuzzy mean.

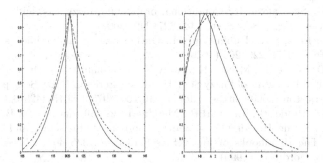

Fig. 5. Market bid-ask (crisp interval), "smart" fuzzy prices (solid), "arithmetic mean" fuzzy prices (dashed) for two examples of CALL options traded on October 5, 2016 with expiration in one month and strike price K = 2040 (left panel) and K = 2210 (right panel). The $S\&P500$ value is $S_0 = 2159.7$.

5   Conclusion

In this paper we propose SMART fuzzy operators to aggregate sources of information and models in order to elicit a unique fuzzy number for some parameter. The operators can be applied to merge any number of fuzzy numbers in a disjunctive or conjunctive way. Such behaviors are obtained through slight modifications of the fuzzy mean that is usually adopted to merge fuzzy estimations of the same quantity obtained in different ways. The definition of the proposed operators is based on an adapted Marzullo's algorithm to properly measure the weight assigned to overlapping intervals in each of the α-cuts of the memberships to be merged. The whole methodology is illustrated within the problem of estimating stock volatility in financial markets to obtain fuzzy option prices, computed with the CRR model. Preliminary results are promising since the proposed merging procedure produces fuzzy option prices which are closer to bid-ask prices with respect to the ones obtained by applying the classical fuzzy mean, whenever closeness is measured by fuzzy similarity as defined in [18]. A full and systematic analysis is demanded to a forthcoming contribution, as well as the adoption of different similarities indexes more apt to compare crisp interval (the market bid-ask spread) with fuzzy numbers (the fuzzy option prices).

References

1. Capotorti, A., Figá-Talamanca, G.: On an implicit assessment of fuzzy volatility in the Black and Scholes environment. Fuzzy Sets Syst. **223**, 59–71 (2013)
2. Capotorti, A., Figá-Talamanca, G.: Smart fuzzy weighted averages of information elicited through fuzzy numbers. In: Laurent, A., Strauss, O., Bouchon-Meunier, B., Yager, R.R. (eds.) IPMU 2014. CCIS, vol. 442, pp. 466–475. Springer, Cham (2014). https://doi.org/10.1007/978-3-319-08795-5_48
3. Coletti, G., Scozzafava, R.: Conditional probability, fuzzy sets, and possibility: a unifying view. Fuzzy Sets Syst. **144**, 227–249 (2004)
4. Cox, J.C., Ross, S.A., Rubinstein, M.: Option pricing: a simplified approach. J. Financ. Econ. **3**(7), 229–263 (1979)
5. Dubois, D., Kerre, E., Mesiar, R.: Fuzzy interval analysis. In: Dubois, D., Prade, H. (eds.) Fundamentals of Fuzzy Sets, Handbook Series of Fuzzy Sets, pp. 483–582. Springer, Boston (2000). https://doi.org/10.1007/978-1-4615-4429-6_11
6. Dubois, D., Prade, H., Yager, R.R.: Merging fuzzy information. In: Bezdek, J.C., Dubois, D., Prade, H. (eds.) Fuzzy Sets in Approximate Reasoning and Information Systems. The Handbooks of Fuzzy Sets Series, pp. 335–401. Kluwer Academic Publishers, Dordrecht (1999)
7. Fodor, J., Yager, R.: Fuzzy set-theoretic operators and quantifiers. In: Dubois, D., Prade, H. (eds.) Fundamentals of Fuzzy Sets. Handbook Series of Fuzzy Sets, pp. 125–193. Springer, Boston (2000). https://doi.org/10.1007/978-1-4615-4429-6_3
8. Grabisch, M., Marichal, J.-L., Mesiar, R., Pap, E.: Aggregation functions. In: Encyclopedia of Mathematics and its Applications, vol. 127. Cambridge University Press (2009)
9. Grabisch, M., Marichal, J.-L., Mesiar, R., Pap, E.: Aggregation functions: means. Inf. Sci. **181**, 1–22 (2011)
10. Grabisch, M., Marichal, J.-L., Mesiar, R., Pap, E.: Aggregation functions: construction methods, conjunctive, disjunctive and mixed classes. Inf. Sci. **181**, 23–43 (2011)
11. Guerra, M.L., Sorini, L., Stefanini, L.: Parametrized fuzzy numbers for option pricing. In: 2007 IEEE Conference on Fuzzy Systems, pp. 728–733 (2007)
12. Jaulin, L.: Robust set-membership state estimation; application to underwater robotics. Automatica **45**(1), 202–206 (2009)

13. Li, W., Han, L.: The fuzzy binomial option pricing model under Knightian uncertainty. In: 2009 Sixth International Conference on Fuzzy Systems and Knowledge Discovery (2009)
14. Li, H.-X., Yen, V.C.: Fuzzy Sets and Fuzzy Decision-making. CDC Press, Boca Raton (1995)
15. Marzullo, K.: Tolerating failures of continuous-valued sensors. ACM Trans. Comput. Syst. **8**(4), 284–304 (1990)
16. Muzzioli, S., Torricelli, C.: A multiperiod binomial model for pricing options in a vague world. J. Econ. Dyn. Control **28**, 861–887 (2004)
17. Muzzioli, S., De Baets, B.: Fuzzy approaches to option price modeling. IEEE Trans. Fuzzy Syst. **25**(2), 392–401 (2017)
18. Scozzafava, R., Vantaggi, B.: Fuzzy inclusion and similarity through coherent conditional probability. Fuzzy Sets Syst. **160**(3), 292–305 (2009)
19. The CBOE Volatility Index - VIX. https://www.cboe.com/micro/vix/vixwhite.pdf. Accessed 17 June 18
20. Yager, R.R.: A general approach to the fusion of imprecise information. Int. J. Intell. Syst. **12**, 1–29 (1997)
21. Zadeh, L.A.: The concept of linguistic variable and its application to approximate reasoning. Inf. Sci. **8**, 199–249 (1975)

Towards a Fuzzy Index of Skewness

Silvia Muzzioli[1,2(✉)] [ID], Luca Gambarelli[1],
and Bernard De Baets[3,4] [ID]

[1] Department of Economics, University of Modena and Reggio Emilia,
Viale Berengario 51, 41121 Modena, Italy
{silvia.muzzioli,luca.gambarelli}@unimore.it
[2] CEFIN, University of Modena and Reggio Emilia, Viale Berengario 51,
41121 Modena, Italy
[3] Department of Data Analysis and Mathematical Modelling, Ghent University,
Coupure links 653, 9000 Ghent, Belgium
bernard.debaets@ugent.be
[4] KERMIT, Ghent University, Coupure links 653, 9000 Ghent, Belgium

Abstract. The aim of this paper is to investigate the potential of fuzzy regression methods for computing a measure of skewness for the market. A quadratic version of the Ishibuchi and Nii hybrid fuzzy regression method is used to estimate the third order moment. The obtained fuzzy estimates are compared with the one provided by standard market practice. The proposed approach allows us to cope with the limited availability of data and to use all the information that is present in the market.

In the Italian market, the results suggest that the fuzzy-regression based skewness measure is closer to the subsequently realized measure of skewness than the one provided by the standard methodology. In particular, the upper bound of the Ishibuchi and Nii method provides the best forecast. The results are important for investors and policy makers who can rely on fuzzy regression methods to get a more reliable forecast of skewness.

Keywords: Fuzzy regression · Skewness · Forecasting · Italian market

1 Introduction

Moments of a distribution are of paramount importance in finance for portfolio allocation, risk management, trading strategies. Volatility of financial assets has attracted the interest of researchers and practitioners for decades. Only later, researchers have moved their interest towards higher-order moments of the distribution. The increasing importance of higher-order moments is supported by the introduction of the CBOE SKEW index for the S&P500 stock market, which measures the third order moment of the S&P500 risk-neutral distribution [6]. In the CBOE SKEW index, skewness is obtained from option prices by means of the Bakshi et al. formula [1] and reflects the investors' expectation of the realized third moment in the next thirty days.

The Bakshi et al. formula [1] is based on the strong assumption that a continuum of option prices with strike price ranging from zero to infinity is available. As in the market only a limited number of option prices is traded, it is standard market practice to

© Springer Nature Switzerland AG 2019
R. Fullér et al. (Eds.): WILF 2018, LNAI 11291, pp. 164–175, 2019.
https://doi.org/10.1007/978-3-030-12544-8_13

generate the missing ones by means of an interpolation-extrapolation procedure of the quoted option prices. Moreover, standard statistical techniques are not able to deal with conflicting information. Therefore, when two options yield different implied volatility, standard market practice retains only out-of-the-money[1] ones and averages the two at-the-money implied volatilities, producing both a considerable loss of information and an element of arbitrariness in the estimation.

A few authors explore the potential of fuzzy techniques to estimate volatility from a limited and conflicting number of option prices (for a literature review see e.g. [15]). References [4, 8] explore fuzzy volatility in the Black-Scholes model [3]. Reference [5] extends previous contributions on the elicitation of the fuzzy volatility membership function in option pricing models by exploiting the Cox-Ross-Rubinstein framework for option pricing developed in [19].

In a model-free setting, [16, 17] combine the Bakshi et al. formula [1] with quadratic fuzzy regression methods (introduced in [18]) to obtain more informative volatility measures. Their methodology presents several advantages. First, it embeds in the estimation of the implied volatility smile function all the information coming from both call and put prices and avoids the a priori choice of discarding some option prices as in standard market practice. Second, the use of fuzzy regression methods ensures the convexity of the volatility smile, and, as a consequence, the absence of arbitrage opportunities. Third, empirical results suggest that the volatility estimates obtained through fuzzy regression methods perform better in forecasting future realized volatility than the volatility measures obtained using the standard procedure.

Given the increasing importance of measuring skewness of the return distribution for both investors and policy makers, and the unsolved problems in the implementation with market data of the Bakshi et al. formula [1], this paper represents the first attempt of computing a skewness index in a fuzzy setting. We complement the existing literature by investigating the potential of fuzzy regression methods to compute a fuzzy measure of skewness for the Italian market. The use of fuzzy regression methods is particularly suitable for this type of data (see [14]). Specifically, fuzzy regression methods allow us to cope with the limited availability of data, given that for the Italian market only a little number of pairs of strike prices and implied volatilities are available to be interpolated. Moreover, it allows us to embed the conflicting information coming from both call and put prices. In fact, for at-the-money strike prices, we have both a call and a put option with different implied volatilities, and standard regression techniques are not able to cope with interval values for the inputs.

An empirical analysis performed in the Italian market (see [18]) concludes that the best estimation method for the volatility smile function is the Ishibuchi and Nii regression method [10], with the preferred h-cut at $h = 0.8$. Therefore, we adopt the quadratic extension of the Ishibuchi and Nii fuzzy regression method to estimate the skewness of the Italian market. In order to assess whether the proposed fuzzy regression method outperform the standard market practice in estimating skewness, we adopt a two-step methodology. First, we evaluate the proposed skewness measure with

---

[1] An option is said to be at-the-money, in-the-money, or out-of-the-money if it generates a zero, positive, or negative payoff, respectively, if exercised immediately.

respect to its forecasting power on future realized skewness using the mean squared error (MSE) indicator, which provides robust results in the presence of noise in the proxy of skewness. Second, we perform the Model Confidence Set test (see [9]) on the MSE loss function to find the best forecast for future realized skewness. Third, we adopt a defuzzification procedure in order to condense all the information content of fuzzy skewness estimates (which provides investors with an interval of possible values and a most possible value within the interval) in a unique value.

The results of this paper suggest that the skewness indices obtained using fuzzy regression methods are closer to the subsequently realized measure of skewness than the one provided by the standard methodology. This result is in line with previous findings in [16, 17] for volatility, indicating that the use of fuzzy regression methods in computing skewness of the option implied distribution enhances its predictive power on future realized skewness. In particular, the best estimate of subsequently realized skewness is the one that combines the Bakshi et al. formula [1] with the upper bound of the Ishibuchi and Nii fuzzy regression method [10].

The paper proceeds as follows. In Sect. 2, we discuss the financial problem. In Sect. 3, we describe the procedure adopted to embed all the information coming from both call and put prices in the estimation of skewness. In Sect. 4, we present the results of the empirical application on the Italian market. In Sect. 5 we present the defuzzification procedure. In Sect. 6 we evaluate the goodness of the measures by assessing their forecasting power on future realized skewness. The last section concludes.

2   Skewness Obtained from Option Prices: From the Smile Function to Skewness Estimation

The standard market formula used to extract volatility and higher order moments from a cross-section of option prices is the model-free formula proposed in [1]. This formula is called model-free since it does not rely on any option pricing model, being consistent with many asset price dynamics. According to [1] model-free skewness can be obtained from the following equations:

$$Skewness(t, \tau) \equiv \frac{e^{r\tau} W(t, \tau) - 3e^{r\tau} \mu(t, \tau) V(t, \tau) + 2\mu(t, \tau)^3}{\left[e^{r\tau} V(t, \tau) - \mu(t, \tau)^2 \right]^{3/2}} \tag{1}$$

where $\mu(t, \tau)$, $V(t, \tau)$ and $W(t, \tau)$ are based on the first, second and third moments of the distribution, respectively, and are obtained from call and put prices as follows:

$$\mu(t, \tau) \equiv E^q \ln[S(t+\tau)/S(t)] = e^{r\tau} - 1 - \frac{e^{r\tau}}{2} V(t, \tau) - \frac{e^{r\tau}}{6} W(t, \tau) - \frac{e^{r\tau}}{24} X(t, \tau) \tag{2}$$

$$V(t, \tau) = \int_{S(t)}^{\infty} \frac{2(1 - \ln[K/S(t)])}{K^2} C(t, \tau; K) dK + \int_{0}^{S(t)} \frac{2(1 + \ln[S(t)/K])}{K^2} P(t, \tau; K) dK \tag{3}$$

$$W(t,\tau) = \int\limits_{S(t)}^{\infty} \frac{6\ln[K/S(t)] - 3ln[K/S(t)]^2}{K^2} C(t,\tau;K)dK - \int\limits_{0}^{S(t)} \frac{6\ln[S(t)/K] + 3ln[S(t)/K]^2}{K^2} P(t,\tau;K)dK \quad (4)$$

$$X(t,\tau) = \int\limits_{S(t)}^{\infty} \frac{12\ln[K/S(t)]^2 - 4ln[K/S(t)]^3}{K^2} C(t,\tau;K)dK + \int\limits_{0}^{S(t)} \frac{12\ln[S(t)/K]^2 + 4ln[S(t)/K]^3}{K^2} P(t,\tau;K)dK$$

$$(5)$$

$C(t,\tau;K)$ and $P(t,\tau;K)$ are the prices of a call and a put option at time t with maturity τ and strike K, respectively, and $S(t)$ is the underlying asset price at time t.

In order to compute the integrals in Eqs. (2)–(5), a continuum of option prices with strike price ranging from zero to infinity is required. However, this hypothesis is not fulfilled in the reality of financial markets. In particular, for European peripheral countries, such as the Italian one, only a small number of strike prices is available (around 15 per day) and the strike prices are spaced by a fixed range of basis points (e.g. for the Italian market, 250–500 basis points depending on the maturity). As a consequence, truncation and discretization errors may occur if a finite range of strike prices and a discrete summation are used to approximate the integrals in Eqs. (2)–(5).

A commonly used solution is the one proposed in [11], who suggest to mitigate both truncation and discretization errors by exploiting an interpolation-extrapolation method. Given that standard statistical techniques are not able to cope with conflicting information, standard market practice uses only a subset of available option prices (it retains only at-the-money and out-of-the-money option prices, therefore put options for strikes below and call options for strikes above the current asset price). Moreover, it averages the two at-the-money implied volatilities (when the strike price equals the current asset price) in a single estimate. It is obvious that this technique produces both a considerable loss of information and introduces an element of arbitrariness in volatility and skewness estimation.

3   The Smile Function Obtained Through Fuzzy Regression Methods

In this section we present the approach adopted in order to include all the available information in the market in the smile estimation procedure to obtain more informative skewness estimates. This methodology represent an appealing solution to deal with a framework characterized by conflicting information that needs to be aggregated (e.g. interval values for the inputs).

Following [16–18], we propose to exploit fuzzy regression methods in order to incorporate all the uncertainty embedded in the data in the smile estimation procedure, without losing the information in the original data. Starting from the initial grid of strike prices (x_p) and implied volatilities (y_p), we compute the minimum and the maximum volatility for each strike price x_p, $p = 1,...,n$ as:

$$\sigma_{min}(x_p) = \min(\sigma_C(x_p), \sigma_P(x_p)) \tag{6}$$

$$\sigma_{max}(x_p) = \max(\sigma_C(x_p), \sigma_P(x_p)) \tag{7}$$

where $\sigma_C(x_p)$ and $\sigma_P(x_p)$ are the volatility of the call and the volatility of the put option associated to the strike price x_p. In this way, for a given strike price x_p, we have a range of possible values for volatility y_p given by $y_p = [\sigma_{min}(x_p), \sigma_{max}(x_p)]$. In order to include all the observations in the smile estimation, we resort to fuzzy regression methods, which are capable to deal with interval values for the inputs. Given that the relationship among strike prices and implied volatilities takes the form of a smile, the so-called volatility smile, we adopt a quadratic fuzzy regression model, in order to achieve the best fit to the data.

The quadratic fuzzy regression model takes the following form:

$$\sigma(x) = A_0 + A_1 x + A_2 x^2 \tag{8}$$

where $\sigma(x)$ is the fuzzy output (i.e., the implied volatility associated to each strike price), x is a non-fuzzy input vector of strike prices and A_i, $i = 0,...,2$, are the fuzzy coefficients of the second order polynomial. Since we deal with strictly positive variables, the lower bound ($\sigma^L(x)$), the upper bound ($\sigma^U(x)$), and the central value ($\sigma^C(x)$) of the fuzzy regression model can be rewritten as:

$$\sigma^L(x) = a_0^L + a_1^L x + a_2^L x^2$$
$$\sigma^U(x) = a_0^U + a_1^U x + a_2^U x^2$$
$$\sigma^C(x) = a_0^C + a_1^C x + a_2^C x^2$$

Relying on a previous empirical analysis performed on the Italian market, we adopt the quadratic extension of Ishibuchi and Nii fuzzy regression method proposed in [18] to estimate the volatility smile function. This approach is based a two-step methodology. In the first step, the coefficients a_0^C, a_1^C, a_2^C of the central regression $\sigma^C(x) = a_0^C + a_1^C x + a_2^C x^2$ are derived using the ordinary least squares:

$$\min z = \sum_{p=1}^{m} \left[y_p - (a_0^C + a_1^C x_p + a_2^C x_p^2) \right]^2 \tag{9}$$

where $y_p = (\sigma_{min}(x_p) + \sigma_{max}(x_p))/2$ is the average of the two implied volatilities which is adopted here to facilitate the use of the least squares estimation for the calculation of the central equation.

In the second step, the lower $\sigma^L(x)$ and the upper $\sigma^U(x)$ bounds are derived by means of the following optimization problem:

$$\min z = \sum_{p=1}^{m} \sigma^{U}(x_p) - \sigma^{L}(x_p) \tag{10}$$

where

$$\sigma^{U}(x) = a_0^{U} + a_1^{U}x + a_2^{U}x^2$$
$$\sigma^{L}(x) = a_0^{L} + a_1^{L}x + a_2^{L}x^2,$$

subject to:

$$h\,\sigma^{C}(x_p) + (1-h)\sigma^{L}(x_p) \le y_p = \sigma_{\min}(x_p), \quad p = 1, \ldots, m$$

$$h\,\sigma^{C}(x_p) + (1-h)\sigma^{U}(x_p) \ge y_p = \sigma_{\max}(x_p), \quad p = 1, \ldots, m$$

$$a_i^{L} \le a_i^{C} \le a_i^{U}, \quad i = 0, 1, 2$$

where a^{C} is pre-determined in the first step.

The fuzzy regression output is used to generate call and put prices to plug into Eqs. (2)–(5). In order to have a benchmark for the proposed fuzzy-regression-based measures of skewness, we also compute a skewness measure by applying the standard cubic spline methodology. Moreover, given the importance of having a constant 30-day measure for skewness (most of the risk measures for financial markets are calculated for a reference time horizon equal to 30 days), a linear interpolation procedure is adopted:

$$Skewness_{30} = w\,Skewness_{near} + (1-w)Skewness_{next} \tag{11}$$

with $w = (T_{next} - 30)/(T_{next} - T_{near})$, and T_{near} (T_{next}) the time to expiration of the near (next) term options, $Skewness_{near}$ (resp. $Skewness_{next}$) is the estimate which refers to the near (resp. next) term options. In general, a first option series with a maturity of less than 30 days (near) and a second series with time to maturity greater than 30 days (next) are used.

4 Fuzzy Skewness for the Italian Market

In this section, we present the results for the skewness measures of the Italian market based either on the standard interpolation-extrapolation methodology or the fuzzy regression method. The data set consists of daily closing prices on FTSE MIB-index options (MIBO), recorded from 1 January 2010 to 28 November 2014. The data set for the MIBO is kindly provided by Borsa Italiana S.p.A, while the time series of the FTSE MIB index, the dividend yield and the Euribor rates are obtained from Datastream. Several filters to the option data set are used in order to eliminate arbitrage opportunities and other irregularities in the prices (for a detailed discussion see e.g. [12, 13]).

We perform the procedures described in Sects. 2 and 3 on the option prices that meet the filter constraints and we obtain 1233 daily observations for each of the 10 estimates of skewness (we choose to use the upper and lower bounds of the h-cuts, with $h = 0$, 0.25, 0.5, 0.75, 1 and the standard method). We also compute the subsequently realized measure of skewness (obtained from historical series) using daily FTSE MIB log-returns and a rolling window of 30 calendar days. In this way the physical measure refers to the same time period covered by the measures computed using option prices, which represent the investors' expectation (under the risk-neutral measure) of the former. In Table 1 we report the average value of realized skewness (first column) and the estimates of skewness computed from option prices (columns 2–7). Specifically, the estimate obtained using the standard procedure is reported in column 2. On the other hand, the upper bound and the lower bound for each h-cut, is reported in columns 3–7.

Several observations are noteworthy. First, it is straightforward to note that all the skewness measures obtained from option prices are on average lower than zero, pointing to a negative risk-neutral skewness (i.e. the risk-neutral distribution is skewed to the left). On the other hand, the subsequently realized distribution is almost symmetrical, the measure of skewness estimated from the historical series of the underlying asset being equal to -0.012 on average. Second, the skewness estimate obtained by setting h equal to one is lower than the one obtained using the standard interpolation-extrapolation methodology. Third, the skewness estimate that is the closest to the subsequently realized measure of skewness is the one provided by the upper bound of the Ishibuchi and Nii ($h = 0$) fuzzy regression method.

5   The Defuzzification Procedure

In Sect. 3 we presented the advantages of skewness estimates obtained using fuzzy regression method. In particular, the proposed skewness measures allow to extrapolate further information with respect to the standard methodology since they provide not only a most possible value for skewness, but also an interval of possible values around the most possible one.

However, investors may prefer to condense all the information content of the skewness estimates obtained using the fuzzy regression method into a unique value (crisp output). This objective can be achieved by exploiting a defuzzification procedure. An appealing solution in order to synthesize all the information embedded in the skewness estimates is the one proposed in [19], who suggest that that the best defuzzifier is the scalar that is "closest" to the triangular fuzzy number:

$$x = \frac{a^L + 2a^C + a^U}{4} \tag{12}$$

where a^L, a^C and a^U are the lower, the central and the upper bound of the triangular fuzzy number.

The defuzzification procedure is used to convert, for each strike price, the different fuzzy regression results in the defuzzified volatility level. The obtained values for volatility are subsequently converted in terms of call prices and used as input in Eqs. (2)–(5) in order to obtain a unique skewness estimate.

The result for the defuzzified skewness estimate obtained with the Ishibuchi and Nii method is reported in Table 1 (last column). We can see that the defuzzified skewness estimate (−0.368) is close to the central estimate of the Ishibuchi and Nii fuzzy regression method ($h = 1$). This suggests that the skewness estimate obtained using the Ishibuchi and Nii fuzzy regression method do not show a pronounced asymmetry.

Table 1. Average value of the estimated skewness measures.

RSkew	Std. Meth.	Ishibuchi and Nii					def
		$h = 1$	$h = 0.75$	$h = 0.50$	$h = 0.25$	$h = 0$	
−0.012	−0.387	−0.368	−0.359	−0.352	−0.345	−0.334	−0.368
			−0.383	−0.397	−0.413	−0.434	

We report in the first and second column the average value for daily realized skewness (RSkew) and the skewness estimate obtained using the standard interpolation-extrapolation method (Std. Meth.), respectively. In columns 3–7 we report the average value for daily skewness measures obtained combining the Bakshi et al. skewness formula (Eq. 1) with the Ishibuchi and Nii fuzzy regression method [10]. The results are reported for different values of h. For each value of h we report the upper bound (first row) and the lower bound (second row) estimate of skewness. Finally in the last column we report the average value for the skewness estimate obtained using the defuzzification procedure.

6   The Assessment of the Best Skewness Forecast

We are interested in evaluating whether fuzzy regression methods to estimate skewness enhance the predictive power on future realized skewness. Given the large number of forecasts (11) for skewness proposed in Table 1, we resort to the model confidence set procedure (MCS) to identify the best model, or a smaller set of best models (see [9]). In order to evaluate the forecasting performance of the proposed models, in line with Patton (2011), we adopt the Mean Squared Error (MSE) error indicator, which provides robust results in the presence of noise in the proxy of skewness:

$$\mathrm{MSE} = \frac{1}{m} \sum_{k=1}^{m} (forecast_k - realized_k)^2 \tag{13}$$

where $forecast_k$ and $realized_k$ are the forecasted and realized measures of moments, respectively, and $forecast_k$ is proxied by the different skewness measures obtained using option prices. The average value of the MSE loss functions are reported in Table 2. We can see that the best forecast for future realized skewness is the one provided by the upper bound of the Ishibuchi and Nii ($h = 0$) fuzzy regression method.

Moreover, also the most possible value provided by the Ishibuchi and Nii ($h = 1$) fuzzy regression method yields a lower error than that of the standard methodology. We also evaluate the forecasting performance of the proposed defuzzified skewness measure on future realized skewness by computing the Mean Squared Error (MSE) error indicator (Eq. (13)). The result, reported in Table 2 (last column), indicates that the unique value of skewness obtained using the defuzzification procedure obtains a slightly worse performance with respect to the central estimate of skewness ($h = 1$). However, the defuzzified skewness estimate is still better than the standard methodology in forecasting future realized skewness (MSE is equal to 0.165 and 0.188 for the defuzzified and the standard method, respectively) and the improvement is significant from a statistical point of view (this result is supported by a t-test, where errors are corrected by Newey West, t-stat = -3.47, p-value = 0.00).

Therefore, investors who prefer to have all the information content of the skewness estimates condensed into a unique value could refer to the estimate obtained by means of the defuzzification procedure to have a more reliable forecast of future realized skewness.

Table 2. Forecasting skewness: MSE error indicator.

	Std. Met.	Ishibuchi and Nii					def
		$h = 1$	$h = 0.75$	$h = 0.50$	$h = 0.25$	$h = 0$	
MSE	0.188	0.161	0.153	0.148	0.143	0.138	0.165
			0.174	0.189	0.209	0.238	

The table reports the results of the skewness forecasting exercise performed using the mean squared error (MSE) indicator defined as follows: $\text{MSE} = \frac{1}{m}\sum_{k=1}^{m}(\text{forecast}_k - \text{realized}_k)^2$ where forecast_k and realized_k are the values of option based forecast of skewness and realized skewness, respectively. For a definition of the skewness measures, see Table 1, (upper bounds in the first row and lower bunds in the second row).

The MSE reported in Table 2 are the inputs of the Model Confidence Set test, which is performed using the MCS package for R developed by [2]. The test allows to assess whether the difference in the forecasting power between the proposed models are significant from a statistical point of view (the statistic t_{ij} is used also in the well-known test for comparing two forecasts, see [7, 23]. The confidence level $(1 - \alpha)$ adopted in the test is equal to 0.95, the number of bootstrapped samples used to construct the statistic test is 1000 (B = 1000). The results for the Model Confidence Set test are reported in Table 3.

Table 3. Forecasting skewness: Model Confidence Set.

Superior Set Model created: (10 models are eliminated), indicator used: MSE

	Rank	$T_{\max,M}$	p-value	$T_{R,M}$	p-value	Loss
I&N f^U (0)	1	−6.896	0.000	−6.896	0.000	0.137

The table reports the Model Confidence Set for the skewness estimates obtained using either the standard methodology and the proposed fuzzy regression method. The input for the Model Confidence Set reported is represented by the MSE loss functions reported in Table 2. $T_{\max,M} = \max_{i \in M} t_i, T_{R,M} = \max_{i,j \in M} |t_{ij}|$ are the test statistics proposed in [9]; p-values for the tests are reported sideways in the p-value column, the corresponding rank is reported in the Rank column. The lower the value of T, the higher the rank. In the last column, we report the average loss of the model.

According to the Model Confidence Set test result reported in Table 3, the upper bound of the Ishibuchi and Nii fuzzy regression method ($h = 0$) is the best forecast for future realized skewness. All the other forecasts, included the one based on the standard procedure, are eliminated.

Given the relevance of correctly measuring skewness to assess the riskiness of asset return distribution, this result is very important for investors and regulators, who can rely on fuzzy regression methods in order to get a more reliable forecast for skewness.

7 Conclusions

In this paper we have proposed a method for estimating skewness from option prices by means of fuzzy regression methods. This approach offers several advantages. First, it is possible to incorporate conflicting information coming from both call and put prices, without having to make the a priori choice of discarding some option prices as in standard market practice. Second, fuzzy regression methods are particularly suited when a limited number of option prices is available. Last, fuzzy regression methods yield a more reliable estimate in the form of interval of possible values, containing the most possible one.

We offer an empirical application of the proposed method in the Italian market, during the 2010–2014 time-period. The measures of skewness are computed on a daily basis (closing values of option price are used) using five different level of h-cut: 0, 0.25, 0.50, 0.75, 1. The proposed skewness measures obtained through fuzzy regression are compared with the measure of skewness provided by the standard procedure, which are used as a benchmark. We also adopt a defuzzification procedure in order to condense all the information content of the fuzzy skewness estimate in a unique value.

We get several results. First, the skewness estimates obtained using the fuzzy regression method allow to extrapolate further information with respect to the standard least square regression, since the coefficients of the fuzzy regression model provide not only a most possible value for the coefficient, but also an interval of possible values around the most possible one. Second, the mean squared error (MSE) indicator suggests that the measures of skewness obtained through a fuzzy regression method are closer to the subsequently realized measures than the one obtained using the standard methodology. Third, the Model Confidence Set indicates that the improvement in the forecasting performance attained using fuzzy regression is significant also from a statistical point of view. Similar results are obtained for volatility estimates through fuzzy regression in [16, 17]. Specifically, the best forecast of subsequently realized skewness is the upper bound of the Ishibuchi and Nii fuzzy regression method ($h = 0$).

Since correctly measuring skewness is of paramount importance in finance in order to correctly assess the riskiness of asset return distribution, this result is very important for investors and regulators, who can rely on fuzzy regression methods to get a more reliable forecast of skewness. Future research should evaluate if the use of other fuzzy regression methods (such as Savic and Pedricz [21] and Tanaka et al. [22]) may improve the forecasting power of the fuzzy skewness estimates.

Acknowledgment. We gratefully acknowledge financial support from Fondazione Cassa di Risparmio di Modena, for the project 2015.0333, from the FAR2015 project "A SKEWness index for Europe (EU-SKEW)" and from the FAR2017 project "The role of Asymmetry and Kolmogorov equations in financial Risk Modelling (ARM)". The usual disclaimer applies.

References

1. Bakshi, G., Kapadia, N., Madan, D.: Stock return characteristics, skew laws, and the differential pricing of individual equity options. Rev. Financ. Stud. **16**, 101–143 (2003)
2. Bernardi, M., Catania, L.: The model confidence set package for R. CEIS Working Paper No. 362 (2014). https://arxiv.org/pdf/1410.8504.pdf
3. Black, F., Scholes, M.: The pricing of options and corporate liabilities. J. Polit. Econ. **81**(3), 637–654 (1973)
4. Capotorti, A., Figà-Talamanca, G.: On an implicit assessment of fuzzy volatility in the Black and Scholes environment. Fuzzy Sets Syst. **223**, 59–71 (2013)
5. Capotorti, A., Figà-Talamanca, G.: A Generalized SMART fuzzy disjunction of volatility indicators applied to option pricing in a binomial model. In: Ferraro, M.B., et al. (eds.) Soft Methods for Data Science. AISC, vol. 456, pp. 95–102. Springer, Cham (2017). https://doi.org/10.1007/978-3-319-42972-4_12
6. CBOE: The CBOE Skew Index (2010). https://www.cboe.com/micro/skew/documents/skewwhitepaperjan2011.pdf
7. Diebold, F.X., Mariano, R.S.: Comparing predictive accuracy. J. Bus. Econ. Stat. **20**(1), 134–144 (1995)
8. Guerra, M.L., Sorini, L., Stefanini, L.: Option price sensitivities through fuzzy numbers. Comput. Math. Appl. **61**(3), 515–526 (2011)
9. Hansen, P.R., Lunde, A., Nason, J.M.: The model confidence set. Econometrica **79**(2), 453–497 (2011)
10. Ishibuchi, H., Nii, M.: Fuzzy regression using asymmetric fuzzy coefficients and fuzzified neural networks. Fuzzy Sets Syst. **119**(2), 273–290 (2001)
11. Jiang, G.J., Tian, Y.S.: The model-free implied volatility and its information content. Rev. Financ. Stud. **18**(4), 1305–1342 (2005)

12. Muzzioli, S.: The forecasting performance of corridor implied volatility in the Italian market. Comput. Econ. **41**(3), 359–386 (2013)
13. Muzzioli, S.: The optimal corridor for implied volatility: from calm to turmoil periods. J. Econ. Bus. **81**, 77–94 (2015)
14. Muzzioli, S., De Baets, B.: A comparative assessment of different fuzzy regression methods for volatility forecasting. Fuzzy Optim. Decis. Making **12**(4), 433–450 (2013)
15. Muzzioli, S., De Baets, B.: Fuzzy approaches to option price modelling. IEEE Trans. Fuzzy Syst. **25**(2), 392–401 (2017)
16. Muzzioli, S., Gambarelli, L., De Baets, B.: Towards a fuzzy volatility index for the Italian market. In: Proceedings of the IEEE International Conference on Fuzzy Systems, Naples (2017)
17. Muzzioli, S., Gambarelli, L., De Baets, B.: Indices for financial market volatility obtained through fuzzy regression. Int. J. Inf. Tech. Decis. Forthcoming
18. Muzzioli, S., Ruggeri, A., De Baets, B.: A comparison of fuzzy regression methods for the estimation of the implied volatility smile function. Fuzzy Sets Syst. **266**, 131–143 (2015)
19. Muzzioli, S., Torricelli, C.: A multiperiod binomial model for pricing options in a vague world. J. Econ. Dyn. Control **28**, 861–887 (2004)
20. Patton, A.J.: Volatility forecast comparison using imperfect volatility proxies. J. Econometrics **160**(1), 246–256 (2011)
21. Savic, D.A., Pedrycz, W.: Evaluation of fuzzy linear regression models. Fuzzy Sets Syst. **39**(1), 51–63 (1991)
22. Tanaka, H., Uejima, S., Asai, K.: Linear regression analysis with fuzzy model. IEEE Trans. Syst. Man. Cybern. **12**, 903–907 (1982)
23. West, K.D.: Asymptotic inference about predictive ability. Econometrica **64**, 1067–1084 (1996)

Grading by Committees:
An Axiomatic Approach

Marta Cardin[ID] and Silvio Giove[✉][ID]

Department of Economics, Ca' Foscari University of Venice,
Cannaregio 873, Venezia, Italy
{mcardin,sgiove}@unive.it

Abstract. We deal the problem of aggregation of individual judgments for a global evaluation of a candidate or a product. In our theoretically oriented approach, aggregation operators are compared with each other based on their mathematical properties. We show that any monotone and strategy-proof operators is characterized by a particular collection of decision makers.

Keywords: Aggregation operators · Axioms · Non-manipulable · Sugeno integral

1 Introduction

We study the problem of aggregating grades that are expressed in a common language consisting of a bounded totally order set of grades and we extend some results in [2,3] and [13].

As an example we consider the evaluation of scientific research by peers or experts review, a problem that is often encountered in practice. In this case we look for a global evaluation of a candidate or of a research product given a set of individual judgments. We are interested in aggregation rules satisfying the requirement that no one can even better off by lying about his or her preference and we call these rules *non-manipulable* or *strategy-proof*.

Judges may attempt to manipulate the outcome by misrepresenting their grades and they are likely to do that if the aggregation rule allows for individual manipulation.

We characterize strategy-proof aggregation rules in our framework by considering Sugeno integral which could be considered when qualitative or ordinal information is used. This is due to the ordinal nature of its definition, which uses only lattice operations.

The structure of the paper is as follows. In Sect. 2, we focus on the problem of a global evaluation given individual evaluations, the notion of strategy proof functional is introduced and we present a characterization of strategy proof aggregation functionals. Section 3 finally provides an axiomatic characterization of anonymous grading functionals.

© Springer Nature Switzerland AG 2019
R. Fullér et al. (Eds.): WILF 2018, LNAI 11291, pp. 176–182, 2019.
https://doi.org/10.1007/978-3-030-12544-8_14

2   Aggregation of Ordinal Assessments by Grading Functionals

We introduce and characterize axiomatically a family of aggregation operators that use a set of evaluations and we assume the existence of a common totally ordered value scale for all experts. The common scale could be the zero-one interval of the reals but also an integer-valued scale. Consider the case in which the value of a product is assessed by a set of experts or the case in which a recruitment committee has to decide among a number of applications for a faculty position. In our model we do not assume that all judges should be counted equally. We propose an axiomatic analysis of such a model and we suppose that the evaluation scale is an ordinal structure not necessarily finite as in others cited papers.

Let $N = \{1, \ldots, n\}$ the sets of experts and let X be a linearly ordered set with a least and a greatest element, denoted by 0 and 1, respectively. As usual we denote by $\leqslant$ the total order and by $<$ the asymmetric part of $\leqslant$. Clearly, every linearly ordered set is a distributive lattice and then the cartesian product X^N constitutes a complete, distributive and bounded lattice by defining the lattice operations componentwise. We use $\mathbf{0}$ and $\mathbf{1}$ to denote the least element and greatest element, respectively, of X^N. We denote the elements of X by lower capital letters $x, y, \ldots$ and the elements of X^N by bold face letters $\mathbf{x}, \mathbf{y}, \ldots$ Moreover for each $c \in X$, we denote by $\mathbf{c}$ the constant c map in X^N.

We say that a functional $F \colon X^N \to X$ on a linearly ordered set X is a *grading functional* if it is monotone with respect to componentwise order and it is *unanimous* i.e such that

$$F(\mathbf{c}) = c, \text{ for every constant map } \mathbf{c} \in X^N.$$

These aggregation functionals are considered also in [13] and play a central role in the majority judgment theory proposed by Balinski and Laraki in [2] and [3]. Balinski and Laraki developed a new concept for aggregating preferences in Voting Theory which is based on the ratings of the candidates. It is important to note that they assume a strong form of monotonicity (see [13]). They consider also the non-manipulability of preferences as in the following definition.

We say that a grading functional is *non manipulable* if for any $\mathbf{x}, \mathbf{y} \in X^N$ and any $i \in N$ are satisfied the following properties:

$$x_i > F(\mathbf{x}) \text{ and } x_j = y_j \text{ for every } j \neq i \text{ implies that } F(\mathbf{x}) \geqslant F(\mathbf{y}) \quad (1)$$

$$x_i < F(\mathbf{x}) \text{ and } x_j = y_j \text{ for every } j \neq i \text{ implies that } F(\mathbf{x}) \leqslant F(\mathbf{y}). \quad (2)$$

Conditions 1 and 2 exclude that an expert i can manipulate the final score $F(\mathbf{x})$ of an element with score $\mathbf{x}$. If the grading functional is non manipulable it is a dominant strategy for an expert to honestly assign to a scientific product the grade that he believes is the correct one and there is no incentive to misreport the evaluations.

For $\mathbf{x} = (x_1, \ldots, x_n) \in X^N$ and $y \in X$ we write as in [13]

$$\mathbf{x}/_i\, y = (x_1, \ldots, x_{i-1}, y, x_{i+1}, \ldots, x_n)$$

and then Conditions 1 and 2 can be expressed as

if $x_i > F(\mathbf{x})$ then for every $y \in X$, $F(\mathbf{x}) \geqslant F(\mathbf{x}/_i\, y)$,

if $x_i < F(\mathbf{x})$ then for every $y \in X$, $F(\mathbf{x}) \leqslant F(\mathbf{x}/_i\, y)$.

The following result characterizes non manipulable grading functionals in terms of generalized committee grading functionals.

We are going to consider discrete Sugeno integrals. The Sugeno integral has been widely studied in aggregation theory and has many applications in different fields such as decision theory, economics and finance, data fusion etc (see [8] for a general background).

Sugeno integrals can be defined on a linearly ordered domain X in terms of fuzzy measures (see also [6] and [7]). If we call $\mathcal{P}$ the set of all subsets of N a X-*fuzzy measure* on N is a nondecreasing mappings $m\colon \mathcal{P} \to X$, such that $m(\emptyset) = \mathbf{0}$.

Then a functional $F\colon X^N \to X$ is a *Sugeno integral* if and only if there exists a X-fuzzy measure on N such that

$$F(\mathbf{x}) = S_m(\mathbf{x}) := \bigvee_{A \subseteq N} \left(m(A) \wedge \bigwedge_{i \in A} x_i \right).$$

In our framework if A is an element of $\mathcal{P}$, $m(A)$ can be viewed as a measure for the importance of A.

A collection of sets $\mathcal{C} \subseteq 2^N$ is said to be a *generalized committee* in N if $A \in \mathcal{C}$ and $A \subset B$ implies that $B \in \mathcal{C}$ (see [11] for a similar definition).

A *generalized committee grading functional* is a functional $F\colon X^N \to X$ on linearly ordered set X such that for every $x \in X$ there exists a generalized committee $\mathcal{C}_x$ such that if $x \geqslant y$ then $\mathcal{C}_x \subseteq \mathcal{C}_y$ and

$$F(\mathbf{x}) = \bigvee \{x \in X : x_i \geqslant x \text{ for every } i \in C \text{ and } C \in \mathcal{C}_x\}.$$

A generalized committee grading functional is a step-based evaluation as there is an increasing set of steps that are the elements in X. The final grade is determined by the best step that an element $\mathbf{x}$ achieves. If there is a set of judges $C \in \mathcal{C}_x$ (of importance x) such that $x_i \geqslant x$ then the best step is x. Note that the well known h-index introduced by Hirsch in [9] (see also [1]) is defined in the same way.

Proposition 1. *A functional $F\colon X^N \to X$ on linearly ordered set X is a non manipulable grading functional if and only it is generalized committee grading functional.*

Proof. We first prove that a functional $F\colon X^N \to X$ on linearly ordered set X is a non manipulable grading functional if and only if there exists a X-fuzzy measure on N such that $F(\mathbf{x}) = S_m(\mathbf{x})$.

Since a Sugeno integral is a monotone and homogeneous aggregation functional (see [6] for example) F is a grading functional. Then it is easy to prove that it is a non manipulable aggregation functional.

Conversely we consider a non manipulable grading functional $F\colon X^N \to X$. If this functional is an idempotent, strongly idempotent functional and has a componentwise convex range by Theorem 2 in [5] is a discrete Sugeno integral with respect to a X-fuzzy measure m on N.

The functional F is idempotent since $F(\mathbf{c}) = c$, for every constant map $\mathbf{c} \in X^N$. A functional $F\colon X^N \to X$ is strongly idempotent if for every $\mathbf{x} \in X^N$ and every $i \in N$ we have that $F(\mathbf{x}/_i F(\mathbf{x})) = F(\mathbf{x})$. Now, if $x_i > F(\mathbf{x})$ we can prove that $F(\mathbf{x}) \geqslant F(\mathbf{x}/_i F(\mathbf{x}))$ by monotonicity of the functional F. Now, for the sake of a contradiction, we assume that $F(\mathbf{x}) > F(\mathbf{x}/_i F(\mathbf{x}))$. By Property 1 we have that $F(\mathbf{x}/_i F(\mathbf{x})) \geqslant F(\mathbf{x}/_i x_i) = F(\mathbf{x})$ and so we can conclude that $F(\mathbf{x}/_i F(\mathbf{x})) = F(\mathbf{x})$. Hence by Property 2 we can prove that $F(\mathbf{x}/_i F(\mathbf{x})) = F(\mathbf{x})$ also when $x_i < F(\mathbf{x})$ and so we have proved that F is strongly idempotent.

The functional F has a a componentwise convex range if when $F(\mathbf{x}/_i y_1) < z < F(\mathbf{x}/_i y_2)$ for $\mathbf{x} \in X^N$, $y_1, y_2 \in X$ and $i \in N$ there exists $y_3 \in X$ such that $z = F(\mathbf{x}/_i y_3)$. If $y_1 > F(\mathbf{x}/_i y_1)$ by non-manipulability of F we have that $F(\mathbf{x}/_i y_2) \leqslant F(\mathbf{x}/_i y_1)$ then we can conclude that $y_1 \leqslant F(\mathbf{x}/_i y_1)$. We can also prove that $y_2 \geqslant F(\mathbf{x}/_i y_2)$ since F is a non manipulable grading functional.

If $F(\mathbf{x}/_i z) < z$ being $z \leqslant y_2$ by Property 1 we have $F(\mathbf{x}/_i y_2) = F(\mathbf{x}/_i z)$ thus we can prove that $F(\mathbf{x}/_i z) \leqslant z$. Moreover we get that $F(\mathbf{x}/_i z) \geqslant z$ hence $F(\mathbf{x}/_i z) = z$ and so we prove that for the grading functional F is a Sugeno integral with respect to a X-fuzzy measure m defined on N.

Now we can prove that a functional $F\colon X^N \to X$ on linearly ordered set X is a non manipulable grading functional if and only it is a generalized committee grading functional since a functional is such that $F(\mathbf{x}) = S_m(\mathbf{x})$ for a fuzzy measure X-fuzzy measure m on N if and only if it is a generalized committee grading functional. We can prove this statement by considering the generalized committee $\mathcal{C}_x = \{C \subseteq N : m(C) \geqslant x\}$.

If we denote by $(\cdot)$ a permutation on N so that $x_{(1)} \leqslant x_{(2)} \leqslant \cdots \leqslant x_{(n)}$ for $\mathbf{x} = (x_1, \ldots, x_n) \in X^N$ by the proof of Proposition 1 a non manipulable grading function $F\colon X^N \to X$. can be written in the following form:

$$F(\mathbf{x}) = S_m(\mathbf{x}) = \bigvee_{i \in N} \left(m(\{(i), \ldots, (n)\}) \wedge x_{(i)} \right)$$

with respect to a X-fuzzy measure m defined on N.

It is important to note that we consider the grade $x_{(i)}$ and also the weight of the set of experts that give a grade greater or equal to $x_{(i)}$.

A linearly ordered set X is a lattice and so we can define in X the ternary *median function* by

$$\text{Med}\,(x_1, x_2, x_3) = (x_1 \wedge x_2) \vee (x_2 \wedge x_3) \vee (x_3 \wedge x_1)$$
$$= (x_1 \vee x_2) \wedge (x_2 \vee x_3) \wedge (x_3 \vee x_1).$$

By Theorem 2 in [5] we can prove the following corollary of Proposition 1 that present a median representation of a grading functional as in [13].

Proposition 2. *A functional $F\colon X^N \to X$ on linearly ordered set X is a non manipulable grading functional if and only for every $\mathbf{x} \in X^N$ and every $i \in N$,*

$$F(\mathbf{x}) = \text{Med}\,(\mathbf{x}/_i\,0, x_i, \mathbf{x}/_i\,1).$$

3   Anonymous Grading Functionals

A grading functional $F\colon X^N \to X$ on linearly ordered set X is said to be *anonymous* if

$$F(x_1, \ldots, x_n) = (x_{\pi(1)}, \ldots, x_{\pi(n)})$$

for any $\mathbf{x} \in X^N$ and any permutation π on N. If the aggregation rule is anonymous the name of the experts does not matter.

It can be easily proved that a grading functional is anonymous if and only it is a generalized committee grading functional with respect to a generalized committee such that for every $x \in X$ there exists a natural number k_x and the elements of $\mathcal{C}_x$ are all the subsets of N of cardinality k_x. In this case the final grade of an element $\mathbf{x}$ is x If there is a set of judges $C \in \mathcal{C}_x$ of cardinality k_x such that $x_i \geqslant x$.

Now we define a class of anonymous grading functional where the importance of a grade is not associated with a specific argument but with the place that a grade occupies in the ordered sequence of grades. We refer to the class of weighted maximum introduced studied in [7] in the more general case of distributive lattice.

A functional $F\colon X^N \to X$ is called a *positional grading functional* if

$$F(\mathbf{x}) = \bigvee_{i \in N} (\omega_i \wedge x_{(i)})$$

where $\omega_1, \ldots, \omega_n$ are elements of X and

$$1 = \omega_1 \geqslant \ldots \geqslant \omega_n.$$

It is easy to prove that a positional grading functional is increasing, unanimous and anonymous functional but we can also prove the following result.

Proposition 3. *A functional $F\colon X^N \to X$ on linearly ordered set X is a non manipulable anonymous grading functional if and only it is a positional grading functional.*

Proof. The result follows directly by Theorem 27 in [7] where we consider a lattice that is a linearly ordered set. In fact a positional grading functional can be represented as a Sugeno integral with respect to a symmetric measure and thus is a non manipulable functional. Conversely a Sugeno integral that is a symmetric functions is an ordered weighted maximum.

Also in this case we have a median-based representation. In fact if we consider an element $\mathbf{x} \in X^{2n-1}$ and we define $\text{Med}_{n-1}(\mathbf{x}) = x_{(n)}$ by Corollary 28 (ii) of [7] we can prove the following result.

Proposition 4. *A functional $F \colon X^N \to X$ on linearly ordered set X is a non manipulable anonymous grading functional if and only it there exist $a_1, \ldots, a_{n-1}$ elements of X where $a_1 \geqslant \ldots, \geqslant, a_{n-1}$ such that*

$$F(\mathbf{x}) = Med_{n-1}(x_1, \ldots, x_n, a_1, \ldots, a_{n-1}).$$

We consider the strong monotonicity property as in [13] and we characterize strongly monotone and non manipulable grading functionals. We say that a grading functional $F \colon X^N \to X$ on a linearly ordered set X is *strongly monotone* if $\mathbf{x}, \mathbf{y} \in X^N$ and $x_i < y_i$ for every $i \in N$ then $F(\mathbf{x}) < F(\mathbf{y})$. We say that a grading functional $F \colon X^N \to X$ is an *order functional* if there exists $i \in N$ such that $F(\mathbf{x}) = x_{(i)}$.

Proposition 5. *A functional $F \colon X^N \to X$ on linearly ordered set X is a non manipulable anonymous and strongly monotone grading functional if and only it is an order functional.*

Proof. If $F \colon X^N \to X$ on linearly ordered set X is a non manipulable anonymous and strongly monotone grading functional then it is a positional grading functional and so $F(\mathbf{x}) = \bigvee_{i \in N}(\omega_i \wedge x_{(i)})$ where $\omega_1, \ldots, \omega_n$ are elements of X and $1 = \omega_1 \geqslant \ldots \geqslant \omega_n$. We can prove that ω_i is 0 or 1 for every $i \in N$. In fact if as an example $\omega_1 = 1$, $0 < \omega_2 = \omega < 1$ and $\omega_i = 0$ for every $i > 2$ we can consider the two elements $\mathbf{x} = (0, \omega, \omega, \ldots, \omega)$ and $\mathbf{y} = (\omega, 1, 1, \ldots, 1)$ and $F(\mathbf{x}) = \omega = F(\mathbf{y})$ that is a impossible since F is strongly monotone.

4   Conclusions

We have studied evaluation structures with different characteristics but in any case we have considered a complete order between scores. Preferences or evaluations are not always representable via complete order and sometimes is more natural to consider the presence of incomparable objects. We intend to extend our approach to explicitly consider this case in a future extension.

References

1. Alonso, S., Cabrerizo, F.J., Herrera-Viedmac, E., Herrera, F.H.: h-index: a review focused in its variants, computation and standardization for different scientific fields. J. Informetr. **3**, 273–1289 (2009)
2. Balinski, M., Laraki, R.: A theory of measuring, electing and ranking. Proc. National Acad. Sci. U.S.A. **104**, 8720–8725 (2007)
3. Balinski, M., Laraki, R.: Majority Judgement: Measuring, Ranking and Electing. MIT Press, Cambridge (2010)
4. Balinski, M., Laraki, R.: Judge: don't vote! Ecole Polytechnique Centre National de la Recherche Scientifique, Cahier n 2010–27 (2010)
5. Couceiro, M., Marichal, J.-L.: Characterizations of discrete Sugeno integrals as lattice polynomial functions. In: Proceedings of the 30th Linz Seminar on Fuzzy Set Theory (LINZ 2009), pp. 17–20 (2009)
6. Couceiro, M., Marichal, J.-L.: Representations and characterizations of polynomial functions on chains. J. Mult.-Valued Log. Soft Comput. **16**(1–2), 65–86 (2010)
7. Couceiro, M., Marichal, J.-L.: Characterizations of discrete Sugeno integrals as polynomial functions over distributive lattices. Fuzzy Sets Syst. **161**, 694–707 (2010)
8. Grabisch, M., Marichal, J.-L., Mesiar, R., Pap, E.: Aggregation Functions: Encyclopedia of Mathematics and Its Applications. Cambridge University Press, Cambridge (2009)
9. Hirsch, J.E.: An index to quantify an individual's scientific research output. Proc. National Acad. Sci. U.S.A. **102**, 16569–16572 (2005)
10. Marchant, T.: Score-based bibliometric rankings of authors. J. Am. Soc. Inf. Sci. Technol. **60**, 1132–1137 (2009)
11. Savaglio, E., Vannucci, S.: Strategy-proofness and single peakedness in bounded distributive lattices, mimeo. Cornell University Library (2014)
12. Sugeno, M.: Theory of fuzzy integrals and its applications. Ph.D. thesis, Tokyo Institute of Technology, Tokyo (1974)
13. Yamamoto, Y., Zhou, Y.: Characterization of anonymous, weakly monotonic and strategy-proof aggregation function. Discussion Paper 1310. University of Tsukuba, Department of Social Systems and Management (2013)

Minimum of Constrained OWA Aggregation Problem with a Single Constraint

Lucian Coroianu[1](✉)(iD) and Robert Fullér[2](iD)

[1] Department of Mathematics and Informatics, University of Oradea,
Oradea, Romania
lcoroianu@uoradea.ro
[2] Department of Informatics, Széchenyi István University, Győr, Hungary
robert.fuller@nik.uni-obuda.hu

Abstract. In a recent paper we found an analytical formula for the constrained ordered weighted aggregation problem (OWA) when we need to maximize the objective function. In this note we prove that the method works in the case when we need to minimize the objective function. If in the case of the maximization problem we need to rearrange the coefficients in the constrained in nondecreasing order, for the nontrivial minimization problem, it suffice to rearrange them in nonincreasing order.

Keywords: OWA operators · Constrained optimization ·
Constrained OWA aggregation

1 Introduction

The OWA operators (ordered weighted average operators) were introduced by Yagger in paper [8]. Since then, OWA operators were successfully used in research fields that belong in broad sense to decision making. One interesting problem is to optimize the OWA operator. This type of investigation started with paper [7] and since then it became a challenging problem for researchers. The issue is that we lack an analytical formula for the solution function. In order to avoid repetition, we refer to our recent paper [2] where the problem is discussed in detail. Then, we refer to the surveys [3] and [4] where the reader can find about many optimization problems related to the OWA operators. Our interest in this topic is to find those types of optimization problems where we can find an analytical expression for the solution function. In paper [7] the idea was to transform the problem into a mixed integer linear problem. As the number of variables increases significantly an some of them are restricted to be integers, it seems hard to find an analytical expression for the solution function in general. The first such concrete result can be found in paper [1] in the special case when we have a single constraint and all coefficients are equal to one. This result was

© Springer Nature Switzerland AG 2019
R. Fullér et al. (Eds.): WILF 2018, LNAI 11291, pp. 183–192, 2019.
https://doi.org/10.1007/978-3-030-12544-8_15

generalized recently in paper [2] where the coefficients are arbitrary this time. The method used in this paper to obtain the analytical expression of the solution function in the case when we maximize the objective function can be adapted in order to find the analytical solution function when we need to minimize the objective function. This is what we will do in this note. A common feature in solving all these problems is that the constrained OWA aggregation problems are transformed into linear programs and the analytical expression of the solution function is obtained using the dual of these linear programs. It is important to mention that there are other works too where one uses the dual of linear programs in order to obtain the solution of certain type of constrained OWA aggregation problems (see papers [5,6]).

The paper is organized as follows. In Sect. 2 we recall the basic theory on the constrained OWA aggregation problem and we also recall our result from the recent paper [2] where we found the analytical expression of the solution function when we have a single constraint with arbitrary coefficients and the objective function needs to be maximized. In Sect. 3, this time we will need to minimize the objective function. Again, we will have a single constraint with arbitrary coefficients. If in the case of the maximization problem we need to rearrange the coefficients in nondecreasing order, for the minimization problem it suffice to rearrange them in nonincreasing order. This similar approach is a consequence of an inequality (often referred as Chebyshev inequality) on finite sequences of reals. There are some differences considering the two types of problems but the cases when the coefficients are positive give a similar type of solution function. It is important to note that in the case of the minimization problem it is not indicated to transform the problem into a maximization problem by considering the opposite of the objective function. In this case, we lose the positiveness of the weights and the solving becomes more complicated. What is more, we cannot use the formulae obtained in paper [2] because there the positiveness of the weights is essential. Indeed, as we said, the solution of this problem is obtained by using the solution of the dual of a linear program. But this solution needs to have positive components and this does not hold if instead of positive weights we consider they opposite values. Section 4 presents an example where both problems, maximization and minimization, are solved according to the expressions of the solution function. The paper ends with conclusions where the main results are discussed and further research on the topic is addressed.

2   Optimization of OWA Operators

Suppose we have the nonnegative weights $w_1, ..., w_n$ such that $w_1 + ... + w_n = 1$ and define a mapping $F : \mathbb{R}^n \to [0, 1]$,

$$F(x_1, ..., x_n) = \sum_{i=1}^{n} w_i y_i,$$

where y_i is the i-th largest element of the sample $x_1, ..., x_n$. This is called an OWA operator associated to the weights $w_1, ..., w_n$ (see [8]). Then consider a matrix A of type (m, n) with real entries and a vector $b \in \mathbb{R}^m$. A constrained OWA aggregation problem corresponding to the above data, is the problem

$$\max F(x_1, ..., x_n)$$

subject to

$$Ax \le b, \, x \ge 0.$$

This problem was proposed by Yagger in [7]. A difficult task is to find an exact analytical solution to this problem. Yagger used a method based on mixed integer linear programming problem which employes the use of auxiliary integer variables and therefore, this method is not always effective. In the special case where we have the single constraint $x_1 + ... + x_n = 1$, the first analytical solution for the constrained OWA aggregation problem is given in paper [1]. This result has been generalized recently in paper [2] where the coefficients in the constraint are arbitrary. This problem can be formulated as

$$\max F(x_1, ..., x_n) \tag{1}$$

subject to

$$\alpha_1 x_1 + ... + \alpha_n x_n \le 1,$$
$$x \ge 0$$

Let us recall this result in the case when we can provide a nontrivial solution (these cases were solved in Propositions 1–2 in [2]). In what follows, S_n denotes the set of permutations of the set $\{1, ..., n\}$.

Theorem 1. *Consider problem (1). Then:*

(i) *if there exists $i_0 \in \{1, ..., n\}$ such that $\alpha_{i_0} \le 0$, then F is unbounded on the feasible set and its supremum over the feasible set is ∞;*

(ii) *if $\alpha_i > 0$, $i \in \{1, ..., n\}$, then taking (any) $\sigma \in S_n$ with the property that $\alpha_{\sigma_1} \le \alpha_{\sigma_2} \le ... \le \alpha_{\sigma_n}$, and $k^* \in \{1, ..., n\}$, such that*

$$\frac{w_1 + ... + w_{k^*}}{\alpha_{\sigma_1} + ... + \alpha_{\sigma_{k^*}}} = \max \left\{ \frac{w_1 + ... + w_k}{\alpha_{\sigma_1} + ... + \alpha_{\sigma_k}} : k \in \{1, ..., n\} \right\},$$

then $(x_1^*, ..., x_n^*)$ is an optimal solution of problem (1), where

$$x_{\sigma_1}^* = ... = x_{\sigma_{k^*}}^* = \frac{1}{\alpha_{\sigma_1} + ... + \alpha_{\sigma_{k^*}}},$$
$$x_{\sigma_{k^*+1}}^* = ... = x_{\sigma_n}^* = 0.$$

In particular, if $0 < \alpha_1 \le \alpha_2 \le ... \le \alpha_n$, and $k^* \in \{1, ..., n\}$ is such that

$$\frac{w_1 + ... + w_{k^*}}{\alpha_1 + ... + \alpha_{k^*}} = \max \left\{ \frac{w_1 + ... + w_k}{\alpha_1 + ... + \alpha_k} : k \in \{1, ..., n\} \right\},$$

then $(x_1^*, ..., x_n^*)$ is a solution of (1), where

$$x_1^* = ... = x_{k^*}^* = \frac{1}{\alpha_1 + ... + \alpha_{k^*}},$$
$$x_{k^*+1}^* = ... = x_n^* = 0.$$

3   Minimizing the Objective Function

In this section we discuss the case when we search for the minimum in the objective function. It seems that we can apply a similar approach as in the case when the objective function is maximized. The general form is

$$\min F(x_1, ..., x_n) \tag{2}$$

subject to

$$\alpha_1 x_1 + \cdots + \alpha_n x_n \leq \beta,$$
$$x \geq 0.$$

Again, we will consider one restriction but in general form. Obviously it suffices to consider only the following three problems (any other problem is reduced to one of them)

$$\min F(x_1, ..., x_n) \tag{3}$$

subject to

$$\alpha_1 x_1 + \cdots + \alpha_n x_n \leq 0,$$
$$x \geq 0$$

$$\min F(x_1, ..., x_n) \tag{4}$$

subject to

$$\alpha_1 x_1 + \cdots + \alpha_n x_n \leq 1,$$
$$x \geq 0$$

and

$$\min F(x_1, ..., x_n) \tag{5}$$

subject to

$$\alpha_1 x_1 + \cdots + \alpha_n x_n \geq 1,$$
$$x \geq 0.$$

The solving of the first two problems is trivial. We observe that in both cases we have the unique solution $(0, 0, ..., 0)$, hence the minimum is 0 for both problems.

Let us discuss now the more interesting problem (5). The first result proves that in searching for the solution, in the case of positive weights it suffices to consider equality in the constraint.

Proposition 1. *Consider problem (5) If* $(x_1^*, ..., x_n^*)$ *is a solution of problem (5) and* $w_i > 0$, $i = 1, ..., n$, *then* $\alpha_1 x_1^* + ... + \alpha_n x_n^* = 1$.

Proof. If the conclusion were false, then we would have $\alpha_1 x_1^* + \cdots + \alpha_n x_n^* > 1$. Obviously, there exists at least one strictly greater than zero component in $(x_1^*, ..., x_n^*)$. Suppose these components are $x_{k_1}^*, ... x_{k_l}^*$. Then, there exists $\varepsilon > 0$ sufficiently small such that $\alpha_1 y_1^* + \cdots + \alpha_n y_n^* > 1$, where $y_{k_i}^* = x_{k_i}^* - \varepsilon > 0$, $i = 1, ..., l$, and all the other components are equal to 0 . Clearly this implies that $(y_1^*, ..., y_n^*)$ is feasible to our problem. What is more, we easily notice that $F(y_1^*, ..., y_n^*) < F(x_1^*, ..., x_n^*)$, which again, contradicts the minimality of $(x_1^*, ..., x_n^*)$.

Let us now discuss on the coefficients of the first constraint. If $\alpha_i \leq 0$ for all $i \in \{1, ..., n\}$ then we have no solution since the feasible set is empty. Next, suppose that there exists $i \in \{1, ..., n\}$ such that $\alpha_i \leq 0$. If $(x_1^*, ..., x_n^*)$ is a solution of problem (5) then it is sufficient to take $x_i^* = 0$ because otherwise, if $x_i^* > 0$, then it is really easy to prove that $(y_1^*, ..., y_n^*)$, where $x_j^* = y_j^*$ if $i \neq j$ and $y_i^* = 0$, belongs to the feasible set of problem (5) and $F(x_1^*, ..., x_n^*) \geq F(y_1^*, ..., y_n^*)$, hence $(y_1^*, ..., y_n^*)$ too, is a solution for (5). It means that if in problem (5) we have nonpositive coefficients in the restriction of problem (5), then we can reduce this problem to a problem where all coefficients are strictly greater than zero (we just eliminate the nonpositive coefficients and the weights from bottom, for example, if only $\alpha_1 \leq 0$ and $\alpha_2 \leq 0$, then in the new problem we eliminate these coefficients and the weights w_{n-1} and w_n) and a solution of the initial problem will be obtained by completing with zeros on the positions where the nonpositive coefficients were standing. For example, if only $\alpha_1 \leq 0$ and $\alpha_2 \leq 0$ then, if $(x_1^*, ..., x_{n-2}^*)$ is a solution of the problem where the coefficients α_1, α_2 and the last two weights are eliminated, then $(0, 0, x_1^*, ..., x_{n-2}^*)$ is a solution of the initial problem. It is important to mention that if the weights are positive and $\alpha_i \leq 0$ for some $i \in \{1, ..., n\}$, then it necessarily follows that $x_i^* = 0$. Indeed reasoning as above, this time we would get $F(x_1^*, ..., x_n^*) > F(y_1^*, ..., y_n^*)$, and this contradicts the minimality of $(x_1^*, ..., x_n^*)$.

Therefore, it will not be at all a limitation for the general case if in all that follows we assume that in problem (5) we have $\alpha_i > 0$, $i = 1, ..., n$. We start with the special case when $\alpha_1 \geq \alpha_2 \geq \cdots \geq \alpha_n$, and only after we shall discuss the general case. If $(x_1^*, ..., x_n^*)$ is a solution of the problem then let $\sigma \in S_n$ be any permutation such that $x_{\sigma_1}^* \geq x_{\sigma_2}^* \geq \cdots \geq x_{\sigma_n}^*$. It is well known that if $a_1 \geq a_2 \geq \cdots \geq a_n$ and $b_1 \geq b_2 \geq \cdots \geq b_n$ then $\sum_{i=1}^{n} a_i b_i \geq \sum_{i=1}^{n} a_i b_{\tau_i}$ for any $\tau \in S_n$. This implies that $\sum_{i=1}^{n} \alpha_i x_{\sigma_i}^* \geq \sum_{i=1}^{n} \alpha_i x_i^*$ and hence $\sum_{i=1}^{n} \alpha_i x_{\sigma_i}^* \geq 1$. It means that $(x_{\sigma_1}^*, ..., x_{\sigma_n}^*)$ is feasible and on the other hand, clearly we have $F(x_1^*, ..., x_n^*) = F(x_{\sigma_1}^*, ..., x_{\sigma_n}^*)$, which means that $(x_{\sigma_1}^*, ..., x_{\sigma_n}^*)$ is a solution of problem (5) as well (in the case when the weights are positive, By Proposition 1 it also means that $\sum_{i=1}^{n} \alpha_i x_{\sigma_i}^* = 1$). But, this implies that $(x_{\sigma_1}^*, ..., x_{\sigma_n}^*)$ is in addition a solution of the linear programming problem

$$\min \sum_{i=1}^{n} w_i x_i \qquad (6)$$

subject to

$$\alpha_1 x_1 + \ldots + \alpha_n x_n \geq 1,$$
$$x_1 \geq x_2 \ldots \geq x_n \geq 0.$$

Indeed, it suffices to notice that the feasible set of this problem is included in the feasible set of problem (5) which combined with the fact that $F(x_1^*, \ldots, x_n^*) = F(x_{\sigma_1}^*, \ldots, x_{\sigma_n}^*)$ and $(x_1^*, \ldots, x_n^*)$ solves (5) while $(x_{\sigma_1}^*, \ldots, x_{\sigma_n}^*)$ is feasible to problem (6), all these imply that $(x_{\sigma_1}^*, \ldots, x_{\sigma_n}^*)$ is a solution of (6).

In view of the above discussion, we start by providing an analytical solution to problem (6). The reasoning is similar to those used in papers [1] and [2] in the cases when we have maximum instead of minimum in the objective function. It is convenient to write the dual of problem (6), which is

$$\max t_1 \qquad (7)$$

subject to

$$\alpha_1 t_1 + t_2 \leq w_1,$$
$$\alpha_2 t_1 - t_2 + t_3 \leq w_2,$$

$$\cdot$$
$$\cdot$$

$$\alpha_{n-1} t_1 - t_{n-1} + t_n \leq w_{n-1},$$
$$\alpha_n t_1 - t_n \leq w_n,$$
$$t \geq 0.$$

Summing up the first k inequalities from above, $k = \overline{1,n}$, we get

$$t_1 \leq \frac{w_1 + \ldots + w_k - t_{k+1}}{\alpha_1 + \ldots + \alpha_k}, \; k = \overline{1, n-1},$$
$$t_1 \leq \frac{w_1 + \ldots + w_n}{\alpha_1 + \ldots + \alpha_n}.$$

We easily notice that $t_1 \leq \frac{w_1 + \ldots + w_{k^*}}{\alpha_1 + \ldots + \alpha_{k^*}}$, where $k^* \in \{1, \ldots, n\}$ satisfies

$$\frac{w_1 + \ldots + w_{k^*}}{\alpha_1 + \ldots + \alpha_{k^*}} = \min \left\{ \frac{w_1 + \ldots + w_k}{\alpha_1 + \ldots + \alpha_k} : k \in \{1, \ldots, n\} \right\}.$$

It means that $(t_1^*, \ldots, t_n^*)$ is a solution of (7), where

$$t_1^* = \frac{w_1 + \ldots + w_{k^*}}{\alpha_1 + \ldots + \alpha_{k^*}},$$
$$t_{k+1}^* = \left(\frac{w_1 + \ldots + w_k}{\alpha_1 + \ldots + \alpha_k} - t_1^* \right) (\alpha_1 + \ldots + \alpha_k), \, k = \overline{1, n-1}.$$

From the duality theorem, if there exists $(x_1^*, ..., x_n^*)$ in the feasible set of problem (6), such that $\sum_{i=1}^{n} w_i x_i^* = t_1^*$, then $(x_1^*, ..., x_n^*)$ is a solution of problem (6). Obviously this solution exists since we can take

$$x_1^* = ... = x_{k^*}^* = \frac{1}{\alpha_1 + ... + \alpha_{k^*}},$$
$$x_{k^*+1}^* = ... = x_n^* = 0.$$

We are now in position to present an analytical solution for the general case of problem (5). We will just need to rearrange the order of the coefficients and variables in order to use the formula from above. We reiterate again the fact that it is not a limitation to assume that the coefficients are positive.

Theorem 2. *Consider problem (5). If $\alpha_i > 0$, $i \in \{1, ..., n\}$, then taking $\sigma \in S_n$ (it is possible to have multiple choices for σ) with the property that $\alpha_{\sigma_1} \geq \alpha_{\sigma_2} \geq \cdots \geq \alpha_{\sigma_n}$, and $k^* \in \{1, ..., n\}$, such that*

$$\frac{w_1 + \cdots + w_{k^*}}{\alpha_{\sigma_1} + \cdots + \alpha_{\sigma_{k^*}}} = \min\left\{\frac{w_1 + \cdots + w_k}{\alpha_{\sigma_1} + \cdots + \alpha_{\sigma_k}} : k \in \{1, ..., n\}\right\},$$

then $(x_1^*, ..., x_n^*)$ is an optimal solution of problem (5), where

$$x_{\sigma_1}^* = ... = x_{\sigma_{k^*}}^* = \frac{1}{\alpha_{\sigma_1} + ... + \alpha_{\sigma_{k^*}}},$$
$$x_{\sigma_{k^*+1}}^* = ... = x_{\sigma_n}^* = 0.$$

In particular, if $\alpha_1 \geq \alpha_2 \geq \cdots \geq \alpha_n$, and $k^* \in \{1, ..., n\}$ is such that

$$\frac{w_1 + \cdots + w_{k^*}}{\alpha_1 + \cdots + \alpha_{k^*}} = \min\left\{\frac{w_1 + \cdots + w_k}{\alpha_1 + \cdots + \alpha_k} : k \in \{1, ..., n\}\right\},$$

then $(x_1^*, ..., x_n^*)$ is a solution of (5), where

$$x_1^* = ... = x_{k^*}^* = \frac{1}{\alpha_1 + \cdots + \alpha_{k^*}},$$
$$x_{k^*+1}^* = ... = x_n^* = 0.$$

As we said in the introduction, transforming problem (5) into a maximization problem, that is, considering $\max -F(x_1, ..., x_n)$ instead of $\min F(x_1, ..., x_n)$, will not lead to a simpler method to find the solution because we loose the positiveness of the weights which is essential in finding the solution of the dual problem that leads to the solution given in Theorem 1. There is another possibility to transform problem (5) into a maximization problem, but in this case too, we do not get an easier method. First, let us discuss the special case when all the coefficients in the constraint are equal to one, that is, we consider problem

$$\min F(x_1, ..., x_n)$$

subject to

$$x_1 + \cdots + x_n \geq 1,$$
$$x \geq 0.$$

As we know, without any loss of generality we may assume that the constraint is $x_1 + \cdots + x_n = 1$. Suppose that $(x_1^*, ..., x_n^*)$ is a solution of the problem from above. Denoting with $(y_1^*, ..., y_n^*)$ the vector that rearranges $(x_1^*, ..., x_n^*)$ in nondecreasing order, then using the substitutions $z_i^* = 1 - x_i^*$ and $y_i^* = 1 - t_i^*$, $i = \overline{1, n}$, we get

$$F(x_1^*, ..., x_n^*)$$
$$= w_1 y_1^* + \cdots + w_n y_n^*$$
$$= w_1 (1 - t_1^*) + \cdots + w_n (1 - t_n^*)$$
$$= \sum_{i=1}^{n} w_i - \sum_{i=1}^{n} w_i t_i^*$$

and

$$x_1^* + \cdots + x_n^*$$
$$= n - \sum_{i=1}^{n} z_i^*.$$

This easily implies that $(1 - x_1^*, ..., 1 - x_n^*)$ and any of its permutations is a feasible solution for the problem

$$\max \overline{F}(z_1, ..., z_n)$$

subject to

$$z_1 + \cdots + z_n = n - 1,$$
$$z \geq 0,$$

where

$$\overline{F}(z_1, ..., z_n)$$
$$= \overline{w}_1 t_1 + \cdots + \overline{w}_n t_n,$$

$\overline{w}_i = w_{n-i}$ and t_i is the i-th largest element from the sequence $z_1, ..., z_n$. Obviously, this later problem is a constrained OWA aggregation problem and the solution is immediate by applying Theorem 1, (ii). Unfortunately, $(1 - x_1^*, ..., 1 - x_n^*)$ it is not necessarily optimal since in general, the solution can have components strictly larger than 1. Actually, one can easily prove that if $(z_1^*, ..., z_n^*)$ is an optimal solution of problem

$$\max \overline{F}(z_1, ..., z_n)$$

subject to

$$z_1 + \cdots + z_n = n - 1,$$
$$z \geq 0,$$
$$z \leq 1$$

then $(1 - z_1^*, ..., 1 - z_n^*)$ and any of its permutations is an optimal solution of problem (5). Clearly, this problem in general is not of type (1). Now, considering the case of arbitrary coefficients we will arrive to a similar construction, that is, a more complex maximization problem having additional constraints.

Comparing Theorems 1 and 2, respectively, we can easily solve both problems (maximum and minimum) in the case of a single constraint. For the maximum problem we just need to rearrange the coefficients in nondecreasing order and in the case of the minimum problem, we need to rearrange them in nonincreasing order.

4   An Example for Both Maximization and Minimization Problems

Example 1. Suppose that $F(x_1, x_2, x_3, x_4) = \frac{1}{3}y_1 + \frac{1}{8}y_2 + \frac{1}{2}y_3 + \frac{1}{24}y_4$ and consider the constraint $x_1 + 4x_2 + 2x_3 + 3x_4 = 1$. Let us find the maximum point of F. Obviously, the minimum points are exactly the same if the constraint would be $x_1 + 4x_2 + 2x_3 + 3x_4 \leq 1$. Therefore, we can apply the conclusion of Theorem 1. We need a permutation of $\{1, ..., 4\}$ which would rearrange the coefficients in nondecreasing order. Such a permutation is

$$\sigma = \begin{pmatrix} 1\,2\,3\,4 \\ 1\,3\,4\,2 \end{pmatrix}$$

and by simple inspection, we get that

$$\max \left\{ \frac{w_1 + ... + w_k}{\alpha_{\sigma_1} + ... + \alpha_{\sigma_k}} : k \in \{1, ..., 4\} \right\}$$

is achieved for $k^* = 1$. Applying the conclusion of Theorem 1, we get that $(x_1^*, ..., x_4^*)$, $x_1^* = 1$, $x_2^* = x_3^* = x_4^* = 0$, is a solution of our problem. We also notice that the maximum value is $F(x_1^*, ..., x_4^*) = \frac{1}{3}$.

Let us find now the minimum of F under the same constraint. Obviously, we have the same solutions if the constraint would be $x_1 + 4x_2 + 2x_3 + 3x_4 \geq 1$. It means that we can apply Theorem 2 for this problem. This time we need a permutation of $\{1, ..., 4\}$ which would rearrange the coefficients in nonincreasing order. Such a permutation is

$$\tau = \begin{pmatrix} 1\,2\,3\,4 \\ 2\,4\,3\,1 \end{pmatrix}$$

and by simple inspection, we get that

$$\min\left\{\frac{w_1 + \ldots + w_k}{\alpha_{\sigma_1} + \ldots + \alpha_{\sigma_k}} : k \in \{1, \ldots, 4\}\right\}$$

is achieved for $k^* = 2$. Applying the conclusion of Theorem 2, we get that $(\overline{x}_1, \ldots, \overline{x}_4)$, $\overline{x}_1 = \overline{x}_3 = 0$, $\overline{x}_2 = \overline{x}_4 = \frac{1}{7}$, is a solution of our problem. We also notice that the minimum value is $F(\overline{x}_1, \ldots, \overline{x}_4) = \frac{11}{168}$.

5   Conclusions

In this note we completed the work in paper [2], as this time we found the analytical expression of the solution function in the case of minimization of OWA aggregation operators with single constraint. In the future, we are interested to extend the results in the case when we have more constraints. Although in general it seems that the method used in this research and in paper [2] cannot be generalized as we cannot find a single permutation to rearrange monotonically the coefficients in all constraints, some important special cases could be investigated. In the case of two constraints we have an ongoing research and results are promising. Another important problem would be to find the solution of the minimum problem from the solution of a derived maximum problem. This would ease on the computer implementation. This problem as well seems to be quite difficult since even in the simplest case when we have a single constraint with all coefficients equal to 1, we obtained a maximization problem that has additional constraints.

Acknowledgement. The contribution of Lucian Coroianu was supported by a grant of Ministry of Research and Innovation, CNCS-UEFISCDI, project number PN-III-P1-1.1-PD-2016-1416, within PNCDI III.

References

1. Carlsson, C., Fullér, R., Majlender, P.: A note on constrained OWA aggregation. Fuzzy Sets Syst. **139**, 543–546 (2003)
2. Coroianu, L., Fullér, R.: On the constrained OWA aggregation problem with single constraint. Fuzzy Sets Syst. **332**, 37–43 (2018)
3. Emrouznejad, A., Marra, M.: Ordered weighted averaging operators 1988–2014: a citation-based literature survey. Int. J. Intell. Syst. **29**, 994–1014 (2014)
4. Nguyen, T.H.: Maximal entropy and minimal variability OWA operators weights: a short survey of recent developments. arXiv preprint arXiv:1804.06331 (2018). arxiv.org
5. Ogryczak, W., Śliwiński, T.: On efficient WOWA optimization for decision support under risk. Int. J. Approximate Reasoning **50**, 915–928 (2009)
6. Ogryczak, W., Śliwiński, T.: On solving linear programs with the ordered weighted averaging objective. Eur. J. Oper. Res. **148**, 80–91 (2003)
7. Yagger, R.R.: Constrained OWA aggregation. Fuzzy Sets Syst. **81**, 89–101 (1996)
8. Yagger, R.R.: On ordered weighted averaging aggregation operators in multicriteria decisionmaking. IEEE Trans. Syst. Man Cybern. **18**, 183–190 (1988)

An Alpha-Cut Evaluation
of Interval-Valued Fuzzy Sets
for Application in Decision Making

Luca Anzilli[✉] and Gisella Facchinetti

Department of Management, Economics, Mathematics and Statistics,
University of Salento, Lecce, Italy
`luca.anzilli@unisalento.it`

Abstract. In this paper we deal with the problem of evaluating an interval-valued fuzzy set, that is a fuzzy quantity delimited by two (lower and upper) membership functions. The problem of associating this type of set with a real number has been dealt with in different ways. Karnik and Mendel proposed an algorithm for computing the mean of centroids of membership functions that lie within the area delimited by the lower and upper memberships. Nie and Tan choose a simpler way by calculating the centroid of the average of the lower and upper membership functions. In both cases, the value obtained is useful not only in ranking problems but also as a value of defuzzification if the set is the final output of a fuzzy inference system. Since in this last case the obtained set is usually not normal and not convex, the centroid seems to be the only useful defuzzifier. Our purpose is to show that other methods based on alpha-cuts, usually applied in convex type-1 case, can also provide useful answers.

Keywords: Fuzzy sets · Fuzzy quantities · Interval-valued fuzzy sets · Evaluation · Decision making

1 Introduction

Type-1 fuzzy sets (T1 FSs) generalize the concept of a classical set allowing its membership function to assume not only the two values zero and one, as in the ordinary sets theory, but all the values between zero and one. This approach lets to face real problems in which, due to the uncertainty that is inherent in information, it is difficult to decide if something belongs or not to a specific class. Either in theoretical framework or in industrial applications and decision making problems, T1 fuzzy inference systems (T1 FISs) have obtained a lot of success. The uncertainty, we find at this step, concerns not only on fuzzy rules but even on inputs-outputs fuzzification. Sometimes, T1 FSs use does not seem the best. We refer to cases in which the fuzzification of inputs derives from choices made by experts. In these cases it is often difficult to reach an agreement

R. Fullér et al. (Eds.): WILF 2018, LNAI 11291, pp. 193–211, 2019.
https://doi.org/10.1007/978-3-030-12544-8_16

on membership fuctions that describe the individual linguistic attributes of the inputs-outputs. These problems carried out the research in the interval-valued fuzzy sets (IVFSs) and interval-valued fuzzy logic systems (IVFLS) [1–6]. IVFSs are particular cases of interval type-2 fuzzy sets (IT2 FSs), which in turn are special cases of type-2 fuzzy sets (T2 FSs) [7–11]. Professor Zadeh introduced the concept of T2 FS [12], but only later this notion has obtained a new life by the work of professor J. Mendel and his group of research. In several book they presented a complete description of Type-2 fuzzy systems [13,14]. An IT2 fuzzy set is a collection of infinite T1 fuzzy sets that define a two-dimensional domain that is called footprint of uncertainty (FOU). The FOU is completely described by its two bounding functions: a lower membership function (LMF)and an upper membership function (UMF). The addition of this new dimension of fuzziness has produced a lot of benefits in dealing with uncertainty, but has generated some problems either in theory or in calculation [15–19]. One of this problem is the defuzzification step of an IT2 fuzzy logic system. The set we have to evaluate, is not connected with a unique membership function, but is a two dimensional set in which are embedded infinite membership functions. The new procedure that faces this problem is called Type Reduction step and one of the most used method is Karnik and Mendel (KM) iterative algorithm introduced for the general case of T2 FSs [20]. In [21] this method is translated for IT2 FLS. It converts the output, which is a IT2 fuzzy set, into a finite set of T1 fuzzy sets embedded in the output FOU zone. Then it evaluates the centroid of all these fuzzy sets. These data lie in an interval and the defuzzified value of IT2 fuzzy sets is its middle point. This procedure use centroid of a fuzzy set as evaluation method. Following the same line, Nie and Tan (NT) [22] have proposed a more simple and useful method for reducing the KM algorithm computational cost. Their proposal of defuzzified value is the centroid of the average of LMF and UMF, that describe the FOU zone of IT2 fuzzy set. Mendel and Liu [23] either show a comparison between the two methods and propose an NT implementation to the continuous case. The literature is therefore all directed to the use of centroid as the unique method of defuzzification. The motivation is clearly linked to the structure of the Fuzzy Sets that are obtained in the defuzzification step. These sets may be non-normal and/or non-convex and while the literature is full of defuzzification methods for fuzzy numbers [24–33] it is very poor for generic fuzzy sets [34–39]. As in the past we have been interested in this problem for T1FSs and we have shown how also methods that seem closely related to the convexity of the fuzzy set, such as those related to the α-cuts can produce interesting results, we have decided to face the type 2 case. In the case of IT2 fuzzy sets the techniques we propose have been defined having in mind that the attribution of a real value to a fuzzy set could be inserted in different fields of research as decision theory in which the final choice of the decision maker may need an "area" of freedom. To give space to this area, the methodology proposes a parametric formulation. We have followed this idea guided by the words: "We note that parametrized classes of methods for ordering fuzzy numbers are particularly useful, in that they allow us to train amethodology to satisfy the user. Parametrized classes are often suggested by a process in which we try to unify and connect already existing approaches" [39]. We underline that our choice produces two different results: the first is to unify the two procedures, one that works on x-axis (centroid) and the other on y-axis (α-cuts); the second is connected with the freedom that the parametric vision offers to the decision maker. We start with an α-cut definition proposed by Hamrawi [40,41]. He presents one different definition of α-cut for IT2 FSs respect to the previous ones, taking the α-cuts of its LMF and UMF at the same level. This definition is coherent withusual definition of T1FS α-cut when the UMF collapses over LMF. Then we consider the family of all α-cuts of IVT2FS and we find the nearest interval to this family with respect to a suitable distance. The midpoint of the solution is the evaluation we propose and its spread is what we call "level of uncertainty". The two results depend on two different types of parameters: the first acts horizontally as an α-cut weight, that is for every alpha fixed it takes into account of all the intervals that form the single α-cut; the second acts vertically as an α-level weight. As a consequence, the parametric view leaves the decision maker to choice the weights in a different manner depending on the context and his opinions. For a particular choice of these parameters we reacquire the NT formula (centroid) in the continuous version. The new formulation of NT method gives more information than the original one since it shows the two families of parameters involved. In this case, as we will show, either the horizontal or the vertical weight depends only on the length of α-cuts. This is an important information for the decision maker as this type of choice produces a particular meaning of his point of view. The freedom and the transparency that the method offers may give more awareness for decision making.

2   Preliminaries and Notation

A type-1 fuzzy set (T1 FS) A of the universe of discourse X is defined by a membership function $\mu_A : X \to [0, 1]$ which assigns to each element of X a grade of membership to the set A. The height of A is $h_A = height\, A = \sup_{x \in X} \mu_A(x)$. The support and the core of A are the crisp sets $supp(A) = \{x \in X; \mu_A(x) > 0\}$ and $core(A) = \{x \in X; \mu_A(x) = 1\}$. A fuzzy set A is normal if its core is nonempty. The intersection (resp. union) of two fuzzy sets A and B is the fuzzy set $A \cap B$ (resp. $A \cup B$) defined by $\mu_{A \cap B}(x) = \min\{\mu_A(x), \mu_B(x)\}$ (resp. $\mu_{A \cup B}(x) = \max\{\mu_A(x), \mu_B(x)\}$). The α-cut of a T1 FS A, with $0 \leq \alpha \leq 1$, is defined as the crisp set $A_\alpha = \{x \in X; \mu_A(x) \geq \alpha\}$ if $0 < \alpha \leq 1$ and as the closure of the support if $\alpha = 0$. We say that $A \subseteq B$ if $\mu_A(x) \leq \mu_B(x)$ for each $x \in X$. Note that $A \subseteq B \iff A_\alpha \subseteq B_\alpha \, \forall \alpha$. A fuzzy set is called convex if each α-cut is a closed interval $A_\alpha = [a_L(\alpha), a_R(\alpha)]$, where $a_L(\alpha) = \inf A_\alpha$ and $a_R(\alpha) = \sup A_\alpha$.

Definition 1. [35,42] *Let N be a positive integer and let $a_1, a_2, \ldots, a_{4N}$ be real numbers with $a_1 < a_2 \leq a_3 < a_4 \leq a_5 < a_6 \leq a_7 < a_8 \leq a_9 < \cdots < a_{4N-2} \leq a_{4N-1} < a_{4N}$. We call type-1 fuzzy quantity (T1 FQ)*

$$A = (a_1, a_2, \ldots, a_{4N};\ h_1, h_2, \ldots, h_N,\ h_{1,2}, h_{2,3}, \ldots, h_{N-1,N}) \tag{1}$$

where $0 < h_j \leq 1$ for $j = 1, \ldots, N$ and $0 \leq h_{j,j+1} < \min\{h_j, h_{j+1}\}$ for $j = 1, \ldots, N - 1$, the type-1 fuzzy set defined by a continuous membership function $\mu : \mathbb{R} \to [0,1]$, with $\mu(x) = 0$ for $x \leq a_1$ or $x \geq a_{4N}$, such that

(i) for $j = 1, 2, \ldots, N$: μ is strictly increasing in $[a_{4j-3}, a_{4j-2}]$, with $\mu(a_{4j-3}) = h_{j-1,j}$ and $\mu(a_{4j-2}) = h_j$; μ is constant in $[a_{4j-2}, a_{4j-1}]$, with $\mu \equiv h_j$; μ is strictly decreasing in $[a_{4j-1}, a_{4j}]$, with $\mu(a_{4j-1}) = h_j$ and $\mu(a_{4j}) = h_{j,j+1}$;
(ii) for $j = 1, 2, \ldots, N - 1$: μ is constant in $[a_{4j}, a_{4j+1}]$, with $\mu \equiv h_{j,j+1}$;

where $h_{0,1} = h_{N,N+1} = 0$. Thus the height of A is $h_A = \max_{j=1,\ldots,N} h_j$.

We observe that in the case $N = 1$ the T1 FQ defined in (1) is fuzzy convex, that is every α-cut A_α is a closed interval. If $N \geq 2$ the T1 FQ defined in (1) is a non-convex fuzzy set with N humps and height $h_A = \max_{j=1,\ldots,N} h_j$.

Figure 1 shows an example of piecewise linear T1 FQ with $N = 2$.

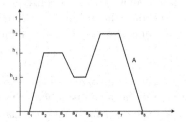

Fig. 1. Piecewise linear T1 FQ ($N = 2$)

If A is a T1 FQ with height h_A then for each $\alpha \in [0, h_A]$ there exist an integer n_α^A, with $1 \leq n_\alpha^A \leq N$, and $A_1^\alpha, \ldots, A_{n_\alpha^A}^\alpha$ disjoint closed intervals such that

$$A_\alpha = \bigcup_{i=1}^{n_\alpha^A} A_i^\alpha = \bigcup_{i=1}^{n_\alpha} [a_i^L(\alpha), a_i^R(\alpha)], \tag{2}$$

where we have denoted $A_i^\alpha = [a_i^L(\alpha), a_i^R(\alpha)]$, with $A_i^\alpha < A_{i+1}^\alpha$ (that is $a_i^R(\alpha) < a_{i+1}^L(\alpha)$). Thus n_α^A is the number of intervals producing the α-cut A_α.

From decomposition theorem [43] for T1 FSs and using previous result, we get the representation

$$A = \bigcup_{\alpha \in [0,h_A]} \alpha A_\alpha = \bigcup_{\alpha \in [0,h_A]} \alpha \bigcup_{i=1}^{n_\alpha^A} A_i^\alpha = \bigcup_{\alpha \in [0,h_A]} \bigcup_{i=1}^{n_\alpha^A} \alpha A_i^\alpha. \tag{3}$$

Definition 2. [21] *A T2 FS $\tilde{A}$ in the universe of discourse X is characterized by a type-2 membership function $\mu_{\tilde{A}}(x, u)$ where $x \in X$ and $u \in J_x \subseteq [0,1]$, i.e. $\tilde{A} = \{((x, u), \mu_{\tilde{A}}(x, u)); \ x \in X, \ u \in J_x \subseteq [0,1]\}$ in which $0 \leq \mu_{\tilde{A}}(x, u) \leq 1$.*

If all $\mu_{\tilde{A}}(x, u) = 1$ then $\tilde{A}$ is called an IT2 FS. An IT2 FS $\tilde{A}$ can be considered as a special case of a T2 FS. J_x is called the primary membership of x. The footprint of uncertainty (FOU) of an IT2 FS $\tilde{A}$ is defined by FOU$(\tilde{A}) = \{(x, u); \ x \in X, \ u \in J_x\}$. The FOU is a complete description of an IT2 FS. The lower membership function (LMF) $\mu_{\tilde{A}}^L$ and the upper membership function (UMF) $\mu_{\tilde{A}}^U$ of an IT2 FS $\tilde{A}$ are defined as the two type-1 membership functions that bound the FOU. An IT2 FS $\tilde{A}$ is also denoted by $\tilde{A} = (A^L, A^U)$ where A^L and A^U are the T1 FSs with membership functions $\mu_{A^L} = \mu_{\tilde{A}}^L$ and $\mu_{A^U} = \mu_{\tilde{A}}^U$, respectively. For the sake of notation simplicity, in the following an IT2 FS $\tilde{A} = (A^L, A^U)$ will be denoted by $\tilde{A} = (B, C)$. Thus, the LMF of $\tilde{A}$ is the membership function μ_B of the T1 FS B and the UMF of $\tilde{A}$ is the membership function μ_C of the T1 FS C. The intersection and the union of two IT2 FSs $\tilde{A}_1 = (B_1, C_1)$ and $\tilde{A}_2 = (B_2, C_2)$ are defined, respectively, by the IT2 FSs

$$\tilde{A}_1 \cap \tilde{A}_2 = (B_1 \cap B_2, C_1 \cap C_2), \qquad \tilde{A}_1 \cup \tilde{A}_2 = (B_1 \cup B_2, C_1 \cup C_2). \qquad (4)$$

Let us denote by $L([0, 1])$ the set of all closed subintervals of $[0, 1]$, that is $L([0, 1]) = \{[a, b]; \ a, b \in [0, 1], \ a \leq b\}$.

Definition 3. [4,7,12] *An interval-valued fuzzy set (IVFS) $\tilde{A}$ on the universe of discourse X is a mapping $\tilde{A} : X \to L([0, 1])$ such that the membership degree of $x \in X$ is given by $\tilde{A}(x) = [A^L(x), A^U(x)] \in L([0, 1])$, where $A^L : X \to [0, 1]$ and $A^R : X \to [0, 1]$ are mappings defining the lower and the upper bound of the membership interval $\tilde{A}(x)$, respectively.*

We may look at IVFSs as special cases of IT2 FSs, when J_x is a closed interval of real numbers [7]. Indeed, in this case, $J_x = [\mu_{\tilde{A}}^L(x), \mu_{\tilde{A}}^U(x)] = [A^L(x), A^U(x)] = \tilde{A}(x)$ for all $x \in X$. In the following an IVFS $\tilde{A} = (A^L, A^U)$ will be denoted by $\tilde{A} = (B, C)$.

3   Evaluation of Interval-Valued Fuzzy Quantities

We now introduce the concept of interval-valued fuzzy quantitity and present an α-cuts decomposition result.

Definition 4. *We call interval-valued fuzzy quantitity (IVFQ) an IVFS $\tilde{A} = (B, C)$, with $B \subseteq C$, such that B and C are T1 FQs. An example of IVFQ is shown in Fig. 2(a).*

Several papers deal with the definition of T2 FS α-cuts [11,40,44–46]. In this study we follow definition given in [41] that preserves the α-cuts representation for a T1 FS obtained by a IT2 FS in which UMF is collapsed to LMF.

Definition 5. [40,41] *The α-cut at level α of an IVFQ $\tilde{A} = (B, C)$ is the IVFS $\tilde{A}_\alpha$ defined by the α-cuts of the LMF and the UMF at the same level (see Fig. 2(b)), that is $\tilde{A}_\alpha = (B_\alpha, C_\alpha)$.*

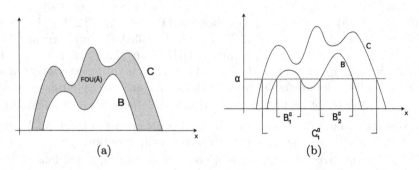

Fig. 2. IVFQ $\tilde{A} = (B, C)$ (a) and IVFQ $\tilde{A} = (B, C)$ with its α-cuts (b)

If we let $\alpha \tilde{A}_\alpha$ be the IVFS given by $\alpha \tilde{A}_\alpha = \alpha(B_\alpha, C_\alpha) = (\alpha B_\alpha, \alpha C_\alpha)$, the following decomposition holds

$$\tilde{A} = \bigcup_\alpha \alpha \tilde{A}_\alpha. \tag{5}$$

Indeed $\bigcup_\alpha \alpha \tilde{A}_\alpha = \bigcup_\alpha \alpha(B_\alpha, C_\alpha) = \bigcup_\alpha (\alpha B_\alpha, \alpha C_\alpha) = (\bigcup_\alpha \alpha B_\alpha, \bigcup_\alpha \alpha C_\alpha) = (B, C) = \tilde{A}$ where the third equality follows from (4) and the fourth equality from decomposition theorem for T1 FSs.

Let us consider an IVFQ $\tilde{A} = (B, C)$. From decomposition (5), all the information contained in $\tilde{A}$ is described by its α-cuts. Furthermore, observing that, from (2), the α-cuts of T1 FQs B and C can be decomposed as $B_\alpha = \bigcup_{i=1}^{n_\alpha^B} B_i^\alpha$, and $C_\alpha = \bigcup_{j=1}^{n_\alpha^C} C_j^\alpha$, from (5) we obtain $\tilde{A} = \bigcup_\alpha \alpha \left(\bigcup_{i=1}^{n_\alpha^B} B_i^\alpha, \bigcup_{j=1}^{n_\alpha^C} C_j^\alpha \right)$. Such decomposition enables us to identify the IVFQ $\tilde{A}$ with the family $\tilde{\mathcal{A}}$ of all the closed intervals B_i^α, C_j^α, that is $\tilde{\mathcal{A}} = \{B_1^\alpha, \dots, B_{n_\alpha^B}^\alpha, C_1^\alpha, \dots, C_{n_\alpha^C}^\alpha; \ 0 \le \alpha \le h\}$ where $h = \max\{h_B, h_C\} = h_C$. For convenience, by defining

$$A_i^\alpha = \begin{cases} B_i^\alpha & i = 1, \dots, n_\alpha^B \\ C_{i-n_\alpha^B}^\alpha & i = n_\alpha^B + 1, \dots, \tilde{n}_\alpha \end{cases} \qquad \alpha \in [0, h] \tag{6}$$

where

$$\tilde{n}_\alpha = n_\alpha^B + n_\alpha^C, \tag{7}$$

we can represent $\tilde{\mathcal{A}}$ as the family of all the closed intervals A_i^α, that is

$$\tilde{\mathcal{A}} = \{A_i^\alpha; \ i = 1 \dots, \tilde{n}_\alpha, \ 0 \le \alpha \le h\}. \tag{8}$$

We observe that each interval $I = [a, b]$ can represented as a point in $\mathbb{R}^2$ by the pair (mid, spr) where $mid(I) = (a + b)/2$ and $spr(I) = (b - a)/2$ are, respectively, the middle point and the spread of the interval. In this way the IVFQ $\tilde{A}$ can be viwed as a $\mathbb{R}^2$ cloud (cluster) of points corresponding to all the α-cuts of the IVFQ. Our idea is to evaluate the IVFQ $\tilde{A}$ by the centre C of the cloud, that can be found by solving a minimum distance problem.

This point C is an interval and we propose the midpoint of this interval as IVFQ $\tilde{A}$ evaluation and its spread as its grade of uncertainty. We employ the distance between two closed intervals I_1, I_2 (see [47]) defined by $d_{\overline{\theta}}(I_1, I_2) = \sqrt{(mid(I_1) - mid(I_2))^2 + \overline{\theta}(spr(I_1) - spr(I_2))^2}$. The parameter $\overline{\theta} \in]0, 1]$ indicates the relative importance of the spreads against the mids [31]. In this paper we consider the parameter θ as a function of α having in mind that the relative importance of the spreads against the mids may depend on the level of uncertainty.

Let $\tilde{A}$ be an IVFQ and $\tilde{\mathcal{A}}$ be the family of closed intervals defined in (8). Let us consider a 3-tuple (p, f, θ) such that, for each level α, the weights $p(\alpha) = (p_i(\alpha))_{i=1,...,\tilde{n}_\alpha}$ satisfy the properties

$$p_i(\alpha) \geq 0 \qquad \sum_{i=1}^{\tilde{n}_\alpha} p_i(\alpha) = 1, \tag{9}$$

the weight function $f : [0, 1] \to [0, +\infty[$ fulfil the condition

$$\int_0^h f(\alpha)\, d\alpha = 1 \tag{10}$$

and $\theta : [0, 1] \to]0, 1]$ is a function.

We are looking for the nearest interval to $\tilde{\mathcal{A}}$ with respect to (p, f, θ), that is the interval $K^*(\tilde{A}) = K^*(\tilde{A}; p, f, \theta)$ which minimizes the weighted mean of the squared distances

$$\mathcal{D}^{(2)}(K; \tilde{A}) = \int_0^h \sum_{i=1}^{\tilde{n}_\alpha} d_{\theta(\alpha)}^2(K, A_i^\alpha)\, p_i(\alpha)\, f(\alpha)\, d\alpha$$

$$= \int_0^h \sum_{i=1}^{\tilde{n}_\alpha} [(mid(K) - mid(A_i^\alpha))^2 + \theta(\alpha)\,(spr(K) - spr(A_i^\alpha))^2] p_i(\alpha)\, f(\alpha) d\alpha \tag{11}$$

among all the intervals K. Note that the function $\theta : [0, 1] \to]0, 1]$ indicates the relative importance of the spreads against the mids. The weights we have introduced work in a different manner: $p(\alpha) = (p_i(\alpha))_{i=1,...,\tilde{n}_\alpha}$ gives the possibility to evaluate in a different way the several intervals that produce the α-cut, the weighting function f offers the possibility to give different importance to each α-level.

Theorem 1. *The interval $K^*(\tilde{A}) = K^*(\tilde{A}; p, f, \theta)$ which minimizes (11) with respect to (p, f, θ) is given by*

$$mid(K^*(\tilde{A})) = \int_0^h \sum_{i=1}^{\tilde{n}_\alpha} mid(A_i^\alpha)\, p_i(\alpha)\, f(\alpha)\, d\alpha$$

$$spr(K^*(\tilde{A})) = \frac{\int_0^h \sum_{i=1}^{\tilde{n}_\alpha} spr(A_i^\alpha)\, p_i(\alpha)\, f(\alpha)\, \theta(\alpha)\, d\alpha}{\int_0^h f(\alpha)\, \theta(\alpha)\, d\alpha}. \tag{12}$$

Proof. By denoting $m_k = mid(K)$ and $s_K = spr(K)$, we have to minimize the function $g(m_K, s_K) = \int_0^h \sum_{i=1}^{\tilde{n}_\alpha} [(m_K - mid(A_i^\alpha))^2 + \theta(\alpha)(s_K - spr(A_i^\alpha))^2] p_i(\alpha) f(\alpha) d\alpha$ with respect to m_K and s_K, where $s_K \geq 0$. We get

$$\frac{\partial g}{\partial m_K}(m_K, s_K) = 2 \int_0^h \sum_{i=1}^{\tilde{n}_\alpha} (m_K - mid(A_i^\alpha)) p_i(\alpha) f(\alpha) d\alpha$$

$$\frac{\partial g}{\partial s_K}(m_K, s_K) = 2 \int_0^h \theta(\alpha) \sum_{i=1}^{\tilde{n}_\alpha} (s_K - spr(A_i^\alpha)) p_i(\alpha) f(\alpha) d\alpha.$$

By solving $\frac{\partial g}{\partial m_K}(m_K, s_K) = \frac{\partial g}{\partial s_K}(m_K, s_K) = 0$, taking into account that p and f satisfy conditions (9) and (10), we easily obtain that the solution (m_K^*, s_K^*) is given by (12). Moreover, by calculation, $\frac{\partial^2 g}{\partial m_K^2}(m_K, s_K) = 2 \int_0^h f(\alpha) d\alpha = 2$, $\frac{\partial^2 g}{\partial s_K^2}(m_K, s_K) = 2 \int_0^h f(\alpha) \theta(\alpha) d\alpha$, $\frac{\partial^2 g}{\partial s_K \partial m_K}(m_K, s_K) = \frac{\partial^2 g}{\partial m_K \partial s_K}(m_K, s_K) = 0$, $\det \begin{bmatrix} \frac{\partial^2 g}{\partial m_K^2}(m_K, s_K) & \frac{\partial^2 g}{\partial s_K \partial m_K}(m_K, s_K) \\ \frac{\partial^2 g}{\partial m_K \partial s_K}(m_K, s_K) & \frac{\partial^2 g}{\partial s_K^2}(m_K, s_K) \end{bmatrix} = 4 \int_0^h f(\alpha) \theta(\alpha) d\alpha > 0$ and $\frac{\partial^2 g}{\partial m_K^2}(m_K, s_K) > 0$. Then (m_K^*, s_K^*) minimizes $g(m_K, s_K)$. $\square$

Remark 1. We observe that $mid(K^*(\tilde{A}))$ doesn't depend on θ. Furthermore, when the function θ is constant, that is $\theta(\alpha) = \bar{\theta}$ for any α, with $\bar{\theta} \in]0, 1]$, we get from (12) and (10)

$$spr(K^*(\tilde{A})) = \int_0^h \sum_{i=1}^{\tilde{n}_\alpha} spr(A_i^\alpha) p_i(\alpha) f(\alpha) d\alpha \qquad (13)$$

and thus even $spr(K^*(\tilde{A}))$ doesn't depend on θ.

Definition 6. *We call evaluation of the IVFQ $\tilde{A} = (B, C)$ with respect to (p, f, θ) the real number $V(\tilde{A}) = V(\tilde{A}; p, f)$ defined by $V(\tilde{A}) = mid(K^*(\tilde{A}))$, that is, from (12),*

$$V(\tilde{A}) = \int_0^h \sum_{i=1}^{\tilde{n}_\alpha} mid(A_i^\alpha) p_i(\alpha) f(\alpha) d\alpha. \qquad (14)$$

As previously observed, this evaluation doesn't depend on θ.

At this point we have reached our aim: for every IVFQ $\tilde{A}$ we may associate two values, one is its evaluation $V(\tilde{A})$ and the second is the spread of its uncertainty interval $spr(K^*(\tilde{A}))$.

4   Application

We observe that the evaluation (14) with respect to (p, f) of an IVFQ $\tilde{A}$ with α-cuts $\tilde{A}_\alpha$ can be expressed, in a more general way, by

$$V(\tilde{A}) = \int_0^h Val_p(\tilde{A}_\alpha) f(\alpha) \, d\alpha$$

where $Val_p(\tilde{A}_\alpha) = \sum_{i=1}^{\tilde{n}_\alpha} mid(A_i^\alpha) p_i(\alpha)$ is the evaluation of α-cut $\tilde{A}_\alpha$ with respect to (vector) weights p.

For instance, we may consider the parametric family of weighting functions

$$f(\alpha) = \frac{(n+1)\alpha^n}{h^{n+1}} \qquad n \geq 0 \tag{15}$$

By varying parameter n we obtain different evaluations of $\tilde{A}$. This choice is motivated by the desire to make decision maker able of assigning different weights to different α-levels by tuning parameter n. Using (15) the evaluation (14) is

$$V(\tilde{A}) = \frac{n+1}{h^{n+1}} \int_0^h Val_p(\tilde{A}_\alpha) \alpha^n \, d\alpha.$$

Furthermore, if in (14) we use for each level α the constant weights

$$p_i(\alpha) = \frac{1}{\tilde{n}_\alpha} \qquad i = 1, \dots, \tilde{n}_\alpha, \tag{16}$$

we obtain the evaluation

$$V_1(\tilde{A}) = \frac{n+1}{h^{n+1}} \int_0^h \left(\frac{1}{\tilde{n}_\alpha} \sum_{i=1}^{\tilde{n}_\alpha} mid(A_i^\alpha) \right) \alpha^n \, d\alpha. \tag{17}$$

Observe that in this case $Val_p(\tilde{A}_\alpha)$ is the simple average of interval midpoints.

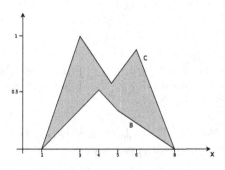

Fig. 3. IVFQ $\tilde{A} = (B, C)$

4.1   A Numerical Example

Let us consider the IVFQ $\tilde{A} = (B, C)$, shown in Fig. 3, adopted from [23, 44]. The T1 FQs B and C have membership

$$\mu_B(x) = \begin{cases} (x-1)/6 & 1 \leq x \leq 4 \\ (7-x)/6 & 4 \leq x \leq 5 \\ (8-x)/9 & 5 \leq x \leq 8 \\ 0 & \text{otherwise} \end{cases} \qquad \mu_C(x) = \begin{cases} (x-1)/2 & 1 \leq x \leq 3 \\ (7-x)/4 & 3 \leq x \leq 43/9 \\ (x-2)/5 & 43/9 \leq x \leq 6 \\ (16-2x)/5 & 6 \leq x \leq 8 \\ 0 & \text{otherwise.} \end{cases}$$

We now compute the α-cuts of B and C. For the T1 FQ B we have $n_\alpha^B = 1$ and α-cuts

$$B_\alpha = \begin{cases} [1+6\alpha, 8-9\alpha] & 0 \leq \alpha \leq 1/3 \\ [1+6\alpha, 7-6\alpha] & 1/3 < \alpha \leq 1/2. \end{cases}$$

For the T1 FQ C we have

$$n_\alpha^C = \begin{cases} 1 & 0 < \alpha \leq 5/9 \\ 2 & 5/9 < \alpha \leq 4/5 \\ 1 & 4/5 < \alpha \leq 1 \end{cases}$$

and α-cuts given by

$$C_\alpha = \begin{cases} [1+2\alpha, 8-\frac{5}{2}\alpha] & 0 < \alpha \leq 5/9 \\ [1+2\alpha, 7-4\alpha] \cup [2+5\alpha, 8-\frac{5}{2}\alpha] & 5/9 < \alpha \leq 4/5 \\ [1+2\alpha, 7-4\alpha] & 4/5 < \alpha \leq 1. \end{cases}$$

Then for the IVFQ $\tilde{A} = (B, C)$ we have (noting that $h = \max\{h_B, h_C\} = 1$)

$$\tilde{n}_\alpha = n_\alpha^B + n_\alpha^C = \begin{cases} 2 & 0 \leq \alpha \leq 1/2 \\ 1 & 1/2 < \alpha \leq 5/9 \\ 2 & 5/9 < \alpha \leq 4/5 \\ 1 & 4/5 < \alpha \leq 1 \end{cases}$$

and, from (6),

$$A_1^\alpha = \begin{cases} [1+6\alpha, 8-9\alpha] & 0 \leq \alpha \leq 1/3 \\ [1+6\alpha, 7-6\alpha] & 1/3 < \alpha \leq 1/2 \\ [1+2\alpha, 8-\frac{5}{2}\alpha] & 1/2 < \alpha \leq 5/9 \\ [1+2\alpha, 7-4\alpha] & 5/9 < \alpha \leq 1 \end{cases} \qquad A_2^\alpha = \begin{cases} [1+2\alpha, 8-\frac{5}{2}\alpha] & 0 \leq \alpha \leq 1/2 \\ [2+5\alpha, 8-\frac{5}{2}\alpha] & 5/9 < \alpha \leq 4/5. \end{cases}$$

Thus from (14) we have $V(\tilde{A}) = \int_0^1 mid(A_1^\alpha) \, p_1(\alpha) \, f(\alpha) \, d\alpha + \int_0^{1/2} mid(A_2^\alpha) \, p_2(\alpha) \, f(\alpha) \, d\alpha + \int_{5/9}^{4/5} mid(A_2^\alpha) \, p_2(\alpha) \, f(\alpha) \, d\alpha$.

In Fig. 4 we have plotted the evaluation $V_1(\tilde{A})$ defined in (17) as a function of parameter n.

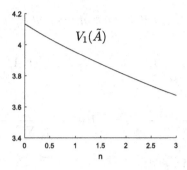

Fig. 4. Evaluation $V_1(\tilde{A})$ as function of parameter n

5   Comparison with Nie-Tan Method

In [22] Nie and Tan (NT) proposed to evaluate an IT2 FS by the center of gravity (COG) of the average of the LMF and the UMF. They presented a discrete version of this formula and underlined that important advantages of the proposed scheme are its low computational cost and that it is expressed in a closed form so it may be possible to analyse IT2 FLSs theoretically.

We show that the NT method, in its continuous formulation [23], may be obtained by the evaluation we propose choosing particular values of the parameters involved. It is interesting to see that we obtain either the NT evaluation or the interval of uncertainty in which it lies. The continuous form of the NT method is given as $V_{NT}(\tilde{A}) = \frac{\int_{-\infty}^{+\infty} x\,\bar{\mu}(x)\,dx}{\int_{-\infty}^{+\infty} \bar{\mu}(x)\,dx}$ where $\bar{\mu}(x) = (\mu_B(x) + \mu_C(x))/2$ is the average of the LMF and the UMF of $\tilde{A} = (B, C)$. Equivalently,

$$V_{NT}(\tilde{A}) = \frac{\int_{-\infty}^{+\infty} x\,\mu_B(x)\,dx + \int_{-\infty}^{+\infty} x\,\mu_C(x)\,dx}{\int_{-\infty}^{+\infty} \mu_B(x)\,dx + \int_{-\infty}^{+\infty} \mu_C(x)\,dx}. \tag{18}$$

Our aim is to show that our evaluation (14) contains, as a particular case, the NT one (18). In order to achieve this, we recall a previous result [35, Proposition 9.3] for T1 FQs.

Lemma 1. *Let A be a T1 FQ as defined in (1) with membership function μ_A, height h_A and α-cuts given by (2). Then for $t \geq 0$ we have $\int_{-\infty}^{+\infty} x^t\,\mu_A(x)\,dx = \frac{1}{t+1}\int_0^{h_A}\sum_{i=1}^{n_\alpha^A}\left(a_i^R(\alpha)^{t+1} - a_i^L(\alpha)^{t+1}\right)\,d\alpha$. In particular, for $t = 0$*

$$\int_{-\infty}^{+\infty} \mu_A(x)\,dx = \int_0^{h_A}\sum_{i=1}^{n_\alpha^A}|A_i^\alpha|\,d\alpha = \int_0^{h_A}|A_\alpha|\,d\alpha \tag{19}$$

and, for $t = 1$

$$\int_{-\infty}^{+\infty} x\,\mu_A(x)\,dx = \int_0^{h_A}\sum_{i=1}^{n_\alpha^A} mid(A_i^\alpha)\,|A_i^\alpha|\,d\alpha, \tag{20}$$

where $|A_i^\alpha| = a_i^R(\alpha) - a_i^L(\alpha)$ *is the length of interval* A_i^α *and* $|A_\alpha|$ *is the Lebesgue measure of* A_α.

Proposition 1. *Let* $\tilde{A} = (B, C)$ *be an IVFQ. Let* A_i^α *be the closed intervals defined in* (6). *If we choose*

$$p_i(\alpha) = \frac{|A_i^\alpha|}{\sum_{j=1}^{\tilde{n}_\alpha} |A_j^\alpha|}, \qquad f(\alpha) = \frac{\sum_{j=1}^{\tilde{n}_\alpha} |A_j^\alpha|}{\int_0^h \sum_{j=1}^{\tilde{n}_\alpha} |A_j^\alpha| \, d\alpha} \qquad (21)$$

then we obtain $V(\tilde{A}) = V_{NT}(\tilde{A})$.

Proof. Substituting the weights (p, f) given in (21) in the expression of $V(\tilde{A})$ (14) we obtain $V(\tilde{A}) = \int_0^h \sum_{i=1}^{\tilde{n}_\alpha} mid(A_i^\alpha) \, p_i(\alpha) \, f(\alpha) \, d\alpha = \frac{\int_0^h \sum_{i=1}^{\tilde{n}_\alpha} mid(A_i^\alpha) |A_i^\alpha| \, d\alpha}{\int_0^h \sum_{i=1}^{\tilde{n}_\alpha} |A_i^\alpha| \, d\alpha}$. Thus from (6) and (7) we get

$$\begin{aligned}
V(\tilde{A}) &= \frac{\int_0^{h_B} \sum_{i=1}^{n_\alpha^B} mid(B_i^\alpha)|B_i^\alpha| \, d\alpha + \int_0^{h_C} \sum_{j=1}^{n_\alpha^C} mid(C_j^\alpha)|C_j^\alpha| \, d\alpha}{\int_0^{h_B} \sum_{i=1}^{n_\alpha^B} |B_i^\alpha| \, d\alpha + \int_0^{h_C} \sum_{j=1}^{n_\alpha^C} |C_j^\alpha| \, d\alpha} \\
&= \frac{\int_{-\infty}^{+\infty} x \, \mu_B(x) \, dx + \int_{-\infty}^{+\infty} x \, \mu_C(x) \, dx}{\int_{-\infty}^{+\infty} \mu_B(x) \, dx + \int_{-\infty}^{+\infty} \mu_C(x) \, dx} = V_{NT}(\tilde{A})
\end{aligned}$$

where in the second equality we have applied (19) and (20) to the T1 FQs B and C and the last equality follows from (18). □

We have shown that the NT method is a particular case of our proposal for suitable values of the parameters. Now we try to analyse how the two procedures work, finding differences and similarities.

We observe that NT formula is the centroid abscissa of the average of the two membership functions μ_B and μ_C that surround the FOU zone. It starts working with a vertical dimension by a function $x \mapsto J_x = [\mu_B(x), \mu_C(x)]$ that assigns to every $x \in X$ the interval $J_x = [\mu_B(x), \mu_C(x)]$. The second step is the evaluation of J_x as its middle point, that is $v(J_x) = (\mu_B(x) + \mu_C(x))/2$. Third step is a horizontal aggregation by $V_{NT}(\tilde{A}) = \frac{\int_X x \, v(J_x) \, dx}{\int_X v(J_x) \, dx}$, that is the centroid abscissa of this new function.

Our proposal starts with a horizontal step by a function $\alpha \mapsto \tilde{A}_\alpha = (B_\alpha, C_\alpha)$ where $\tilde{A}_\alpha$ is the α-cut of the IVFQ $\tilde{A}$. It is identified by the couple of sets B_α and C_α which are the α-cuts of T1 FQs B and C. The evaluation we propose for $\tilde{A}_\alpha$ is a convex combination of the two evaluations of B_α and C_α, that is $v(\tilde{A}_\alpha) = v(B_\alpha) q_B(\alpha) + v(C_\alpha) q_C(\alpha)$ with $q_B, q_C \geq 0$, $q_B + q_C = 1$. As B_α and C_α are union of intervals, we may write them $B_\alpha = \bigcup_{i=1}^{n_\alpha^B} B_i^\alpha$ and $C_\alpha = \bigcup_{j=1}^{n_\alpha^C} C_j^\alpha$, their evaluation is defined as a weighted average of the midpoints of every subinterval, that is $v(B_\alpha) = \sum_{i=1}^{n_\alpha^B} mid(B_i^\alpha) p_i'(\alpha)$ and $v(C_\alpha) = \sum_{j=1}^{n_\alpha^C} mid(C_j^\alpha) p_j''(\alpha)$, with $\sum_{i=1}^{n_\alpha^B} p_i'(\alpha) = \sum_{j=1}^{n_\alpha^C} p_j''(\alpha) = 1$. Therefore, the evaluation $v(\tilde{A}_\alpha)$ can be written as

$$v(\tilde{A}_\alpha) = \sum_{i=1}^{n_\alpha^B} mid(B_i^\alpha)p_i'(\alpha)\,q_B(\alpha) + \sum_{j=1}^{n_\alpha^C} mid(C_j^\alpha)p_j''(\alpha)\,q_C(\alpha) = \sum_{i=1}^{\tilde{n}_\alpha} mid(A_i^\alpha)\,p_i(\alpha)$$

where the intervals A_i^α are defined in (6) and the weights

$$p_i(\alpha) = \begin{cases} p_i'(\alpha)\,q_B(\alpha) & i = 1,\ldots,n_\alpha^B \\ p_{i-n_\alpha^B}''(\alpha)\,q_C(\alpha) & i = n_\alpha^B + 1,\ldots,\tilde{n}_\alpha \end{cases}$$

are such that $\sum_{i=1}^{\tilde{n}_\alpha} p_i(\alpha) = 1$. The evaluation (14) is then obtained by means of a vertical aggregation, using a weighting function f, as $V(\tilde{A}) = \int_0^h v(\tilde{A}_\alpha)\,f(\alpha)\,d\alpha$.

The differences and similarities between the two methods are now more evident. Either the NT method or our are given in a closed form leaving the possibility to have more theoretical information. The NT one works starting with the vertical dimension and then with the horizontal one. The aggregation chosen is the simple average. Our method works on the two axis too. Indeed at horizontal level the weights p_i depend on the α-level and are different respectively for T1 FSs B and C that surround the FOU zone leaving to the decision maker the possibility to attribute different importance to B_α and C_α and even to their subintervals. For example, taking into account that weights may depend on $\tilde{A}$, the decision maker may give more importance to the intervals with relatively large length or, in a pessimistic (optimistic) perspective, he may put more (less) weight on $B_\alpha = \{x;\ \inf J_x \geq \alpha\}$ rather than $C_\alpha = \{x;\ \sup J_x \geq \alpha\}$. In the vertical level a new weight function appears. Its presence may assign different importance to different values of α. We conclude that our approach allows the decision maker to select an evaluation method according to his own criteria.

5.1 Modified Nie-Tan Evaluation

We now consider a modified version of Nie-Tan method by introducing a pessimistic/optimistic parameter $\lambda \in [0,1]$. We define

$$v_\lambda(J_x) = (1-\lambda)\mu_B(x) + \lambda\mu_C(x)$$

and

$$V_{NT}^\lambda(\tilde{A}) = \frac{\int_X x\,v_\lambda(J_x)\,dx}{\int_X v_\lambda(J_x)\,dx}.$$

Here $\lambda = 0$ reflects a pessimistic point of view of decision maker and $\lambda = 1$ an optimistic perspective. This evaluation can be expressed as

$$V_{NT}^\lambda(\tilde{A}) = \frac{(1-\lambda)\int_{-\infty}^{+\infty} x\,\mu_B(x)\,dx + \lambda\int_{-\infty}^{+\infty} x\,\mu_C(x)\,dx}{(1-\lambda)\int_{-\infty}^{+\infty} \mu_B(x)\,dx + \lambda\int_{-\infty}^{+\infty} \mu_C(x)\,dx}.$$

In the next result we show that our evaluation (14) contains, as a special case, the evaluation V_{NT}^λ.

Proposition 2. *Let $\tilde{A} = (B, C)$ be an IVFQ. Let A_i^α be the closed intervals defined in (6). If we choose*

$$
p_i(\alpha) = \begin{cases} \dfrac{(1-\lambda)|A_i^\alpha|}{(1-\lambda)\sum_{j=1}^{n_\alpha^B}|A_j^\alpha|+\lambda\sum_{j=n_\alpha^B+1}^{\tilde{n}_\alpha}|A_j^\alpha|} & i = 1, \ldots, n_\alpha^B \\[4mm] \dfrac{\lambda|A_i^\alpha|}{(1-\lambda)\sum_{j=1}^{n_\alpha^B}|A_j^\alpha|+\lambda\sum_{j=n_\alpha^B+1}^{\tilde{n}_\alpha}|A_j^\alpha|} & i = n_\alpha^B+1, \ldots, \tilde{n}_\alpha \end{cases}
$$

and

$$
f(\alpha) = \frac{(1-\lambda)\sum_{j=1}^{n_\alpha^B}|A_j^\alpha| + \lambda\sum_{j=n_\alpha^B+1}^{\tilde{n}_\alpha}|A_j^\alpha|}{(1-\lambda)\int_0^h\sum_{j=1}^{n_\alpha^B}|A_j^\alpha|\,d\alpha + \lambda\int_0^h\sum_{j=n_\alpha^B+1}^{\tilde{n}_\alpha}|A_j^\alpha|\,d\alpha}
$$

then we obtain $V(\tilde{A}) = V_{NT}^\lambda(\tilde{A})$.

6   A Numerical Example

Let us consider the IVFQ $\tilde{A} = (B, C)$, shown in Fig. 3. We will perform the evaluation of $\tilde{A}$ using four pairs of parameters. The fourth choice corresponds to NT method.

From (14) we have $V(\tilde{A}) = \int_0^1 mid(A_1^\alpha)\,p_1(\alpha)\,f(\alpha)\,d\alpha + \int_0^{1/2} mid(A_2^\alpha)\,p_2(\alpha)$ $f(\alpha)\,d\alpha + \int_{5/9}^{4/5} mid(A_2^\alpha)\,p_2(\alpha)\,f(\alpha)\,d\alpha$. Similarly, assuming θ constant, the spread of interval $K^*(\tilde{A})$ given in (13) can be expressed as $spr(K^*(\tilde{A})) = \int_0^1 spr(A_1^\alpha)$ $p_1(\alpha)\,f(\alpha)\,d\alpha + \int_0^{1/2} spr(A_2^\alpha)\,p_2(\alpha)\,f(\alpha)\,d\alpha + \int_{5/9}^{4/5} spr(A_2^\alpha)\,p_2(\alpha)\,f(\alpha)\,d\alpha$. We now compute the evaluation of $\tilde{A}$ and the spread of interval $K^*(\tilde{A})$ for different proposals.

(I) If we choose the weights

$$
p_i(\alpha) = \frac{1}{\tilde{n}_\alpha}, \qquad f(\alpha) = \frac{\tilde{n}_\alpha}{\int_0^h \tilde{n}_\alpha\,d\alpha} \tag{22}
$$

we get $V(\tilde{A}) = 4.25$ and $spr(K^*(\tilde{A})) = 1.63$. This evaluation is obtained, for the horizontal step, by a simple average of the of the midpoints of its intervals which produce B_α and C_α. The vertical aggregation is obtained as a weighted average of α-cuts values, where the weights are connected with the number of intervals producing every α-cut, as we can see substituting weights (22) in (14)

$$
V(\tilde{A}) = \frac{1}{\int_0^h \tilde{n}_\alpha\,d\alpha} \int_0^h \left(\frac{1}{\tilde{n}_\alpha}\sum_{i=1}^{\tilde{n}_\alpha} mid(A_i^\alpha)\right)\tilde{n}_\alpha\,d\alpha.
$$

(II) If we choose the weights

$$
p_i(\alpha) = \frac{|A_i^\alpha|}{\sum_{i=1}^{\tilde{n}_\alpha}|A_i^\alpha|}, \qquad f(\alpha) = \frac{1}{h} \tag{23}
$$

we get $V(\tilde{A}) = 4.03$ and $spr(K^*(\tilde{A})) = 1.70$. This evaluation is obtained, for the horizontal step, by a weighted average of the midpoints of α-cut intervals, where the weights are connected with the interval lengths. The vertical aggregation is obtained as the arithmetic mean of α-cuts evaluations viewed in the horizontal step

$$V(\tilde{A}) = \frac{1}{h} \int_0^h \frac{\sum_{i=1}^{\tilde{n}_\alpha} mid(A_i^\alpha)|A_i^\alpha|}{\sum_{i=1}^{\tilde{n}_\alpha} |A_i^\alpha|} \, d\alpha.$$

(III) If we choose the weights

$$p_i(\alpha) = \frac{|A_i^\alpha|}{\sum_{i=1}^{\tilde{n}_\alpha} |A_i^\alpha|}, \qquad f(\alpha) = \frac{\tilde{n}_\alpha}{\int_0^h \tilde{n}_\alpha \, d\alpha} \qquad (24)$$

we obtain $V(\tilde{A}) = 4.13$ and $spr(K^*(\tilde{A})) = 1.84$. This evaluation is obtained, for the horizontal step, by a weighted average of the midpoints of α-cut intervals, where the weights are connected with the interval spreads. The vertical aggregation is obtained as a weighted average of α-cuts evaluations, where the weights are connected with the number of intervals producing every α-cut

$$V(\tilde{A}) = \frac{1}{\int_0^h \tilde{n}_\alpha \, d\alpha} \int_0^h \left(\frac{\sum_{i=1}^{\tilde{n}_\alpha} mid(A_i^\alpha)|A_i^\alpha|}{\sum_{i=1}^{\tilde{n}_\alpha} |A_i^\alpha|} \right) \tilde{n}_\alpha \, d\alpha.$$

(IV) If we choose the weights (p, f) given by (21)

$$p_i(\alpha) = \frac{|A_i^\alpha|}{\sum_{j=1}^{\tilde{n}_\alpha} |A_j^\alpha|}, \qquad f(\alpha) = \frac{\sum_{j=1}^{\tilde{n}_\alpha} |A_j^\alpha|}{\int_0^h \sum_{j=1}^{\tilde{n}_\alpha} |A_j^\alpha| \, d\alpha},$$

we obtain $V(\tilde{A}) = 4.32 = V_{NT}(\tilde{A})$ and $spr(K^*(\tilde{A})) = 2.43$. This evaluation is obtained, for the horizontal step, by a weighted average of the midpoints of α-cut intervals, where the weights are connected with the interval lengths. The vertical aggregation is obtained as a weighted average of α-cuts evaluations, where the weights are connected with the sum of the lengths of every interval producing the α-cut $\tilde{A}_\alpha$

$$V(\tilde{A}) = \frac{1}{\int_0^h \sum_{i=1}^{\tilde{n}_\alpha} |A_i^\alpha| \, d\alpha} \int_0^h \left(\frac{1}{\sum_{j=1}^{\tilde{n}_\alpha} |A_j^\alpha|} \sum_{i=1}^{\tilde{n}_\alpha} mid(A_i^\alpha)|A_i^\alpha| \right) \sum_{j=1}^{\tilde{n}_\alpha} |A_j^\alpha| \, d\alpha.$$

It is interesting to note that NT method viewed in this new perspective, offers more information than in the centroid version. Indeed, even in this case, the richness of the parameters, we have found, shows that the decision maker may enter in the evaluation selecting the parameters involved, while in the centroid version this richness it is impossible to look.

Remark 2. If $\tilde{A}$ is a T1 FQ, that is $B = C$, then we recover the uncertainty interval and the evaluation for a T1 FQ given in [35]. We recall that this type-1 evaluation includes, for suitable choices of parameters, other evaluations. In particular using the parameters proposed in (I) we obtain the method proposed by Fortemps and Roubens [37,42], using the parameters proposed in (II) we obtain the method proposed by Yager and Filev [38,39], at the same way using the parameters proposed in (III) we obtain what we have proposed in [42] and using the (IV)-th parameters we obtain the Center of Gravity (COG).

7   A Different Proposal

In Eq. (17) we have defined the evaluation $V_1(\tilde{A})$ using the parametric weighting function f introduced in (15) and the weights p given by (16).

Motivated by previous results, we now propose an alternative evaluation to $V_1(\tilde{A})$ using in (14) the weights p given in (21) and (24). We define

$$V_2(\tilde{A}) = \frac{n+1}{h^{n+1}} \int_0^h \left(\frac{\sum_{i=1}^{\tilde{n}_\alpha} mid(A_i^\alpha)|A_i^\alpha|}{\sum_{i=1}^{\tilde{n}_\alpha} |A_i^\alpha|} \right) \alpha^n \, d\alpha. \tag{25}$$

We observe that for $n = 0$ the evaluation $V_2(\tilde{A})$ agrees with the evaluation $V(\tilde{A})$ obtained using (23).

In Fig. 5 we have plotted the evaluations $V_1(\tilde{A})$ and $V_2(\tilde{A})$ as functions of parameter n.

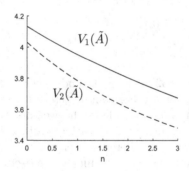

Fig. 5. Evaluations V_1 and V_2 as functions of parameter n

8   Conclusion

In this paper we present a parametric way to evaluate a particular type of IVFS, called interval-valued fuzzy quantity (IVFQ), in which LMF and UMF are continuous and, in general, non-convex and/or non-normal. This set is the typical output of IVFLS. This formula seems useful as it offers either an interval that gives information about uncertainty or an evaluation of an IVFQ. The contribution of this work can be highlighted by two aspects. The first relates to the possibility of using other defuzzification methods than centroid. The second is related to the parametrization of the proposal. The presence of these two types of parameters shows that both in cases of fuzzy control system and in cases of decision theory as in the case of optimization of fuzzy-value functions, different results can be obtained depending on the problem and the decision-maker. This form of freedom that acts both horizontally and vertically provides a wealth that the centroid can not offer. But remember that the centroid is a special case and in this new perspective you can see what it means to use the centroid. There is another interesting factor that is the result of our further line of investigation. Referring to what has been said in Sects. 4 and 7, the idea of using a power weighting function f in the evaluation recalls, in the case of fuzzy numbers, the evaluation proposed in [27] in which it was noted that for powers greater (less) than one this is equivalent to apply a "concentration" ("dilation") operator to the considered fuzzy number.

References

1. Bustince, H.: Interval-valued fuzzy sets in soft computing. Int. J. Comput. Intell. Syst. **3**(2), 215–222 (2010)
2. Gorzałczany, M.B.: A method of inference in approximate reasoning based on interval-valued fuzzy sets. Fuzzy Sets Syst. **21**(1), 1–17 (1987)
3. Gorzałczany, M.B.: An interval-valued fuzzy inference method-some basic properties. Fuzzy Sets Syst. **31**(2), 243–251 (1989)
4. Sambuc, R.: Fonctions *Phi*-floues: application a l'aide au diagnostic en pathologie thyroidienne. Faculté de Médecine de Marseille (1975)
5. Turksen, I.B., Zhong, Z.: An approximate analogical reasoning schema based on similarity measures and interval-valued fuzzy sets. Fuzzy Sets Syst. **34**(3), 323–346 (1990)
6. Wu, D., Mendel, J.M., Coupland, S.: Enhanced interval approach for encoding words into interval type-2 fuzzy sets and its convergence analysis. IEEE Trans. Fuzzy Syst. **20**(3), 499–513 (2012)
7. Bustince Sola, H., Fernandez, J., Hagras, H., Herrera, F., Pagola, M., Barrenechea, E.: Interval type-2 fuzzy sets are generalization of interval-valued fuzzy sets: towards a wider view on their relationship. IEEE Trans. Fuzzy Syst. **23**, 1876–1882 (2015)
8. John, R.: Type 2 fuzzy sets: an appraisal of theory and applications. Int. J. Uncertain. Fuzziness Knowl.-Based Syst. **6**(06), 563–576 (1998)
9. Deschrijver, G., Cornelis, C.: Representability in interval-valued fuzzy set theory. Int. J. Uncertain. Fuzziness Knowl.-Based Syst. **15**(03), 345–361 (2007)
10. Hwang, C.-M., Yang, M.-S., Hung, W.-L.: On similarity, inclusion measure and entropy between type-2 fuzzy sets. Int. J. Uncertain. Fuzziness Knowl.-Based Syst. **20**(03), 433–449 (2012)
11. Zeng, W., Shi, Y., Li, H.: Representation theorem of interval-valued fuzzy set. Int. J. Uncertain. Fuzziness Knowl.-Based Syst. **14**(03), 259–269 (2006)

12. Zadeh, L.A.: The concept of a linguistic variable and its application to approximate reasoning-I. Inf. Sci. **8**(3), 199–249 (1975)
13. Mendel, J., Hagras, H., Tan, W.-W., Melek, W.W., Ying, H.: Introduction to Type-2 Fuzzy Logic Control: Theory and Applications. Wiley, Hoboken (2014)
14. John, R., Hagras, H., Castillo, O.: Type-2 Fuzzy Logic and Systems. SFSC, vol. 362. Springer, Cham (2018). https://doi.org/10.1007/978-3-319-72892-6
15. Ontiveros-Robles, E., Melin, P., Castillo, O.: Comparative analysis of noise robustness of type 2 fuzzy logic controllers. Kybernetika **54**(1), 175–201 (2018)
16. Castillo, O., Amador-Angulo, L., Castro, J.R., Garcia-Valdez, M.: A comparative study of type-1 fuzzy logic systems, interval type-2 fuzzy logic systems and generalized type-2 fuzzy logic systems in control problems. Inf. Sci. **354**, 257–274 (2016)
17. Ontiveros, E., Melin, P., Castillo, O.: New methodology to approximate type-reduction based on a continuous root-finding karnik mendel algorithm. Algorithms **10**(3), 77 (2017)
18. Sanchez, M.A., Castillo, O., Castro, J.R.: Information granule formation via the concept of uncertainty-based information with interval type-2 fuzzy sets representation and takagi-sugeno-kang consequents optimized with cuckoo search. Appl. Soft Comput. **27**, 602–609 (2015)
19. Castillo, O., Melin, P.: Intelligent systems with interval type-2 fuzzy logic. Int. J. Innov. Comput. Inf. Control **4**(4), 771–783 (2008)
20. Karnik, N.N., Mendel, J.M.: Centroid of a type-2 fuzzy set. Inf. Sci. **132**(1), 195–220 (2001)
21. Mendel, J.M., John, R.I., Liu, F.: Interval type-2 fuzzy logic systems made simple. IEEE Trans. Fuzzy Syst. **14**(6), 808–821 (2006)
22. Nie, M., Tan, W.W.: Towards an efficient type-reduction method for interval type-2 fuzzy logic systems. In: IEEE International Conference on Fuzzy Systems, FUZZ-IEEE 2008, IEEE World Congress on Computational Intelligence, pp. 1425–1432, June 2008
23. Mendel, J.M., Liu, X.: Simplified interval type-2 fuzzy logic systems. IEEE Transactions on Fuzzy Systems **21**(6), 1056–1069 (2013)
24. Carlsson, C., Fullér, R.: On possibilistic mean value and variance of fuzzy numbers. Fuzzy Sets Syst. **122**(2), 315–326 (2001)
25. Chanas, S.: On the interval approximation of a fuzzy number. Fuzzy Sets Syst. **122**(2), 353–356 (2001)
26. de Campos Ibáñez, L.M., Muñoz, A.G.: A subjective approach for ranking fuzzy numbers. Fuzzy Sets Syst. **29**(2), 145–153 (1989)
27. González, A.: A study of the ranking function approach through mean values. Fuzzy Sets Syst. **35**(1), 29–41 (1990)
28. Dubois, D., Prade, H.: The mean value of a fuzzy number. Fuzzy Sets Syst. **24**(3), 279–300 (1987)
29. Facchinetti, G.: Ranking functions induced by weighted average of fuzzy numbers. Fuzzy Optim. Decis. Making **1**(3), 313–327 (2002)
30. Fullér, R., Majlender, P.: On weighted possibilistic mean and variance of fuzzy numbers. Fuzzy Sets Syst. **136**(3), 363–374 (2003)
31. Grzegorzewski, P.: On the interval approximation of fuzzy numbers. In: Greco, S., Bouchon-Meunier, B., Coletti, G., Fedrizzi, M., Matarazzo, B., Yager, R.R. (eds.) IPMU 2012. CCIS, vol. 299, pp. 59–68. Springer, Heidelberg (2012). https://doi.org/10.1007/978-3-642-31718-7_7
32. Heilpern, S.: The expected value of a fuzzy number. Fuzzy Sets Syst. **47**(1), 81–86 (1992)

33. Liou, T.-S., Wang, M.-J.J.: Ranking fuzzy numbers with integral value. Fuzzy Sets Syst. **50**(3), 247–255 (1992)
34. Anzilli, L., Facchinetti, G.: The total variation of bounded variation functions to evaluate and rank fuzzy quantities. Int. J. Intell. Syst. **28**(10), 927–956 (2013)
35. Anzilli, L., Facchinetti, G., Mastroleo, G.: A parametric approach to evaluate fuzzy quantities. Fuzzy Sets Syst. **250**, 110–133 (2014)
36. Facchinetti, G., Pacchiarotti, N.: Evaluations of fuzzy quantities. Fuzzy Sets Syst. **157**(7), 892–903 (2006)
37. Fortemps, P., Roubens, M.: Ranking and defuzzification methods based on area compensation. Fuzzy Sets Syst. **82**(3), 319–330 (1996)
38. Yager, R.R.: A procedure for ordering fuzzy subsets of the unit interval. Inf. Sci. **24**(2), 143–161 (1981)
39. Yager, R.R., Filev, D.: On ranking fuzzy numbers using valuations. Int. J. Intell. Syst. **14**(12), 1249–1268 (1999)
40. Hamrawi, H., Coupland, S., John, R.: A novel alpha-cut representation for type-2 fuzzy sets. In: 2010 IEEE International Conference on Fuzzy Systems (FUZZ), pp. 1–8. IEEE (2010)
41. Hamrawi, H.: Type-2 Fuzzy Alpha-cuts. Ph.D. thesis, De Montfort University (2011)
42. Anzilli, L., Facchinetti, G., Mastroleo, G.: Evaluation and interval approximation of fuzzy quantities. In: 8th Conference of the European Society for Fuzzy Logic and Technology (EUSFLAT 2013), pp. 180–186 (2013)
43. Bede, B.: Mathematics of Fuzzy Sets and Fuzzy Logic. Springer, Heidelberg (2013). https://doi.org/10.1007/978-3-642-35221-8
44. Liu, F.: An efficient centroid type-reduction strategy for general type-2 fuzzy logic system. Inf. Sci. **178**(9), 2224–2236 (2008)
45. Mendel, J.M., Liu, F., Zhai, D.: α-plane representation for type-2 fuzzy sets: theory and applications. IEEE Trans. Fuzzy Syst. **17**(5), 1189–1207 (2009)
46. Yager, R.R.: Level sets and the extension principle for interval valued fuzzy sets and its application to uncertainty measures. Inf. Sci. **178**(18), 3565–3576 (2008)
47. Trutschnig, W., González-Rodríguez, G., Colubi, A., Gil, M.Á.: A new family of metrics for compact, convex (fuzzy) sets based on a generalized concept of mid and spread. Inf. Sci. **179**(23), 3964–3972 (2009)

Tutorial

Paving the Way to Explainable Artificial Intelligence with Fuzzy Modeling
Tutorial

Corrado Mencar[1]([⊠])[iD] and José M. Alonso[2][iD]

[1] Dipartimento di Informatica, Università degli Studi di Bari Aldo Moro, Bari, Italy
corrado.mencar@uniba.it
[2] Centro Singular de Investigación en Tecnoloxías da Información (CiTIUS),
Universidade de Santiago de Compostela, Santiago de Compostela, Spain
josemaria.alonso.moral@usc.es

Abstract. Explainable Artificial Intelligence (XAI) is a relatively new approach to AI with special emphasis to the ability of machines to give sound motivations about their decisions and behavior. Since XAI is human-centered, it has tight connections with Granular Computing (GrC) in general, and Fuzzy Modeling (FM) in particular. However, although FM has been originally conceived to provide easily understandable models to users, this property cannot be taken for grant but it requires careful design choices. Furthermore, full integration of FM into XAI requires further processing, such as Natural Language Generation (NLG), which is a matter of current research.

1 Introduction

Explainable Artificial Intelligence (XAI) is gaining consensus among researchers and engineers in Computer Science, as an alternative approach to current AI methods that show great learning capabilities but are relatively ineffective in explaining the reasons of the produced outputs in a human-intelligible way. Fuzzy Modeling has a huge potential for the development of advanced XAI systems, provided that some methodological requirements are fulfilled. The aim of this tutorial is to give a short overview of XAI and the way to reach it through Fuzzy Modeling in particular. After a brief introduction to XAI (Sect. 2), the role of Granular Computing is highlighted as the theoretical background that motivates the adoption of Fuzzy Modeling for XAI (Sect. 3). In particular, interpretability in Fuzzy Modeling is a key requirement for XAI, which is outlined in the subsequent Sect. 4. The next step toward XAI is the generation of natural language expressions to explain the decisions of a fuzzy (rule-based) model; NLG is briefly described in Sect. 5. Finally, some notes of possible future developments conclude this paper.

2 Towards Explainable Artificial Intelligence

In 2013, Eric Loomis was found driving a car that had been used in a crime. The judge sentenced him six-year of prison, which was determined in part by his

© Springer Nature Switzerland AG 2019
R. Fullér et al. (Eds.): WILF 2018, LNAI 11291, pp. 215–227, 2019.
https://doi.org/10.1007/978-3-030-12544-8_17

score on the COMPAS scale, an algorithmically determined assessment used to predict an individual's risk of recidivism. COMPAS is a proprietary algorithm, and its risk assessment procedure is opaque to the public. Loomis appealed against the sentence by objecting that the use of a predictive algorithm violated the principle of a due process but the Wisconsin Supreme Court ruled against Mr. Loomis because he would have gotten the same sentence based solely on the usual factors, including his crime and his criminal history [31][1].

Loomis' case is perhaps one of the first and most apparent examples of AI used to determine the course of a person's life. More and more cases accumulated in recent years, in very disparate situations, including autonomous vehicles, robot-assisted surgery, health-care, warfare, etc. AI is preponderantly entering our life and we must ask ourselves if we want this new presence and at which conditions.

The scientific community already recognized this new trend and began to react accordingly. In 2017, ACM issued a Statement on Algorithmic Transparency and Accountability which, by recognizing that computer algorithms have far-reaching impacts, their use may consciously or unconsciously result in harmful discrimination[2]. Accordingly, ACM recommends to use the same standards as institutions where humans have traditionally made decisions and outlines a set of principles, including the ability of explanation (a.k.a. *explainability*) which encourages to produce explanations regarding both the procedures followed by an algorithm and the specific decisions that are made.

From a political standpoint, the importance of data and their processing has recently been recognized and regulated. The General Data Protection Regulation (GDPR) is a EU regulation, emanated in 2016 and implemented in 2018, for the protection of natural persons with regard to the processing of personal data and on the free movement of such data[3]. GDPR is motivated, among other things by the right to obtain an explanation of the decision reached after any assessment provided by automatic procedures[4].

Explainable Artificial Intelligence (XAI) is a new approach to AI where the ability to explain the decisions provided by algorithms is the primary objective. The XAI program was firstly defined by the Defense Advanced Research Projects Agency (DARPA), with the objective of creating machine learning techniques that produce more explainable models, while maintaining a high level of prediction accuracy and «enable human users to understand, appropriately trust, and effectively manage the emerging generation of artificially intelligent partners»[5]. Figure 1 illustrates the differences between current AI (mainly based on Machine

---

[1] The full history has been reported by The New York Times, on May 2, 2017, p. A22. See https://nyti.ms/2qoe8FC.

[2] https://www.acm.org/binaries/content/assets/public-policy/2017_joint_statement_algorithms.pdf.

[3] https://eur-lex.europa.eu/legal-content/EN/ALL/?uri=CELEX:32016R0679.

[4] See note (71) in the preamble of GDPR. Actually, GDPR is quite timid in affirming the right of explanation [36], thence the need of more precise regulations on the subject in future.

[5] https://www.darpa.mil/program/explainable-artificial-intelligence.

Learning) and XAI according to DARPA: the learned function is replaced by an explainable model and an explainable interface for helping users understanding the results of a machine learning process.

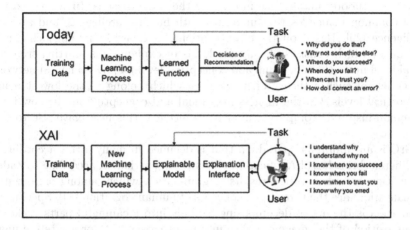

Fig. 1. XAI according to DARPA. Source: see footnote (5)

The importance of XAI is outstanding for several reasons, including: (i) the possibility of integrating machine and human knowledge in a simple way that is accessible by non-technical users; (ii) the possibility of interaction between users and machines in order to tackle complex problems; (iii) the ability of users to validate the functionality of an intelligent machine with respect to criteria of performance, ethics, safety, causality, etc.; (iv) the possibility of *trusting* machines for mission-critical applications [14].

XAI is growing widespread and reaching new frontiers on both scientific and technological sides. In this tutorial we will highlight the role of Granular Computing in general, and Fuzzy Modeling in particular, to the development of XAI.

3   Granular Computing

Granular Computing (GrC) is a computing paradigm where the object of processing is the *information granule*, i.e. a clump of objects kept together by some relations of indistinguishability, similarity, functionality or alike [44]. GrC is motivated by the need to approach AI through *human-centric information processing* [9], thence its central role in XAI.

GrC moves from some long-stated considerations concerning the apparent difficulty in developing common-sense reasoning in computers, while it seems so natural in human beings [28, Sect. 2.5]. These considerations led to the development of highly challenging branches of Informatics, such as Brain Informatics and

Cognitive Informatics, which aim at understanding the informational nature of human brain and mind by using the techniques provided by Informatics [37, 45]. In particular, understanding the human brain and mind from the point of view of Informatics brought to a couple of fundamental assumptions: (i) brains and computers embody intelligence mostly for the same reasons; (ii) there exists a set of common principles that underlies both human intelligence and artificial intelligence [29]. Based on these assumptions, a theory of intelligence can be envisioned, which consists of multiple levels of explanations starting from the neural level up to the functional and conceptual level [38]. Deep neural networks are models of such a theory of intelligence which belong to the lowest neural and cortical levels. On the highest functional and conceptual levels, new forms of "human-inspired" computing models are needed; this need gave rise to GrC [38].

GrC is an "umbrella" paradigm that is declined in many forms according to the different branches of Artificial Intelligence. In particular, according to Zadeh, information granules are the results of granulation which, among organization and causation, are the three basic concepts of human cognition [44]. Specifically, granulation is the act of decomposing a whole into meaningful parts – like the decomposition of the image of a face into mouth, eyes, etc., or a satellite image into terrain, rivers, lakes, and so on.

Independent on the specific formal theories that can be developed under the paradigm of GrC, there are two common principles that are generally preserved: the *multilevel* and the *multiview* principles [39]. According to the multilevel principle, granulation yields a hierarchical granular structure, with levels in the hierarchy corresponding to different degrees of abstraction; on the other hand, each granular structure offers just a partial view of a phenomenon, therefore different granular structures (i.e. multiple views) may be used to provide a more complete understanding of the reality that is modeled. (A handy example is the scientific publishing model: title-abstract-content is a multilevel granular structure that is represented in a paper, and more papers are usually published on a subject to highlight the methodology, the application, the implementation, etc.).

Information granules at one level are treated as primitives for the higher level of a granular structure. Therefore, each information granule is informally defined as a collection of objects (i.e., information granules of the lower level) related together by some relation that makes objects indistinguishable at the higher level. Similarity, spatial proximity, functionality are examples of such relations.

Many concepts in the human mind are formed through an act of *perception*, i.e. the organization, identification and interpretation of a sensation in order to form a mental representation [32, Chap. 4]. Since what is perceived belongs to a continuous Reality and concepts are formed through perceptions, it is straightforward to assume that such concepts reflect the continuity of perceptions. Information granules are used to represent and process concepts as conceived by human minds, therefore information granules should be defined in order to preserve the continuity of perception-based concepts. Fuzzy Set

Theory (FST) offers a suitable mathematical underpinning to define this kind of information granules [43]. In other words, «fuzziness of information granules is a direct consequence of fuzziness of the concepts of indistinguishability, similarity, proximity and functionality» [40].

Very often, perception-based concepts are designated by labels forming our Natural Language [42]. Therefore, FST can be used for *Computing With Words* [41]: propositions in natural language are translated into fuzzy constraints on the involved variables; inference is carried out through the machinery offered by FST; the results of inference are eventually expressed in natural language. FST is a promising approach for defining the theoretical background to represent perception-based information granules, which are designated by linguistic terms drawn from natural language. Thus, FST is a natural candidate for designing models in XAI. In the next Section, we'll look at the opportunities and challenges deriving from the use of FST in XAI.

4   Interpretability in Fuzzy Modeling

Fuzzy Modeling (FM) is a methodology oriented toward the design of explanatory and predictive models using FST. FM is long-standing, with pioneering works dated in the seventies. The original intent of FM was to develop knowledge-based models capable of both representing highly non-linear relations between inputs and outputs, and at the same time offering an intelligible view of such relations through the use of a simplified natural language [22]. This was accomplished by "fuzzy rules"; nowadays, fuzzy rule-based models are common practice in FM.

In the eighties FST met Machine Learning [33,34], and since then several methods for automatically deriving fuzzy rule-based models from data arose. As a result, such fuzzy models were mainly designed for accuracy, while the original intent of FST to represent perception-based knowledge became of secondary relevance. But fuzzy rule-based models that are not *interpretable* are akin to black-box models, like neural networks, for which an armamentarium of powerful learning techniques already exist and are continuously refined. Interpretability is a property of fuzzy rule-based models which can be roughly defined as the capability of reading and understanding the knowledge-base of a (fuzzy) model. Interpretability is not given from grant by the mere use of FST but it requires a methodology that is still in development.

The definition of interpretability cannot be formulated in strict mathematical sense because it involves the human factor which is hard, if not impossible, to formalize. However, the basic principle underlying interpretability can be found in Michalski's Comprehensibility Postulate, which parallels the results of a learning algorithm with the description that a human expert might produce by observing the same entities [27]. Roughly speaking, the perception-based concepts acquired by a human should be *co-intensive* with the information granules that are automatically generated by a learning algorithm, provided that the same objects are observed [23]. In particular, since we use symbolic terms drawn from natural language to communicate knowledge, then the implicit semantics that a term conveys when used in a context should overlap with the explicit semantics a term is given by interpreting it with a fuzzy set (see Fig. 2).

It is possible to illustrate the concept of co-intension with an explanatory example involving two actors, Alice and Bob [24], where Alice is a scholar communicating some piece of information to Bob by adopting an appropriate language that is capable to represent her own knowledge. In order be understood by Bob, Alice chooses linguistic terms whose meanings are supposed to be shared by Bob. (It is not necessary that Alice's and Bob's meanings are exactly the same, but they should be overlapping enough in order to understand each other.) This is possible if Alice and Bob share similar environment, language, culture, experiences, etc. Therefore, co-intension can be achieved if Alice and Bob share similar conceptualizations of the piece of reality they are talking about. This illustrative scenario is very common among humans as it enables communication of information and knowledge. The comprehensibility Postulate tries to extend this principle to the communication of knowledge acquired by machines to humans.

Interpretability calls for both semantic and structural requirements, whereas the semantic facet is related to the co-intension of information granules with perception-based concepts, and the structural facet is needed to cope with the limited capabilities of the human brain in processing information [7]. In order to achieve an effective definition of interpretability, a collection of interpretability constraints and criteria can be adopted. This collection is not standardized, because different constraints can be selected according to the needs of the designer. As a consequence, there is not a unique computational definition of interpretability. It is common to organize interpretability constraints according to the level of modeling. Therefore, there are interpretability constraints for fuzzy sets, for linguistic variables, for multi-dimensional information granules, for fuzzy rules and for entire fuzzy models [25]. Assessment of interpretability is aimed at formalizing measures that quantify the degree of fulfillment of interpretability constraints by any model component. Also in the case of assessment, both structural and semantic measures are used and eventually aggregated to define a global evaluation of interpretability [16]. As an alternative approach, interview-based experiments can be used to evaluate the interpretability of a fuzzy rule-based model in a holistic way [5].

Designing interpretable fuzzy models requires some additional steps with respect to the usual modeling stages. In particular, the source of knowledge may be twofold: the available data and the expert's knowledge. The way of considering these two sources is critical for an effective model. An iterative approach is recommended to integrate induced knowledge with expert rules in order to drive the design toward a model that is balanced in terms of predictive and explanatory capabilities [6]. Also, several modeling approaches are available, which may favor interpretability over accuracy or the converse; other approaches recognize that interpretability and accuracy are conflicting objectives and adopt multi-objective techniques to achieve a Pareto front of solutions [12]. Alternatively,

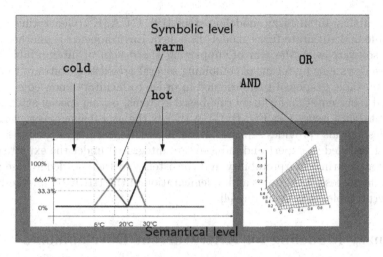

Fig. 2. Interpretation of symbols with fuzzy sets. Interpretability is assured only if the implicit semantics conveyed by each linguistic term is co-intensive with the explicit semantics determined by its interpretation.

ad-hoc algorithms may be used to incorporating interpretability constraints within the algorithms that induce fuzzy rules from data [13].

Most methods for modeling interpretable fuzzy systems adopt type-1 fuzzy sets, i.e. fuzzy sets that map objects of a universe of discourse into a scalar degree of membership. There is, however, a large corpus of literature concerning the use of type-2 fuzzy sets in fuzzy modeling. Type-2 fuzzy sets map elements of a universe of discourse into a type-1 fuzzy set defined on the domain of membership degrees. Type-2 fuzzy sets are justified by the assumption that «words mean different things to different people», therefore the uncertainty, related to the membership degree an object has to a set modeling a word, can only be represented by another level of uncertainty, thus giving rise to type-2 fuzzy sets [26]. Type-2 fuzzy sets gained attention in the last 15 years, not without complicacies and misconceptions [20]. For example, set operations on type-2 fuzzy sets can be defined in different ways, leading to very different theories [11]. Also, type-2 fuzzy sets may have different interpretations (e.g. in terms of intuitionistic or bipolar information). Therefore, type-2 fuzzy sets have a potential usefulness in modeling the meaning of words, but their manipulation and interpretation requires a full understanding of the subject of modeling. The authors' position is to favor type-1 fuzzy sets to model the knowledge base of a specific agent, while type-2 fuzzy sets are more suitable to model a kind of "social knowledge" that is shared among different agents. This is, however, matter of future research.

There are not many software tools to support designers in developing interpretable fuzzy models [1]. FISPRO[6] is an open-source software that facilitates

---

[6] https://www7.inra.fr/mia/M/fispro/fispro2013_en.html.

interpretability in all fuzzy modeling steps [19]. GUAJE[7] (Generating Understandable and Accurate fuzzy models in a Java Environment) is another open-source software with the aim of supporting the design of interpretable fuzzy rule-based systems by means of combining several preexisting software tools [2]. It is a portable graphical tool designed in order to facilitate knowledge extraction and representation for fuzzy rule-based systems, paying special attention to interpretability issues (see Fig. 3). GUAJE lets the user define expert variables and rules, but also provides supervised and automatic learning capabilities. Both types of knowledge, expert and induced, are integrated under the expert supervision for ensuring interpretability and consistency of the knowledge base along the whole process. The tool is an implementation of the HILK++ methodology for interpretable fuzzy modeling [4].

5   Fuzzy Modeling for XAI: Current Developments

Interpretability in fuzzy modeling is a requirement that leads to the development of methods and techniques to generate fuzzy models—mostly fuzzy rule-based systems—whose knowledge bases can be read and understood by users. In order to develop XAI, a step forward must be done, since in this case the new requirement is to explain the decision provided by a system. An interpretable fuzzy system gives the necessary information, but the explanation of a decision needs further processing.

XAI is a flourishing research direction is Artificial Intelligence, particularly in Machine Learning [10,18]; in Fuzzy Logic, research is gradually including the

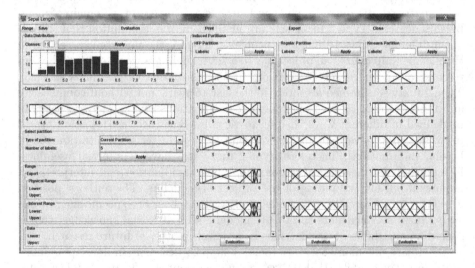

Fig. 3. A screen-shot of GUAJE.

---

[7] https://sourceforge.net/projects/guajefuzzy/.

results in interpretability to develop explainable models based on fuzzy models
[15]. A promising methodology that drives interpretable fuzzy modeling toward
XAI is Natural Language Generation (NLG). NLG enables the generation of
text from other data sources and finds application in state-of-art systems such
as speech recognition, machine translation and conversational systems among
others [17]. A specific branch of NLG is the so-called "data-to-text" (D2T-NLG),
whereas linguistic descriptions are automatically generated from a complex of
data. A particular approach for D2T-NLG is based on Linguistic Description
of Complex Phenomena (LDCP), a method for NLG that produces a Granular
Linguistic Model of a Phenomenon (GLMP), i.e. a network of processing units
called "perception mappings", each of them representing a computational perception or an aggregation thereof [35]. A computational perception is a unit of
meaning for the phenomenon under analysis and is identified by a set of linguistic expressions and their corresponding validity values given a situation (e.g.
an input sample). Perception mappings aggregate computational perceptions by
means of aggregation functions, which could be implemented in form of fuzzy
rules, and generate appropriate text by an algorithm. The output of a GLMP is
a linguistic description that explains a possibly complex situation, thanks to use
of one or more underlying interpretable fuzzy models that are distributed among
the perception mappings [3]. Figure 4 illustrates an example of GLMP used for
generating an explanation of the inference carried out by a fuzzy rule-based
classifier and the corresponding explanation for a given input sample.

A challenge in LDCP is to explain a phenomenon involving correlated data,
whereas this relation has been learned by some inductive algorithm. A typical
example is given by a Machine Learning algorithm that is used for learning a
classification function: this algorithm could be highly accurate but it may hardly
explain why a class label has been assigned to a given input. A possible approach
is to use the classification algorithm as an oracle and a collection of interpretable

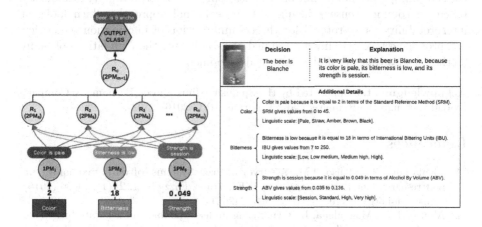

Fig. 4. Example of GLMP for explaining the classification of beers (left); textual explanation of a classification (right) [8].

models (including fuzzy models) as candidates for generating an explanation. Given an input sample, the simplest interpretable model in accordance with the oracle is used to generate an explanation through LDCP [8].

6   Future Developments

NLG is a promising way for developing XAI systems that generate textual descriptions concerning their inferences. Fuzzy sets seem appropriate models of the meaning of words, therefore fuzzy modeling is a promising approach for NLG, as exemplified by LDCP. Current works are still in the introductory stage and shed light on new research opportunities in the field. In particular, the interaction with deep neural networks is a mid-term objective since it could offer the best of two worlds: the outstanding learning abilities of deep neural networks with the human-centrality of conceptual models like those generated by LDCP.

From the point of view of interpretability in fuzzy modeling, future developments will be focused on representational issues: flat rule-based models are quite standard nowadays but suffer structural limits that could be overcome by more structured representations of knowledge. There are some tentative approaches in this sense by hierarchical fuzzy systems [30] but they are not exempt from criticism [21]. A tighter integration of fuzzy models with explanation models like GLMP may reconcile the need of interpretability of acquired knowledge with the requirement of providing explanation in complex scenarios.

Interpretability itself is matter of ongoing research, in order to cope with current challenges resulting from the higher complexity of data that is used to acquire knowledge. The use of incremental inductive algorithms, for example, is welcome to cope with stream data; nevertheless, these algorithms should take into account the requirement of interpretability of both the resulting knowledge and its historical evolution.

Finally, it must be noticed that the interpretability constraints and criteria, used for an operational definition and assessment of interpretability, are mostly based on common-sense principles. A more formal approach, which looks at interpretability as a protocol for the communication of information semantics, is a promising research direction aimed at establishing the foundations of many methodologies that are under current development.

Acknowledgments. Supported by the Spanish "Ministerio de Economía y Competitividad" through the Ramón y Cajal Program (RYC-2016-19802).

References

1. Alcala-Fdez, J., Alonso, J.M.: A survey of fuzzy systems software: taxonomy, current research trends, and prospects. IEEE Trans. Fuzzy Syst. **24**(1), 40–56 (2016). https://doi.org/10.1109/TFUZZ.2015.2426212
2. Alonso, J.M., Magdalena, L.: Generating understandable and accurate fuzzy rule-based systems in a Java environment. In: Fanelli, A.M., Pedrycz, W., Petrosino, A. (eds.) WILF 2011. LNCS (LNAI), vol. 6857, pp. 212–219. Springer, Heidelberg (2011). https://doi.org/10.1007/978-3-642-23713-3_27

3. Alonso, J., Conde-Clemente, P., Trivino, G.: Linguistic description of complex phenomena with the rLDCP R package. In: Proceedings of the 10th International Conference on Natural Language Generation, pp. 243–244 (2017)
4. Alonso, J.M., Magdalena, L.: HILK++: an interpretability-guided fuzzy modeling methodology for learning readable and comprehensible fuzzy rule-based classifiers. Soft Comput. **15**(10), 1959–1980 (2011). https://doi.org/10.1007/s00500-010-0628-5
5. Alonso, J.M., Magdalena, L., González-Rodríguez, G.: Looking for a good fuzzy system interpretability index: an experimental approach. Int. J. Approx. Reason. **51**(1), 115–134 (2009). https://doi.org/10.1016/j.ijar.2009.09.004
6. Alonso, J.M., Magdalena, L., Guillaume, S.: HILK: a new methodology for designing highly interpretable linguistic knowledge bases using the fuzzy logic formalism. Int. J. Intell. Syst. **23**(7), 761–794 (2008). https://doi.org/10.1002/int.20288
7. Alonso, J.M., Castiello, C., Mencar, C.: Interpretability of fuzzy systems: current research trends and prospects. In: Kacprzyk, J., Pedrycz, W. (eds.) Springer Handbook of Computational Intelligence, pp. 219–237. Springer, Heidelberg (2015). https://doi.org/10.1007/978-3-662-43505-2_14
8. Alonso, J.M., Ramos-soto, A., Castiello, C., Mencar, C.: Hybrid data-expert explainable beer style classifier. In: IJCAI/ECAI Workshop on Explainable Artificial Intelligence (XAI 2018), pp. 1–5 (2018). https://www.dropbox.com/s/jgzkfws41ulkzxl/proceedings.pdf?dl=0
9. Bargiela, A., Pedrycz, W.: Human-Centric Information Processing Through Granular Modelling. SCI, vol. 182. Springer, Heidelberg (2009). https://doi.org/10.1007/978-3-540-92916-1
10. Biran, O., Cotton, C.: Explanation and justification in machine learning: a survey. In: Workshop on Explainable AI (XAI), IJCAI 2017, pp. 8–13 (2017). http://www.intelligentrobots.org/files/IJCAI2017/
11. Bustince, H., Barrenechea, E., Fernández, J., Pagola, M., Montero, J.: The origin of fuzzy extensions. In: Kacprzyk, J., Pedrycz, W. (eds.) Springer Handbook of Computational Intelligence, pp. 89–112. Springer, Heidelberg (2015). https://doi.org/10.1007/978-3-662-43505-2_6
12. Casillas, J., Cordón, O., Triguero, F.H., Magdalena, L.: Interpretability Issues in Fuzzy Modeling, vol. 128. Springer, Heidelberg (2013). https://doi.org/10.1007/978-3-540-37057-4
13. Castiello, C., Mencar, C., Lucarelli, M., Rothlauf, F.: Efficiency improvement of DC* through a genetic guidance. In: 2017 IEEE International Conference on Fuzzy Systems (FUZZ-IEEE), pp. 1–6. IEEE, Naples, July 2017. https://doi.org/10.1109/FUZZ-IEEE.2017.8015585
14. Doran, D., Schulz, S., Besold, T.R.: What does explainable AI really mean? A new conceptualization of perspectives. In: Proceedings of the First International Workshop on Comprehensibility and Explanation in AI and ML 2017 co-located with 16th International Conference of the Italian Association for Artificial Intelligence (AI*IA 2017). CEUR Workshop Proceedings, vol. 2071 (2017). http://ceur-ws.org/Vol-2071/CExAIIA_2017_paper_2.pdf
15. Fernandez, A., del Jesus, M.J., Cordon, O., Marcelloni, F., Herrera, F.: Evolutionary fuzzy systems for explainable artificial intelligence: why, when, what for, and where to? IEEE Comput. Intell. Mag., 69–81 (2019). https://doi.org/10.1109/MCI.2018.2881645
16. Gacto, M., Alcalá, R., Herrera, F.: Interpretability of linguistic fuzzy rule-based systems: an overview of interpretability measures. Inf. Sci. **181**(20), 4340–4360 (2011). https://doi.org/10.1016/J.INS.2011.02.021

17. Gatt, A., Krahmer, E.: Survey of the state of the art in natural language generation: core tasks, applications and evaluation. J. Artif. Intell. Res. **61**, 65–170 (2018). https://doi.org/10.1613/jair.5477

18. Guidotti, R., Monreale, A., Ruggieri, S., Turini, F., Giannotti, F., Pedreschi, D.: A survey of methods for explaining black box models. ACM Comput. Surv. **51**(5), 1–42 (2018). https://doi.org/10.1145/3236009

19. Guillaume, S., Charnomordic, B.: Learning interpretable fuzzy inference systems with FisPro. Inf. Sci. **181**(20), 4409–4427 (2011). https://doi.org/10.1016/J.INS.2011.03.025

20. John, R., Coupland, S.: Type-2 fuzzy logic: challenges and misconceptions [discussion forum]. IEEE Comput. Intell. Mag. **7**(3), 48–52 (2012). https://doi.org/10.1109/MCI.2012.2200632

21. Magdalena, L.: Do hierarchical fuzzy systems really improve interpretability? In: Medina, J., et al. (eds.) IPMU 2018. CCIS, vol. 853, pp. 16–26. Springer, Cham (2018). https://doi.org/10.1007/978-3-319-91473-2_2

22. Mamdani, E.H., Assilian, S.: An experiment in linguistic synthesis with a fuzzy logic controller. Int. J. Man-Mach. Stud. (1975). https://doi.org/10.1016/S0020-7373(75)80002-2

23. Mencar, C., Castiello, C., Cannone, R., Fanelli, A.M.: Design of fuzzy rule-based classifiers with semantic cointension. Inf. Sci. **181**(20), 4361–4377 (2011). https://doi.org/10.1016/j.ins.2011.02.014

24. Mencar, C., Castiello, C., Cannone, R., Fanelli, A.M.: Interpretability assessment of fuzzy knowledge bases: a cointension based approach. Int. J. Approx. Reason. **52**(4), 501–518 (2011). https://doi.org/10.1016/j.ijar.2010.11.007

25. Mencar, C., Fanelli, A.M.: Interpretability constraints for fuzzy information granulation. Inf. Sci. **178**(24), 4585–4618 (2008). https://doi.org/10.1016/j.ins.2008.08.015

26. Mendel, J.: Fuzzy sets for words: a new beginning. In: The 12th IEEE International Conference on Fuzzy Systems, FUZZ 2003, vol. 1, pp. 37–42 (2003). https://doi.org/10.1109/FUZZ.2003.1209334

27. Michalski, R.S.: A theory and methodology of inductive learning. Artif. Intell. **20**, 111–161 (1983). https://doi.org/10.1016/0004-3702(83)90016-4

28. Minsky, M.: Society of Mind. Simon and Schuster, New York (1988)

29. Pinker, S.: How the Mind Works, vol. 882. Wiley/Blackwell (10.1111) (1999). https://doi.org/10.1111/j.1749-6632.1999.tb08538.x

30. Razak, T.R., Garibaldi, J.M., Wagner, C., Pourabdollah, A., Soria, D.: Interpretability indices for hierarchical fuzzy systems. In: Proceedings of IEEE International Conference on Fuzzy Systems (FUZZ-IEEE 2017) (2017). https://doi.org/10.1109/FUZZ-IEEE.2017.8015616

31. Revell, T.: Computer says "no comment". New Sci. **238**(3173), 40–43 (2018). https://doi.org/10.1016/S0262-4079(18)30664-X

32. Schacter, D.L., Gilbert, D.T., Wegner, D.M.: Psychology, 2nd edn. Worth, New York (2011)

33. Sugeno, M., Kang, G.: Structure identification of fuzzy model. Fuzzy Sets Syst. **28**(1), 15–33 (1988). https://doi.org/10.1016/0165-0114(88)90113-3

34. Takagi, T., Sugeno, M.: Fuzzy identification of systems and its applications to modeling and control. IEEE Trans. Syst. Man Cybern. **SMC-15**(1), 116–132 (1985). https://doi.org/10.1109/TSMC.1985.6313399

35. Trivino, G., Sugeno, M.: Towards linguistic descriptions of phenomena. Int. J. Approx. Reason. **54**(1), 22–34 (2013). https://doi.org/10.1016/J.IJAR.2012.07.004

36. Wachter, S., Mittelstadt, B., Floridi, L.: Why a right to explanation of automated decision-making does not exist in the general data protection regulation. Int. Data Priv. Law **7**(2), 76–99 (2017). https://doi.org/10.1093/idpl/ipx005
37. Wang, Y.: On cognitive informatics. Brain Mind **4**(2), 151–167 (2003). https://doi.org/10.1023/A:1025401527570
38. Yao, Y.: The rise of granular computing. J. Chongqing Univ. Posts Telecommun. Nat. Sci. Ed. **20**(3), 229–308 (2008)
39. Yao, Y.: A triarchic theory of granular computing. Granul. Comput. **1**(2), 145–157 (2016). https://doi.org/10.1007/s41066-015-0011-0
40. Zadeh, L.A.: Information granulation and its centrality in human and machine intelligence. In: 1997 IEEE International Conference on Systems, Man, and Cybernetics. Computational Cybernetics and Simulation, vol. 1, pp. 486–487, October 1997. https://doi.org/10.1109/ICSMC.1997.625798
41. Zadeh, L.A.: From computing with numbers to computing with words. From manipulation of measurements to manipulation of perceptions. IEEE Trans. Circ. Syst. I: Fundam. Theory Appl. **46**(1), 105–119 (1999). https://doi.org/10.1109/81.739259
42. Zadeh, L.A.: A new direction in AI: toward a computational theory of perceptions. AI Mag. **22**(1), 73–84 (2001). https://doi.org/10.1609/aimag.v22i1.1545
43. Zadeh, L.A.: Is there a need for fuzzy logic? Inf. Sci. **178**(13), 2751–2779 (2008). https://doi.org/10.1016/j.ins.2008.02.012
44. Zadeh, L.A.: Toward a theory of fuzzy information granulation and its centrality in human reasoning and fuzzy logic. Fuzzy Sets Syst. **90**(2), 111–127 (1997). https://doi.org/10.1016/S0165-0114(97)00077-8
45. Zhong, N., et al.: Web intelligence meets brain informatics. In: Zhong, N., Liu, J., Yao, Y., Wu, J., Lu, S., Li, K. (eds.) WImBI 2006. LNCS (LNAI), vol. 4845, pp. 1–31. Springer, Heidelberg (2007). https://doi.org/10.1007/978-3-540-77028-2_1

Round Table: Zadeh and the Future of Fuzzy Logic

The Fuzzy Logic Gambit as a Paradigm
of Lotfi's Proposals

Marco Elio Tabacchi[1,2(✉)] and Settimo Termini[1]

[1] DMI, Università degli Studi di Palermo, Palermo, Italy
{marcoelio.tabacchi,settimo.termini}@unipa.it
[2] Istituto Nazionale di Ricerche Demopolis, Palermo, Italy

Abstract. Lotfi Zadeh, in discussing the future directions the discipline should have taken, has insisted in highlighting what he called 'the Fuzzy Logic Gambit' , whose basic idea is that, when dealing with the solution of a problem through the use of Fuzzy Logic, two different type of precisions exist: "precision in value", which is connected to the ability of measuring reality, and "precision in meaning", which is what we want to attain when dealing with the real world.

While the final goal of Fuzzy Logic is to provide some degree of precision to what is less precise in nature, he has brilliantly suggested that this can be obtained by bartering between precision in value and precision in meaning.

The thesis we present (and argue in favor of) here is that this idea can be seen as a sort of paradigm of many proposals that Zadeh advanced along many decades of research, and that this simple observation lurks in many of the realizations that Fuzzy Logic has developed after becoming mainstream

Keywords: Fuzzy Logic · Fuzzy Logic Gambit

Why "Fuzzy Logic Gambit" Can Be Seen as a Paradigm of Lotfi's Proposals

Trying to contribute to the main question of the round table in a heretical way we shall not focus on specific aspects but on some (creative) internal tensions existing in the development of Fuzzy Logic. We think, in fact, that great progress can be obtained if big problems are faced in a direct way. In the last years of his prolific career, Zadeh, in discussing the future directions the discipline should have taken, has insisted in highlighting what he called 'the Fuzzy Logic Gambit' [1]. The fundamental idea behind the gambit is that, when dealing with solution of a problem through the use of Fuzzy Logic, two different type of precisions exist: precision in value, which is connected to the ability of measuring reality, and precision in meaning, which is what we want to attain when dealing with the real world. It is true that Fuzzy Logic aims at giving some degree of precision to what is less precise in nature, but according to Zadeh this can be obtained by bartering between precision in value and precision in meaning. Fuzzy Logic

© Springer Nature Switzerland AG 2019
R. Fullér et al. (Eds.): WILF 2018, LNAI 11291, pp. 231–235, 2019.
https://doi.org/10.1007/978-3-030-12544-8_18

sacrifices precision in value, a strategy that reduces computational costs and enhances tractability, and enhances precision in meaning, which has a cost but is desirable in order to obtain a solution that makes sense in the real world. The Fuzzy Logic Gambit is exactly the idea of starting earlier with imprecisiation, or what Bělohlávek et al. call 'a purposeful employment of imprecision' [2], to get to precisiation at a later stage.

One thesis we want to present and argue in favor here is that this idea can be seen as a sort of paradigm of many proposals that Zadeh advanced along many, many decades of research, and that this simple observation lurks in many of the realizations that Fuzzy Logic has assumed after becoming mainstream.

The aim of this kind of analysis residing in the hope of focusing a more or less meaningful pattern in his epistemological attitude. In fact, this could be extracted also starting from other crucial observations done by him such as the distinction between Fuzzy Logic in a restricted and general sense (or narrow and wide) [3,4] or his project of "Computing with words" [5] seen as "a system of computation in which the objects of computation are words, phrases, propositions, questions, commands and other types of semantic entities drawn from natural language" as he writes in the Preface to [6]. Another meaningful point is his reference also to 'perceptions'. Something indicating a very brave attitude, since perceptions, traditionally, belong to the realm of 'secondary' qualities which are usually assumed as not directly approachable by the use of scientific methodologies.

We think that a part of Lotfi Zadeh's scientific attitudes has been the one of presenting a number of huge challenges which were impossible to answer and solve within the usual methodologies of science – and for which he himself was also unable to provide convincing reasons to the contrary – counting on the fact that the 'fuzzy' scientific community would provide 'reasonable' answers, although such answers would inevitably 'normalize' and downplay the importance of his original proposals. His typical reaction would be both a subtle restatement of the original idea – in most cases with a reinforcement of the heretical content – and, more often than not, the acceptance of what the community was busy dealing with at the moment (see e.g. [7–10]). The overall result has been, on one side, a progressive (and useful) enlargement of the fuzzy domain as a whole; the other side of the coin was paid at the price of both a methodological fragmentation of what has been done and the complementarity of the results obtained by traditional scientific standards and the reference to 'visionary' ideas. One could say that such a tension is present in every discipline. However, we think that the modality along which this happens in the realm of fuzziness is incomparably stronger.

The Fuzzy Logic Gambit, in our view, represents a good starting point for this kind of reflections since it presents itself as a methodological guide free from other direct implications on the 'scientific production' that the other two topics, mentioned above, are usually burdened with. The idea of sacrificing precision with values but requiring more of it when meanings are involved resonates perfectly well in the world of big data and in the society of information.

The "epistemology" of the Fuzzy Logic Gambit and Some Questions of Big Data

As a very brief and summary indication of the road that can be followed along the line indicated above let us now briefly present a few remarks having just to do with "big data". We refer to the well known and quarrelsome debate on the autonomy of data with respect to the necessity of the construction of a theory.

Big Data proponents have recently posited the "End of Science" as we know it. Not recently indeed, as the first formulation of such an idea dates back to the early nineteens, well before the technology allowed the analysis of petabytes of data. The usual steps of outlining a theory and then try to find proof inside the data will be outdated. We will have (and already have) so much data and means of analysis that we don't need theories anymore. We will just limit ourselves to "let the data speaks for themselves", extracting useful correlations just by the sheer force of Big Data analysis. Small creeps in this façade start showing: a number of authors (see, e.g., [11,12]) point to the correlation/causation problem in findings obtained from big data, sometimes to a humorous extent (see [13]). Not every use of big data suffers from such drawbacks, and in [11] a number of positive results culled from linguistics are presented. But anecdotic evidence is not enough: the pipe dream of post-theory science, fuelled by the money invested in cloud computing, risks crashing against limits that are of a theoretical nature, and imposed by computation theory: the more data you have, the more the correlation that can be found by automatic data analysis are due to random properties of the data itself, and not on real causation effects. Some of such limits are highlighted in the paper by Calude and Longo [12], and based on ergodic theory and Ramsey theorem. There is another central and crucial question that is becoming more and more important. The one of responsibility with respect to a wide use of new technologies. For a long time, humans entertained the utopian idea that the combination of availability of a huge quantity of data produced from sensors and humans, as well as the democratisation of such data, in the shape of publication, accessibility for anyone of analysis software that works in an easy and intuitive way, its sharing and discussion through social networks, would have ushered us in a new era of factualness, scientificity and truthfulness. A number of recent developments (e.g. the Facebook/Cambridge Analytica affair, the alleged meddling by Russian hackers/government agents in the USA presidential elections) have turned the dream into a nightmare: decentralising news and mining the authoritativeness of reliable – if sometimes incoherent – sources are among the causes of the rise of fake news, and among others of a more sociological nature, of populisms in many European (and non-European) countries. It has been postulated [14] that the democratization of data is just apparent: that the control over what is measured, analyzed and discussed is concentrated in few, powerful hands. This brings along new and potentially dangerous consequences. Public opinion can be more easily manipulated, and most decisions are now in the 'hands of algorithms' – whatever this means. The authors maintain that the strength of new technologies and, specifically, their ability in collecting, preserving and analyzing Big Data provides an important advantage for addressing electoral consensus. This happens with the active collaboration of the users (which, in fact, as of today is practically anyone), seemingly willing to offer information that will be used to trick them in exchange for free services and the illusion of inclusion. Other, less glamorous examples (e.g. OpenPolis) are a hint that knowledge is not bad by itself, and that its real power can be harnessed for the good as well as the bad.

All the envisaged dangers are concretely possible. However, just to afford the problems from a correct viewpoint, we must remember that similar situations have occurred in the past. A suitable comparison can help us in forging correct instruments for today's questions. In particular, we must reflect on the following facts: (a) the possible dangers – today - are certainly "quantitatively" different from similar ones feared or occurred in the past. Are they also "qualitatively" different? (b) This last point should be carefully analyzed in order to understand, first, how this difference can be afforded (granted that this would be possible) and, secondly, understand whether the qualitative difference pops up only from phenomena of the type exemplified, e.g., in [15] (transformation of quantity in a qualitative difference) or it is related to original, independent reasons (causes).

We could say that the inspiration behind Fuzzy Logic Gambit would suggest that solutions based on the purely correlations emerging from the analysis of Big Data are to be taken seriously as a first step, to be compared with experimental findings. This check could be useful in providing suggestions for the subsequent construction of a theory. This example as such does not specifically involve fuzzy sets. The questions could be more interesting in case we strongly use fuzzy techniques in the modelling of data.

References

1. Zadeh, L.A.: Is there a need for fuzzy logic? Inf. Sci. **178**, 2751–2779 (2008)
2. Bêlohlávek, R., Dauben, J.W., Klir, G.J.: Fuzzy Logic and Mathematics: A Historical Perspective. Oxford Press, Oxford (2018)
3. Zadeh, L.A.: Foreword. In: Trillas, E., Bonissone, P.P., Magdalena, L., Kacprzyk, J. (eds.) Combining Experimentation and Theory: A Homage to Abe Mamdani. Studies in Fuzziness and Soft Computing, vol. 271. Springer, Heidelberg (2012)
4. Tabacchi, M.E., Termini, S.: Back to "reasoning". In: Ferraro, M.B., et al. (eds.) Soft Methods for Data Science. AISC, vol. 456, pp. 471–478. Springer, Cham (2017). https://doi.org/10.1007/978-3-319-42972-4_58
5. Zadeh, L.A.: Fuzzy Logic = Computing with Words. IEEE Trans. Fuzzy Syst. **4**(2), 103–111 (1996)
6. Zadeh, L.A.: Computing with Words. Studies in Fuzziness and Soft Computing, vol. 277. Springer, Heidelberg (2012). https://doi.org/10.1007/978-3-642-27473-2
7. Zadeh, L.A.: Fuzzy languages and their relation to human and machine intelligence, man and computer. In: Proceedings of the International Conference, Bourdeaux (1970)
8. Blair, B.: Interview with Lotfi Zadeh, creator of fuzzy logic. Azerbaijan Int. **2–4**, 46–57 (1994)
9. Zadeh, L.A.: Foreword. Appl. Soft Comput. **1**(1), 1–2 (2001)

10. Zadeh, L.A.: From computing with numbers to computing with words–from manipulation of measurements to manipulation of perceptions. Int. J. Appl. Math. Comput. Sci. **12**, 307–324 (2002)
11. Calude, A.: Does big data equal big problems? (2018). https://blogs.crikey.com.au/fullysic/2015/11/13/does-big-data-equal-big-problems/
12. Calude, C., Longo, G.: The deluge of spurious correlations in big data. In: Foundations of Science, pp. 1–18, March 2016
13. Vigen, T.: Spurious Correlations. Hachette Books, New York (2015)
14. Bellotti, R.: https://www.eticaeconomia.it/propaganda-e-manipolazione-nelle-elezioni-politiche-il-ruolo-dei-social-network-e-degli-algoritmi-basati-sulla-intelligenza-artificiale/ (2018)
15. Carneiro, R.L.: The transition from quantity to quality: a neglected causal mechanism in accounting for social evolution. PNAS **97**(23), 12926–12931 (2000)

How Best to Design Fuzzy Sets and Systems

In Memory of Prof. Lotfi A. Zadeh

Plamen Angelov[(✉)]

Director LIRA Research Centre, Chair of Intelligent Systems,
School of Computing and Communications, Lancaster University, Lancaster, UK
p.angelov@lancaster.ac.uk
https://www.lancaster.ac.uk/lira

The fundamental shift in dealing with uncertainties [12] and computerised reasoning was made by the late Professor Lotfi Aliasker Zadeh (1921–2017) in 1965 in his seminal paper [1]. For the last over five decades the Fuzzy Sets theory has matured and was applied to a long list of applications spanning from engineering, social sciences, biology to transport, mathematics and many more. One of the developments in which Prof. L. A. Zadeh had a strong personal input is the Fuzzy rule-based (FRB) systems. Perhaps the main specific features characteristic for the fuzzy sets caused a remarkable rethinking of some postulates and established concepts can be narrowed down to the following two;

(a) The partial degree of membership, satisfaction, association;
(b) The duality (and, more generally, the multi-multiplicity) of association.

Throughout the years of these last five decades many problems were solved in a new way thanks to the flexibility the fuzzy sets theory offers. The role of fuzzy sets in making AI (artificial intelligence) more interpretable and explainable is undeniable. Fuzzy sets offered the opportunity to formulate and solve more realistic optimisation, decision support and control problems. They are hard to be replaced in areas such as customer preferences modelling, etc.

However, one particular area of applications attracted my attention, in particular in mid-1980s when I started my research career under the supervision of Dr. Filev, FIEEE, FNAE [2], namely the issue of the design of FRB systems and the closely related to them artificial neural network systems which confluence into so called neuro-fuzzy systems around that time. Another significant milestone was reached somewhat later when Hornik [3] (1990 for neural networks) and Wu and Mendel [4] (1992 for the FRB systems), respectively theoretically proved the property of the respective systems and models to be universal approximators.

The issue of the design of fuzzy sets was traditionally related to the definition of the membership function as its descriptor [5]. This postulate was not questioned so far although in late 1980s and 1990s in addition to the traditional subjective way of designing fuzzy sets (Fig. 1) the so-called *data driven* design method started to be popular and was developed as well (Fig. 2).

The subjective approach has its own very strong rationale in the two way process of:

R. Fullér et al. (Eds.): WILF 2018, LNAI 11291, pp. 236–239, 2019.
https://doi.org/10.1007/978-3-030-12544-8_19

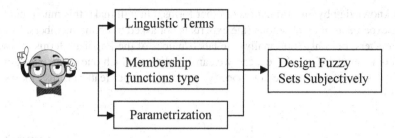

Fig. 1. Subjective design of fuzzy sets.

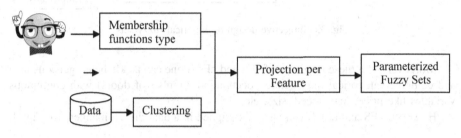

Fig. 2. Data-driven design of fuzzy sets.

(i) extracting expert knowledge and representing it in a mathematical form through the membership functions, and

(ii) the ability to represent and extract form data human intelligible and understandable, transparent linguistic information in the form of IF …THEN rules.

In addition, since mid-1970s (Mamdani [7] or Zadeh-Mamdani [5]) and since mid-1980s (Takagi-Sugeno [8]) FRB systems started to be developed and are now widely applied. Although, there are other types of fuzzy systems (relational [9], etc.) one particular type of FRB systems which we introduced recently with one of the pioneers of fuzzy sets theory, Professor Yager [10] called AnYa offers a great potential, specifically to address the issue of design of the fuzzy sets. While, both Mamdani-Zadeh and Takagi-Sugeno type of FRB share the exact same antecedent part (the IF) and only (although significantly) differ by the consequent (THEN) part, theYa type FRB has a quite different antecedent (IF) part. The main issue in the design of the fuzzy sets and systems is the very fundamental one – the membership function by which they are defined in first place. It is practically very difficult and controversial to define membership functions both form experts and from data. This is also related to the more general issue of assumptions made and handcrafting which machine learning (including statistical methods) are facing and is now hotly researched.

Recently, we proposed a new approach [11], which leads to a new form of fuzzy sets and systems – empirical fuzzy sets and FRB systems (εFS and εFRB, respectively). εFS and εFRB allow preserving the subjective specifics that fuzzy sets and systems are strong with. At the same time, εFS and εFRB can benefit from the vast amount of data that may be available. For example, εFS and εFRB still allow extracting expert knowledge by questionnaires or other forms, but will make this much more easy for the expert and not ambiguous (the experts is not asked to define membership values or parameters, but only (optionally) the labels/names of the linguistic terms, classes (if any). For example, if we chose a car, we can simply say which one we like (or possibly how much), but we do not need to specify why or define per feature (price, max speed, etc.) (Fig. 3).

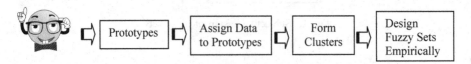

Fig. 3. Subjective design of empirical fuzzy sets

Moreover, with these new type of εFS and εFRB one can tackle heterogeneous data and combine categorical (e.g. gender, occupation, number of doors) with continuous variables like price, max speed, size, etc.

However, εFS and εFRB can also be designed in a data-driven manner, see Fig. 4.

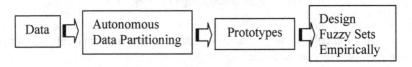

Fig. 4. Data-driven design of empirical fuzzy sets

More details are provided in [6] and [11].

On the basis of εFS and εFRB one can build empirically fuzzy classifiers (εF Classifiers), predictors (εF Predictors), controllers (εF Controllers), recommender systems, etc. Moreover, these can be evolving, not just fixed structure. This will allow studying the dynamic changes in human preferences as well as to build more efficient recommender systems where the only necessary input form the users is the preference ("likes" or "retweets" or "clicks").

References

1. Zadeh, L.A.: Fuzzy Sets. Inf. Control **8**(3), 338–353 (1965)
2. Zhao, Q., Filev, D.: A pioneer in car intellgence. SMC Mag. (2018, to appear)
3. Hornik, K.: Approximation capabilities of multilayer feedforward networks. Neural Netw. **4** (2), 251–257 (1991)
4. Wang, L.X., Mendel, J.M.: Fuzzy basis functions, universal approximation, and orthogonal least-squares learning. IEEE Trans. Neural Netw. **3**(5), 807–814 (1992)
5. Zadeh, L.A.: The concept of linguistic variable and it applications to approximate reasoning – II. Inf. Sci. **8**(4), 301–357 (1975)

6. Angelov, P., Gu, X.: Empirical Approach to Machine Learning. Springer, Heidelberg (2018). Chap. 5

7. Mamdani, E.H., Assilian, S.: An experiment in linguistic synthesis with a fuzzy logic controller. Int. J. Man-Mach. Stud. **7**(1), 1–13 (1975)

8. Takagi, T., Sugeno, M.: Fuzzy identification of systems and its applications to modeling and control. IEEE Trans. Syst. Man Cybern. **15**(1), 116–132 (1985)

9. Pedrycz, W.: An Identification algorithm in fuzzy relational systems. Fuzzy Sets Syst. **13**(2), 153–167 (1984)

10. Angelov, P., Yager, R.: A new type of simplified fuzzy rule-based system. Int. J. Gener. Syst. **41**(2), 163–185 (2011)

11. Angelov, P., Gu, X.: Empirical fuzzy sets. Int. J. Intell. Syst. (2017). https://doi.org/10.1002/int.21935

12. Angelov, P., Sotirov, S.: Imprecision and Uncertainty In Information Representation and Processing: New Tools Based on Intuitionistic Fuzzy Sets and Generalized nets. Springer, Heidelberg (2015). https://doi.org/10.1007/978-3-319-26302-1

Some Thoughts About Appealing Directions for the Future of Fuzzy Theory and Technologies Along the Path Traced by Lotfi Zadeh

Dario Malchiodi[(✉)] [iD]

Dipartimento di Informatica and DSRC, Università degli Studi di Milano, Milan, Italy
malchiodi@di.unimi.it
http://malchiodi.di.unimi.it

Keywords: Fuzzy systems future directions ·
Fuzzy systems and big data · Fuzzy system security

Extended Abstract

> Alas, it is always dangerous to
> prophesy, particularly, as the
> Danish proverb says, about the
> future.
>
> ————————————————
> *Proceedings of the Meeting of the*
> *Royal Statistical Society* [1]

The quoted text is an interesting instance of a fuzzy object: it is currently known in slightly diversified forms, each rather different from the quoted one, which corresponds to the first known appearance in English of this adage[1]. Indeed, most of the times we are used to reading or hearing variations of the sentence "it is difficult to make predictions, particularly about the future". The fuzziness here also extends to the source of the aphorism, which over the years has been attributed, among others, to Mark Twain, Niels Bohr, and even Nostradamus. This induces *a fortiori* further uncertainty of about half a century on the time of introduction of this saying. Actually, the first written evidence of what we could call an academic proverb is found in the autobiography of a Danish politician (published in 1948, by the way, in Danish). Summing up, what we have here is a rather definite concept (a humorous, yet effective warning about the assertion of forecasts) exhibiting several forms of imprecision: in its statement, in its authorship and in its temporal origin. Having this warning in mind, the challenging idea of shaping the future of fuzzy logic and fuzzy

---

[1] According to https://quoteinvestigator.com/2013/10/20/no-predict/, where the subject of tracing the various incarnations of this quote is covered in depth.

R. Fullér et al. (Eds.): WILF 2018, LNAI 11291, pp. 240–243, 2019.
https://doi.org/10.1007/978-3-030-12544-8_20

sets fields one year after Lotfi Zadeh passed away looks like a venturesome and hazardous task. On the one hand, almost any scholar investigating the broad umbrella of soft computing knows the papers originating the rich veins of fuzzy sets and fuzzy logic [22,26]; however, the fact of having extended the basic brick of mathematical architecture (namely, the concept of set, along with the immediate application bringing to the definition of *fuzzy numbers* [8]) allowed Zadeh to publish more than forty years ago quite a number of seminal papers concerning the "fuzzification" of several key fields in mathematics and informatics. The following list shows some interesting example of such fields[2], without any pretence of exhaustiveness:

- the notion of *fuzzy languages* [12], introduced as fuzzy sets defined over the universe of strings induced by a finite alphabet;
- the extension of probability theory characterized by expressing the probabilities of events in terms of the above mentioned fuzzy numbers, giving thus rise to *fuzzy probabilities* [25];
- the concept of *fuzzy random variable*, which over the years has been studied under various interpretations, intended for instance as random variables whose specifications are fuzzy random numbers [11], or identified with random fuzzy sets [17];
- the field of *fuzzy control* [24], which has been widely applied in the industrial domains, also in its *neurofuzzy* variant [16] devoted to the integration of neural and fuzzy technologies;
- the reformulation of algorithms to the fuzzy domain [18,23], for instance through the use of rules based on fuzzy conditional statements.

Since they have been introduced, some of these promising research lines have had less fortune than the widely known ones: for instance, the papers authored by Zadeh and focusing on fuzzy probabilities and on fuzzy random variables have been cited, respectively, two and one order of magnitude fewer times than its paper originating the rich vein of fuzzy sets. Thus for sure there are relatively unexplored fields of the fuzzy universe which are worth studying and whose investigation could even bring to serendipitous research results. This could also help to strengthen the mathematical foundations of fuzzy systems and to establish a tighter connection with other techniques such as deep learning, which recently experienced a tremendous expansion also in terms of industrial applications. Within this thread, big data and security emerge as two critical issues, as briefly explained below.

Big data. The amount of data to be processed in modern applications is more and more characterized by inputs not even fitting the hard disk of a computer. The only option in such cases consists in considering distributed storage systems (see [6,19] for two of the mainly used technologies). Moreover, in several interesting situations data is organized in *streams* requiring each item to be either

---

[2] It is also worth mentioning the contribution of Zadeh in the rise of other specific frameworks dealing with uncertainty, such as those of possibility theory [9] and granular computing [3–5].

processed on the fly, or forgotten forever. Although some efforts to adapt mainstream fuzzy procedures to such a massive scale already exist in the literature (see for instance [14] for an extension of the fuzzy C-means algorithm exploiting the map-reduce distributed computational framework [7]), a lot of work still needs to be done; this process will inevitably bring to new views of the fuzzy world when the available techniques are inherently difficult to scale up with the dimension of processed data (such as when standard optimization techniques are involved, as in [15]). Moreover, several approaches and algorithms born expressly to deal with big data problems might represent interesting grounds for fuzzy techniques as well (just to state an example, [13] and [20] deal with a fuzzy declination of the collaborative filtering procedure at the basis of several modern recommendation systems).

Security. The transition from research domain to industrial and real-world applications brings up the need of ensuring robustness of the proposed methods against unauthorized access. Again, other soft computing realms might inspire new categories of malicious attacks, as well as suggest defensive techniques. An example is constituted by the attacks expressly designed in order to fool deep neural networks on the basis of adversarial examples [10,21]. Analogously, the maturity of fuzzy systems for the business market requires the design and development of specific digital rights management techniques against the unauthorized use of illegal copies of a sold fuzzy artifact; here, once again, the neural networks world might suggest interesting candidate solutions, for instance referring to watermarking techniques [2]. More generally, the security field is highly unexplored as far as fuzzy techniques are concerned.

Summing up, Lotfi Zadeh has unveiled a world which for sure is far from being fully explored and understood. Instead, his work laid the foundations for a thorough management of uncertainty in several fields of mathematics and informatics yet to be completed in its theoretical form.

Moreover, it is worth pointing out that one of the key factors which led to the success of the vast umbrella of fuzzy technologies is related to their early implementation in industrial and electronic devices. In order to keep the pace up in this domain, a big challenge is envisaged in terms of applications, with specific reference to big data and stream processing, as well as of providing a secured access to artifacts based on fuzzy technologies.

References

1. Proceedings of the meeting. J. Roy. Stat. Soc. Ser. A (Gen.) 119(2), 146–149 (1956). http://www.jstor.org/stable/2342881
2. Adi, Y., Baum, C., Cisse, M., Pinkas, B., Keshet, J.: Turning your weakness into a strength: watermarking deep neural networks by backdooring. In: 27th USENIX Security Symposium (USENIX Security 2018), Baltimore, MD, pp. 1615–1631. USENIX Association (2018)
3. Apolloni, B., Bassis, S., Gaito, S., Malchiodi, D.: Bootstrapping complex functions. Nonlinear Anal. Hybrid Syst. **2**(2), 648–664 (2008)

4. Apolloni, B., Bassis, S., Malchiodi, D., Pedrycz, W.: The Puzzle of Granular Computing. Springer, Heidelberg (2008). https://doi.org/10.1007/978-3-540-79864-4
5. Bargiela, A., Pedrycz, W.: Granular computing. In: Handbook on Computational Intelligence. Fuzzy Logic, Systems, Artificial Neural Networks, and Learning Systems, vol. 1, pp. 43–66. World Scientific (2016)
6. Chang, F., et al.: Bigtable: a distributed storage system for structured data. ACM Trans. Comput. Syst. (TOCS) 26(2), 4 (2008)
7. Dean, J., Ghemawat, S.: MapReduce: simplified data processing on large clusters. Commun. ACM 51(1), 107–113 (2008)
8. Dubois, D., Prade, H.: Operations on fuzzy numbers. Int. J. Syst. Sci. 9(6), 613–626 (1978)
9. Dubois, D., Prade, H.: Possibility theory. In: Meyers, R.A. (ed.) Computational Complexity, pp. 2240–2252. Springer, New York (2012). https://doi.org/10.1007/978-1-4614-1800-9
10. Goodfellow, I.J., Shlens, J., Szegedy, C.: Explaining and harnessing adversarial examples (2014). arXiv preprint arXiv:1412.6572
11. Kruse, R., Meyer, K.D.: Statistics with Vague Data, vol. 6. Springer, Netherlands (2012)
12. Lee, E.T., Zadeh, L.A.: Note on fuzzy languages. Inf. Sci. 1(4), 421–434 (1969)
13. Leung, C.W.K., Chan, S.C.F., Chung, F.L.: A collaborative filtering framework based on fuzzy association rules and multiple-level similarity. Knowl. Inf. Syst. 10(3), 357–381 (2006)
14. Ludwig, S.A.: MapReduce-based fuzzy c-means clustering algorithm: implementation and scalability. Int. J. Mach. Learn. Cybern. 6(6), 923–934 (2015)
15. Malchiodi, D., Pedrycz, W.: Learning membership functions for fuzzy sets through modified support vector clustering. In: Masulli, F., Pasi, G., Yager, R. (eds.) WILF 2013. LNCS (LNAI), vol. 8256, pp. 52–59. Springer, Cham (2013). https://doi.org/10.1007/978-3-319-03200-9_6
16. Nauck, D., Klawonn, F., Kruse, R.: Foundations of Neuro-Fuzzy Systems. Wiley, Hoboken (1997)
17. Puri, M.L., Ralescu, D.A., Zadeh, L.: Fuzzy random variables. In: Readings in Fuzzy Sets for Intelligent Systems, pp. 265–271. Elsevier (1993)
18. Santos, E.S.: Fuzzy algorithms. Inf. Control 17(4), 326–339 (1970)
19. Shvachko, K., Kuang, H., Radia, S., Chansler, R.: The hadoop distributed file system. In: 2010 IEEE 26th Symposium on Mass Storage Systems and Technologies (MSST), pp. 1–10. IEEE (2010)
20. Son, L.H.: HU-FCF: a hybrid user-based fuzzy collaborative filtering method in recommender systems. Expert Syst. Appl. Int. J. 41(15), 6861–6870 (2014)
21. Szegedy, C., et al.: Intriguing properties of neural networks. arXiv preprint arXiv:1312.6199 (2013)
22. Zadeh, L.A.: Fuzzy sets. Inf. Control 8(3), 338–353 (1965)
23. Zadeh, L.A.: Fuzzy algorithms. Inf. Control 12(2), 94–102 (1968)
24. Zadeh, L.A.: A rationale for fuzzy control. J. Dyn. Syst. Meas. Control 94(1), 3–4 (1972)
25. Zadeh, L.A.: Fuzzy probabilities. Inf. Process. Manage. 20(3), 363–372 (1984)
26. Zadeh, L.A.: Fuzzy logic = computing with words. IEEE Trans. Fuzzy Syst. 4(2), 103–111 (1996)

From Zadeh's Computing with Words Towards eXplainable Artificial Intelligence

Jose M. Alonso[✉] [iD]

Centro Singular de Investigación en Tecnoloxías da Información (CiTIUS),
Universidade de Santiago de Compostela, Santiago de Compostela, Spain
josemaria.alonso.moral@usc.es
https://citius.usc.es/v/jose-maria-alonso-moral

Keywords: Fuzzy Logic · Explainable AI · Computing with words ·
Computing with perceptions · Cointension ·
Interpretable fuzzy systems

Extended Abstract

The European Commission has identified Artificial Intelligence (AI) as the "most strategic technology of the 21st century" [7]. AI is already part of our everyday life through many successful applications into real-world usage and according to Accenture [16] the economic impact of the automation of knowledge work, robots and self-driving vehicles could reach between 6.5 and 12 €trillion annually by 2025. People are used to buzzwords like smart watch, smart phone, smart home, smart car, smart city, etc. In practice, we are surrounded by smart gadgets, i.e., devices connected to Internet and endowed with some level of autonomy and intelligence thanks to AI systems. The cohabitation of humans and smart gadgets makes society demand the development of a new generation of explainable AI systems, i.e., AI systems ready to explain naturally (as humans do) their automatic decisions.

Thus, the research field on explainable AI is flourishing and attracting more and more attention not only regarding technical but also ethical and legal issues [8]. The ACM Code of Ethics [1] highlighted explanation as a basic principle in the search for "Algorithmic Transparency and Accountability". In addition, Floridi et al. defined the concept of "explicability" in reference to both "intelligibility" and "explainability" and hence captured the need for transparency and for accountability in an ethical framework for AI [10]. Moreover, the new European General Data Protection Regulation (GDPR) [14] refers to the "right to explanation", i.e., GDPR states that European citizens have the right to ask for explanations of decisions affecting them, no matter who (or what AI system) makes such decision.

The term eXplainable Artificial Intelligence (XAI) was coined by the USA Defense Advanced Research Projects Agency (DARPA) [11]. Assuming that

© Springer Nature Switzerland AG 2019
R. Fullér et al. (Eds.): WILF 2018, LNAI 11291, pp. 244–248, 2019.
https://doi.org/10.1007/978-3-030-12544-8_21

"even though current AI systems offer many benefits in many applications, their effectiveness is limited by a lack of explanation ability when interacting with humans" DARPA launched to the research community (including both academy and industry) the challenge of designing new self-explanatory AI systems from 2017 to 2021.

In Europe, there is not any initiative similar to the DARPA challenge on XAI yet. However, the European Commission has already pointed out the convenience of launching a pilot in XAI [7]. In June 2018, the Confederation of Laboratories for Artificial Intelligence Research in Europe (CLAIRE[1]), a novel initiative to create a network of excellence in AI with the most well-recognized universities and R+D centres, emphasized in its European vision for AI the need to search for transparent, explainable, fair and socially compatible intelligent systems. Moreover, The AI4EU[2] H2020 Project is funded by call ICT-26 2018 (grant 825619) with the aim of: (1) to mobilize the entire European AI community to make AI promises real for the European Society and Economy; and (2) to create a leading collaborative AI European platform to nurture economic growth. Explainable Human-centered AI is highlighted as one of the five key research areas to consider and it is present in 5 out of the 8 experimental pilots to be developed.

In the rest of this manuscript we briefly review a selection of outstanding Zadeh's contributions which are likely to have direct impact in the research field of XAI. The paradigm of Computing with Words (CWW) is especially relevant because humans are used to explanations in natural language (NL).

From Prof. Zadeh's seminal ideas on fuzzy sets and systems [21], many key concepts such as linguistic variables and linguistic rules have turned up in the field of Fuzzy Logic (FL). Accordingly, FL has many successful applications [19]. In addition, as it is described in [4], about 30% of publications in XAI come from authors well recognized in the field of FL. This is mainly due to the commitment of the fuzzy community to produce interpretable fuzzy systems [3]. Actually, interpretability is deeply rooted in the fundamentals of FL. However, it is worthy to note that interpretability is not guaranteed only because of applying FL. In practice, producing interpretable fuzzy systems is a matter of careful design [17].

In XAI, interpretability is a key issue but understandability and comprehensibility which are not so deeply considered by the FL community also play a prominent role. Nowadays, a new generation of intelligent systems is expected to provide users with natural explanations. Those explanations should be easy to understand no matter the user background. Since, humans think and compute naturally with words, explanations in NL are likely to be considered as natural explanations. Prof. Zadeh was the first to talk about CWW [22] as an extension of fuzzy sets and systems. Later, Prof. Kacprzyk gave some hints about how to implement CWW [12]. Moreover, he highlighted the need to connect CWW with the paradigm of NL Generation (NLG) [9]. It is worth noting that NLG is a well-known area within the Computational Linguistics and AI research fields.

---

[1] https://claire-ai.org/.
[2] http://ai4eu.org/.

The connection between FL and NLG has been further researched by other authors [2,15].

In addition, Prof. Zadeh was pioneer to introduce a new generation of more natural intelligent systems, ready to compute with perceptions and make approximate reasoning as humans naturally do. Thus, the Computational Theory of Perceptions (CTP) [20,23] was first introduced by Zadeh and later applied by Trivino and Sugeno to automatically generate linguistic descriptions of complex phenomena [18]. The CTP has been successfully applied for example to explain the energy consumption at home [5] or to automatically generate linguistic descriptions associated to the USA census data [6].

In addition, Prof. Zadeh also coined the concept of cointension [24]. The semantic-cointension approach [13] is already applied to assess interpretability of fuzzy systems. Likewise, it can be considered when evaluating the understandability of explanations in XAI. In short, two different concepts referring almost to the same entities are taken as cointensive. Accordingly, an explanation in NL is deemed as comprehensible only when the explicit semantics embedded in it is cointensive with the implicit semantics inferred by the user when reading and processing the given explanation.

To sum up, Prof. Zadeh made many highly valuable contributions to the FL field and beyond. Many of these contributions were pioneer ideas and/or challenging proposals with a lot of potential to be fully developed later by other researchers. Nowadays, XAI is a prominent and fruitful research field where many of Zaden's contributions can become crucial if they are carefully considered and thoroughly developed. For example, two major open challenges for XAI are: (1) how to build conversational agents able to provide humans with semantic grounding, persuasive and trustworthy interactive explanations; and (2) how to measure the effectiveness and naturalness of automatically generated explanations. CWW as well as fuzzy measures and Z-numbers [25] introduced by Zadeh are likely to contribute to successfully address both challenges and achieve valuable results.

Acknowledgments. Jose M. Alonso is *Ramón y Cajal* Researcher (RYC-2016-19802). In addition, this research was funded by the Spanish Ministry of Economy and Competitiveness (grants TIN2014-56633-C3-1-R, TIN2017-90773-REDT and TIN2017-84796-C2-1-R) and the Galician Ministry of Education (grants ED431F 2018/02, GRC2014/030 and "accreditation 2016–2019, ED431G/08"). All grants were co-funded by the European Regional Development Fund (ERDF/FEDER program).

References

1. ACM US Public Policy Council: Statement on Algorithmic Transparency and Accountability (2017). http://www.acm.org/binaries/content/assets/public-policy/2017_usacm_statement_algorithms.pdf
2. Alonso, J.M., Bugarin, A., Reiter, E.: Special issue on natural language generation with computational intelligence. IEEE Comput. Intell. Mag. **12**(3), 8–9 (2017). https://doi.org/10.1109/MCI.2017.2708919

3. Alonso, J.M., Castiello, C., Mencar, C.: Interpretability of fuzzy systems: current research trends and prospects. In: Kacprzyk, J., Pedrycz, W. (eds.) Springer Handbook of Computational Intelligence, pp. 219–237. Springer, Heidelberg (2015). https://doi.org/10.1007/978-3-662-43505-2_14

4. Alonso, J.M., Castiello, C., Mencar, C.: A bibliometric analysis of the explainable artificial intelligence research field. In: Medina, J., et al. (eds.) IPMU 2018. CCIS, vol. 853, pp. 3–15. Springer, Cham (2018). https://doi.org/10.1007/978-3-319-91473-2_1

5. Conde-Clemente, P., Alonso, J.M., Trivino, G.: Towards automatic generation of linguistic advice for saving energy at home. Soft Comput. **22**(2), 345–359 (2018)

6. Conde-Clemente, P., Trivino, G., Alonso, J.M.: Generating automatic linguistic descriptions with big data. Inf. Sci. **380**(2), 12–30 (2017)

7. European Commission: Artificial Intelligence for Europe. Technical report (2018). https://ec.europa.eu/digital-single-market/en/news/communication-artificial-intelligence-europe

8. EU AI HLEG: AI Ethics Guidelines. Technical report (2019). https://ec.europa.eu/digital-single-market/en/news/draft-ethics-guidelines-trustworthy-ai

9. Gatt, A., Krahmer, E.: Survey of the state of the art in natural language generation: core tasks, applications and evaluation. J. Artif. Intell. Res. **61**, 65–170 (2018)

10. Floridi, L., et al.: AI4People - an ethical framework for a good AI society: opportunities, risks, principles, and recommendations. Mind. Mach. **28**(4), 689–707 (2018)

11. Gunning, D.: Explainable artificial intelligence (XAI). Technical report, Defense Advanced Research Projects Agency (DARPA), DARPA-BAA-16-53, Arlington, USA (2016). https://www.darpa.mil/program/explainable-artificial-intelligence

12. Kacprzyk, J.: Computing with words is an implementable paradigm: fuzzy queries, linguistic data summaries, and natural-language generation. IEEE Trans. Fuzzy Syst. **8**, 451–472 (2010)

13. Mencar, C., Castiello, C., Cannone, R., Fanelli, A.M.: Interpretability assessment of fuzzy knowledge bases: a cointension based approach. Int. J. Approximate Reasoning **52**(4), 501–518 (2011)

14. Parliament and Council of the European Union: General data protection regulation (GDPR) (2016). http://data.europa.eu/eli/reg/2016/679/oj

15. Ramos-Soto, A., Bugarin, A., Barro, S.: On the role of linguistic descriptions of data in the building of natural language generation systems. Fuzzy Sets Syst. **285**, 31–51 (2016)

16. Schoeman, W.: Why AI is the future of growth. Accenture technical report (2016). https://www.accenture.com/za-en/company-news-release-why-artificial-intelligence-future-growth

17. Trillas, E., Eciolaza, L.: Fuzzy Logic: An Introductory Course for Engineering Students. SFSC, vol. 320. Springer, Cham (2015). https://doi.org/10.1007/978-3-319-14203-6

18. Trivino, G., Sugeno, M.: Towards linguistic descriptions of phenomena. Int. J. Approximate Reasoning **54**, 22–34 (2013)

19. Yager, R.R., Zadeh, L.A.: An Introduction to Fuzzy Logic Applications in Intelligent Systems. The Springer International Series in Engineering and Computer Science. Springer, New York (2012). https://doi.org/10.1007/978-1-4615-3640-6

20. Zadeh, L.A.: Toward a perception-based theory of probabilistic reasoning with imprecise probabilities. Stat. Plann. Infer. **105**, 233–264 (2002)

21. Zadeh, L.A.: Fuzzy sets. Inf. Control **8**(3), 338–353 (1965)

22. Zadeh, L.A.: From computing with numbers to computing with words - from manipulation of measurements to manipulation of perceptions. IEEE Trans. Circuits Syst.—I: Fundam. Theory Appl. **45**(1), 105–119 (1999)
23. Zadeh, L.A.: A new direction in AI: toward a computational theory of perceptions. Artif. Intell. Mag. **22**(1), 73–84 (2001)
24. Zadeh, L.A.: Is there a need for fuzzy logic? Inf. Sci. **178**, 2751–2779 (2008)
25. Zadeh, L.A.: A note on Z-numbers. Inf. Sci. **181**, 2923–2932 (2011)

Looking at the Branches and Roots

Corrado Mencar(✉)

Department of Informatics and Centro Interdipartimentale di Logica e Applicazioni
(CILA), Università degli Studi di Bari Aldo Moro, Bari, Italy
corrado.mencar@uniba.it

Abstract. When looking at the future of Fuzzy Logic (FL), it is immediate to think at applications where FL could be used to compute with perceptions, possibly expressed in natural language, thus enabling Explainable Artificial Intelligence (XAI). Scholars in FL have been working on
Interpretability, an important part of XAI, for decades. Yet, the research
community in FL seems isolated from other Artificial Intelligence (AI)
communities. There is a gap between FL and AI that due to the relative
youth of FL when compared with the foundational theories underlying
AI. If we want FL growing its branches in XAI as well as in other fields,
we need to develop, both in Research and Education, more robust roots
supporting all the theories, the methodologies and the technologies that
are going to be developed now and in future.

Keywords: Foundations of Fuzzy Logic ·
Explainable Artificial Intelligence

Extended Abstract

We are all happy with the infosphere.[1] We live, strive, fight, love and die for
information. And we want more. Technology, our complacent servant, gives us
more and more opportunities to replace our physical lives with an informational
counterpart, so much that someone started to worry that, in a not so far future,
we may well be Technology's servants. But this is the time of enthusiasm and
we want information technology every-where, every-time. That is why we are
witnessing a bloom of applied science, and applied research. This has a cultural
impact too. Computer Science is more and more oriented towards finding new
applications and teaching students how to create new artifacts that work well
and eventually make profits. So, when looking at the future of FL, the first
thing that comes into mind is: what are the applications where FL could be more
successful? Zadeh was indefatigable in telling us that machine intelligence can be
improved by enabling perception-based computing, and FL is *the* scientific way
to compute with perceptions [14]; therefore outstanding applications are those

---

[1] Infosphere denotes denote the whole informational environment constituted by all
informational entities, their properties, interactions, processes and mutual relations.
See Floridi [5] but also https://en.wikipedia.org/wiki/Infosphere.

R. Fullér et al. (Eds.): WILF 2018, LNAI 11291, pp. 249–252, 2019.
https://doi.org/10.1007/978-3-030-12544-8_22

dealing with perception-based information and knowledge, possibly expressed in natural language.

Explainable Artificial Intelligence (XAI) is an evolution of AI methodologies focusing on the development of agents capable of both (i) generating decisions that a human could understand in a given context, and (ii) explicitly explaining decisions [9]. What is an explanation, what are its function and structure are questions posed in Philosophy, Psychology and Cognitive Science. Interestingly, these are the very same fields where the concept of perception has been formulated and studied. Many concepts in the human mind are formed through an act of perception, i.e. the organization, identification and interpretation of a sensation in order to form a mental representation [11, Ch. 4]. Since what is perceived belongs to a continuous Reality and concepts are formed through perceptions, it is straightforward to assume that such concepts reflect the continuity of perceptions. Therefore, as FL gives a computational account to perception representation and processing, it is arguable that XAI is a field where FL could flourish, especially in the days after the binge of deep learning, when we will eventually realize that black boxes might be fragile [10] or even dangerous [8]. I am pretty sure that XAI will be the right mean for *collaborative* intelligence [4], with machine helping and not replacing humans to tackle more and more complex problems. This would dramatically reduce the risks of fear and opposition to the advancement of AI technologies, which are more and more often seen as competing with humans and menacing well-being. But to achieve collaboration, humans and machines should be able to communicate at the same level: this might be accomplished if machines embody perception-based knowledge as FL promises to provide.

XAI is a relatively new field where a number of research efforts are converging. If we look back in the history of FL, we see that in the last twenty years a great deal of research was around the keyword "interpretability" and, with due distinctions, interpretability of fuzzy systems may be considered part of the XAI program. In fact, research on interpretability moved from the definition of a number of structural constraints aimed at keeping knowledge representation as simple as possible, towards the recognition of a more complex phenomenon embracing both structural and semantical aspects [2]. However, a recent research of ours showed that, within XAI at least, the research community in FL is isolated from other Artificial Intelligence (AI) communities [1]; as a consequence, wheels are often re-discovered and a common language is not matured.

There is a gap between FL and AI. Some scholars between the two worlds already recognized this problem and tried to find the reasons of this separation [7]. FL appears as a growing tree but its roots are still not as robust as in other AI branches. In fact, AI is based on theoretical foundations that have hundreds or thousands of years, while FL challenges some well-established dogmas on the basis of a visionary viewpoint that may not be easily accepted. This is not a problem in principle but in practice. If we want FL growing its branches in XAI as well as in other fields, we need robust roots supporting all the theories, the methodologies and the technologies that are going to be developed now and in future.

We have infinite degrees of freedom in developing models based on FL. This makes modeling very difficult because it may be extremely hard to give a clear rationale behind all the design choices in a FL model: shape of membership functions, type of set operators, type of inference mechanism, defuzzification, etc. Sometimes, all such degrees of freedom are translated into parameterized models that are subject of numerical or evolutionary optimization; but the results are completely opaque and incomprehensible to the final users; as a consequence, with such an approach, the vision of FL to support XAI gets lost. Sometimes it is even hard to see a clear distinction between fuzzy set and membership function (and we find a fuzzy set *exactly* defined as its membership function, although denoted by two different symbols). In short, we see amazing applications without sound foundations. We should not content ourselves with such a partial result, because a beautiful tree with big branches but undersized roots is destined to fall.

This is why the future of fuzzy logic must look at its roots, other than at its branches. Sometimes, the need of a new Mathematics is invoked to develop the foundations of FL; but I do not subscribe to this point of view. Mathematics, as we know it, is a suitable language for a plethora of scientific theories, from Boolean algebra to Quantum Mechanics: there is no need to invent a new one for FL. Instead, within Mathematics, we need to formalize the irrefutable ideas behind FL[2] and, within Informatics, to undoubtedly show that they have the real world as a model. To this end, the recent works of some FL scholars (see, e.g. Trillas [12])—which separate the concept of fuzzy set as a collection of loosely ordered objects, and the concept of membership function as a (possibly approximate) measure of gradualness—are very promising and deserve to be further investigated in future. Also, the seminal works of Dubois and Prade on Possibility Theory and—more in general—on systematizing the semantics of fuzzy sets [3], the theoretical breakthroughs given by Mathematical Fuzzy Logic [6], etc. constitute the roots of FL in the soil of Science, which deserve both development in Research and settlement in Education.

Education in FL should give a strong emphasis on the theoretical foundations of FL, including some basics on Mathematical Fuzzy Logic, Possibility Theory, lattice theory, L-fuzzy sets and everything else that allows students to understand the roots of FL so as to develop applications with a stronger awareness of what they are doing. Current education programs may be too much oriented towards an engineering approach aimed at quickly enabling students to be productive; a step back so as to plunge deeper on theory is however important to preserve the scientific culture from oblivion. (An interesting endeavor is given by Trillas and Eciolaza [13].)

---

[2] Of course, FL in the wide sense is implicitly intended; Mathematical Fuzzy Logic is a well formalized discipline, but sometimes distant from the concepts and ideas of FL in the wide sense. Bridging the gap between narrow and wide FL is another interesting way to look at the future of FL.

Zadeh's genius moved thousands of people to question the rigid structure of theories onto which many classical AI methodologies are founded. Thanks to him, we found the key ingredient for human-centric information processing. Now time is come to make this wonderful vision a hard-rock science.

References

1. Alonso, J.M., Castiello, C., Mencar, C.: A bibliometric analysis of the explainable artificial intelligence research field. In: Medina, J., et al. (eds.) IPMU 2018. CCIS, vol. 853, pp. 3–15. Springer, Cham (2018). https://doi.org/10.1007/978-3-319-91473-2_1
2. Alonso, J.M., Castiello, C., Mencar, C.: Interpretability of fuzzy systems: current research trends and prospects. In: Kacprzyk, J., Pedrycz, W. (eds.) Springer Handbook of Computational Intelligence, pp. 219–237. Springer, Heidelberg (2015). https://doi.org/10.1007/978-3-662-43505-2_14
3. Dubois, D., Prade, H.: The legacy of 50 years of fuzzy sets: a discussion. Fuzzy Sets Syst. **281**, 21–31 (2015). https://doi.org/10.1016/j.fss.2015.09.004
4. Epstein, S.L.: Wanted: collaborative intelligence. Artif. Intell. **221**, 36–45 (2015). https://doi.org/10.1016/j.artint.2014.12.006. ISSN: 0004-3702
5. Floridi, L.: Philosophy and Computing: An Introduction. Routledge, Abingdon (2002)
6. Hájek, P.: What is mathematical fuzzy logic. Fuzzy Sets Syst. **157**(5), 597–603 (2006). https://doi.org/10.1016/j.fss.2005.10.004
7. Hüllermeier, E.: Does machine learning need fuzzy logic? Fuzzy Sets Syst. **281**, 292 (2015). https://doi.org/10.1016/j.fss.2015.09.001
8. Knight, W.: The dark secret at the heart of AI. MIT Technol. Rev. (2017). https://www.technologyreview.com/s/604087/the-dark-secret-at-the-heart-of-ai/
9. Miller, T.: Explanation in artificial intelligence: insights from the social sciences. Artif. Intell. **267**, 1–38 (2019). https://doi.org/10.1016/j.artint.2018.07.007
10. Nguyen, A., Yosinski, J., Clune, J.: Deep neural networks are easily fooled: high confidence predictions for unrecognizable images. In: 2015 IEEE Conference on Computer Vision and Pattern Recognition (CVPR), pp. 427–436. IEEE, June 2015. https://doi.org/10.1109/CVPR.2015.7298640
11. Schacter, D.L., Gilbert, D.T., Wegner, D.M.: Psychology, 2nd edn. Worth, New York (2011)
12. Trillas, E.: On the logos: a naïve view on ordinary reasoning and fuzzy logic. SFSC, vol. 354. Springer, Cham (2017). https://doi.org/10.1007/978-3-319-56053-3
13. Trillas, E., Eciolaza, L.: Fuzzy Logic: An Introductory Course for Engineering Students. SFSC, vol. 320. Springer, Cham (2015). https://doi.org/10.1007/978-3-319-14203-6
14. Zadeh, L.A.: Computing with Words and perceptions—a paradigm shift. In: 2009 IEEE International Conference on Information Reuse Integration, pp. viii–x, August 2009. https://doi.org/10.1109/IRI.2009.5211627

Deep Neural Networks and Explainable Machine Learning

Antonio Maratea[(⊠)] [iD] and Alessio Ferone[iD]

University of Naples "Parthenope", Isola C4, Centro Direzionale, Naples, Italy
{antonio.maratea,alessio.ferone}@uniparthenope.it
http://www.scienzeetecnologie.uniparthenope.it/

Keywords: Deep Neural Networks · XAI · Granular Computing

Extended Abstract

From a general perspective, the most impressive results [1,6] in Machine Learning have been recently obtained via black-box models, being Deep Neural Networks (DNNs) the major player in the game. Nowadays, the same wonder that Eugene Wigner expressed for the *unreasonable effectiveness* of mathematics in describing physical world in the sixties of last century [7], should and is to a certain degree permeating computer scientists concerning the ability that computers show of solving complex tasks, often better than expert humans [3].

The reasons for the success of DNNs are mostly technological, and due firstly to the unprecedented computational power, parallel processing ability and low energy consumption that characterize contemporary computing devices and secondly to the ubiquitous abundance of low-cost sensors that generate every day an unimaginable amount of data. Exploiting computational power to build implicit models when data are abundant and *a priori* knowledge is scarce is the realm of Machine Learning, and nowadays computing devices can handle DNNs with millions of parameters, thousands of input variables and terabytes of training data. By consequence, we are witnessing scientists diving into DNNs from every field of human knowledge, ultimately only confirming the flexibility of powerful computers in solving every sort of task, given enough training data. Machine Learning in form of DNNs is becoming a popular shortcut and is slowly replacing the fight for transparent models, even when data are not abundant and *a priori* knowledge is indeed available. What will likely happen in a few years is that pretrained DNNs will be made available for download or straight use in the cloud by major software companies (as already happens with the IBM watson API[1]), giving a powerful *prêt-à-porter* solution engine within reach of every programmer, disregarding altogether the proper formalization of the tackled problem.

On the one hand, despite their achievements in terms of crude accuracy, DNNs are not a panacea: apart curse of dimensionality, they suffer from domain specificity, catastrophic forgetting, overfitting, and are easily fooled with high

---

[1] https://www.ibm.com/watson/developer-resources/.

R. Fullér et al. (Eds.): WILF 2018, LNAI 11291, pp. 253–256, 2019.
https://doi.org/10.1007/978-3-030-12544-8_23

confidence [5]. While efforts are being made in the scientific literature to over-come each of these problems (for example with biologically inspired neuron consolidation mechanisms [4], dropout, transfer learning, selective learning and robust optimization), the lack of a general principle that justifies their effectiveness and the lack of a structured theory for the choice of architecture and parameters of DNNs are problems that will grab the efforts of the scientific literature for many years to come. We are still far from Human Level Machine Intelligence (HLMI) [9], and even if in many specific tasks (most notably translation) computers now approach or outperform human experts, it is likely that the expectations on the ability of state of the art DNNs to subsume human intelligence are overly optimistic.

On the other hand, such a pervasive and disruptive technological advancement generated many concerns from people and domain experts facing with classification results and decisions made by machine learning algorithms whose rationale is not explainable and that may well be discriminatory, unethical, unacceptable or plainly wrong with respect to the general plot where they are applied—see [8] for what looks like a contemporary resurgence of physiognomy. If the lack of a justification principle for DNNs may be seen as a technical question, the opaqueness of models strongly limits the trust and the spread of Machine Learning, giving rise to the so-called Explainable Artificial Intelligence (XAI) movement [2], that is a necessary step forward to apply confidently autonomous decision systems in sensitive applications as health, military or financial domains.

Once the orgy of successful applications will come to an end and it will be given as granted, raising no wonder, that DNNs reach a close to perfect accuracy in most tasks, what will really become deep is the schism among unexplainable but effective computer based solutions in reach of everyone and less than perfect, explainable transparent models requiring experts, scientific background, efforts, time, funds and conscious choices to be conceived, developed and deployed. If the industry may well be satisfied with the former, academia should defend the profound value of the latter, as its purpose is to pursue, to defend and to advance human knowledge in its highest form. Many scientists have been seduced by the DNNs power and usability, and academia is increasingly interwoven with industry in this contemporary rush where the craved gold is the accuracy, but their long term purpose diverges: pre-trained models that give a 100% accuracy straight out-of-the-box will lead to countless applications where the need for comprehension and skillful modeling will be simply skipped, and ultimately industry will have a marginal interest in funding what will be considered purely theoretical speculation.

Fuzzy Logic (FL) and more in general Granular Computing (GC) allow to build more expressive human-centric models that can process natural language words and include various facets of imprecision, uncertainty, ambiguity and incompleteness into more intelligible, controllable and customizable models operating on information granules; these reasons make them ideal candidate tools to hijack DNNs towards transparent and justifiable deductions. Humans are able to perform deductive and inductive reasoning in the framework of logic exploiting common sense *a priori* knowledge and handling noise, outliers, partial truth, imprecision, ambiguity, uncertainty and missing values, ultimately producing a decision that is evaluable by other humans; machines mostly learn through curve fitting on the base of myriads of examples, with a learning process that mimics at the lowest level what happens in the human brain, but that remains largely obscure and non reproducible for a human being. Fuzziness, roughness and other soft computing theories can and should be injected in all levels of the analysis: in modeling imprecise input or output data, labels, granules, costs, *a priori* knowledge, conclusions. In any of these levels there is an accuracy versus interpretability trade-off, similar to what happens with sets of fuzzy rules, and open issues abound: how to model input data and *a priori* knowledge, how to keep the interpretability while training, how to obtain intelligible conclusions, how to choose the right granularity for the task at hand, how to model concepts and information granules with words.

Most likely, Deep Neuro-Granular Systems (DNGS) that integrate multiple soft computing theories and Natural Language Processing is what will help the scientists to build at least partially justifiable decision systems or semi-transparent autonomous systems, both computationally lighter and more powerful in their generalization ability with respect to the current state of the art; most likely these systems will be based on modules corresponding to words, fuzzy quantifiers and self-learnt ontologies. DNGS is the next step forward towards an XAI that exploits the DNNs power remaining interpretable. The paradigm shift evoked by Zadeh [9] is still far ahead.

References

1. Esteva, A., et al.: Dermatologist-level classification of skin cancer with deep neural networks. Nature **542**(7639), 115 (2017)
2. Gunning, D.: Explainable Artificial Intelligence (XAI). Defense Advanced Research Projects Agency (DARPA), nd Web (2017)
3. He, K., Zhang, X., Ren, S., Sun, J.: Delving deep into rectifiers: surpassing human-level performance on imagenet classification. In: Proceedings of the IEEE International Conference on Computer Vision, pp. 1026–1034 (2015)
4. Kirkpatrick, J., et al.: Overcoming catastrophic forgetting in neural networks. Proc. Natl. Acad. Sci. **114**, 3521–3526 (2017). https://doi.org/10.1073/pnas.1611835114
5. Nguyen, A., Yosinski, J., Clune, J.: Deep neural networks are easily fooled: high confidence predictions for unrecognizable images. In: Proceedings of the IEEE Conference on Computer Vision and Pattern Recognition, pp. 427–436 (2015)
6. Silver, D., et al.: Mastering the game of go without human knowledge. Nature **550**(7676), 354 (2017)
7. Wang, Y., Kosinski, M.: Deep neural networks are more accurate than humans at detecting sexual orientation from facial images. J. Pers. Soc. Psychol. **114**(2), 246–257 (2018). https://doi.org/10.31234/osf.io/hv28a

8. Wigner, E.P.: The unreasonable effectiveness of mathematics in the natural sciences. Richard courant lecture in mathematical sciences delivered at New York university, May 11, 1959. Commun. Pure Appl. Math. **13**(1), 1–14 (1960). https://doi.org/10.1002/cpa.3160130102
9. Zadeh, L.A.: Toward human level machine intelligence - is it achievable? The need for a paradigm shift. IEEE Comput. Intell. Mag. **3**(3), 11–22 (2008). https://doi.org/10.1109/MCI.2008.926583

Fuzzy Models for Big Data Mining

Pietro Ducange[✉][iD]

SMART Engineering Solutions and Technologies (SMARTEST) Research Centre,
eCAMPUS University, Via Isimbardi 10, 22060 Novedrate, Italy
`pietro.ducange@uniecampus.it`

Keywords: Big data · Distributed computing · Fuzzy classifiers

Extended Abstract

We are currently experiencing the *Big Data Era* [20]: large *volume* of information is generated by different sources and may have different formats (*variety*) [5]. Such data are often produced at very high speed and need to be elaborated in almost real time (*velocity*). However, these data represent a very important source of added-*values* in several contexts, such as in marketing strategies [11], industrial applications [29] and Internet of Things [2].

When dealing with big data, due to their volume, diversity and complexity, new techniques, algorithms and analyses are required to extract the *hidden* and *valuable knowledge*. Indeed, classical *data mining* and *machine learning* algorithms, that in the last decades have been successfully adopted for extracting knowledge and value from data [19], cannot directly applied to the big data. In this context, the state-of-the art paradigms for data storage and elaboration are not suitable. With aim of designing and experimenting data mining and machine learning algorithms for big data, researchers moved to new distributed frameworks, such as *Apache Hadoop* and *Apache Spark*. Recent contributions in the field big data mining exploit the *MapReduce* paradigm [9] for implementing distributed versions of clustering algorithms [21,23] and classification algorithms [6,24]. Highlights on the recent advances, challenges and objectives in designing, developing and using data mining and machine learning algorithms for big data can be found in [31].

As regards Fuzzy Models (FMs), recently, several works in the specialized literature have focused on the design, the implementation and the experimentation of classifiers for big data [12,14,17,22,25–28]. As stated in [15], FMs are particularly suitable for handling the variety and veracity of Big Data. This is mainly due to their good capability of coping with vague, imprecise and uncertain concepts. Moreover, the use of overlapped fuzzy labels ensures a good coverage of the problem space. This issue is especially relevant when dealing with very large datasets that may be dived into a number of heterogeneous chunks, such as in the MapReduce programming paradigm. Indeed, the different chunks may influence in a different way the parameters learning process of the classification model.

R. Fullér et al. (Eds.): WILF 2018, LNAI 11291, pp. 257–260, 2019.
https://doi.org/10.1007/978-3-030-12544-8_24

The *Chi-FRBCS-BigData* algorithm, discussed in [26], represents the first attempt of extending fuzzy rule-based classifiers (FRBCs) to the distributed framework. It is designed and developed according to the MapReduce programming model and is based on the well-known Chi et al. algorithm, which was introduced in [8] for generating fuzzy classification rules. The Chi-FRBCS-BigData algorithm have been also experimented in [22] and [16], considering imbalanced classification datasets and analyzing the effects of different granularities of the fuzzy partitions, respectively. Last year, an optimized version of the distributed Chi et al. algorithm has been introduced in [12]. The optimization regards both the scheme for generating the rules and the architecture of the distributed execution scheme. Two completely different fuzzy models for classifying big datasets have been recently proposed in [27] and in [28]: the two models are based on *Fuzzy Associative Classifiers* and on *Fuzzy Decision Trees*, respectively. Moreover, some *Evolutionary-based* methods for learning FMs for Big Data have been also proposed [14,25].

The aforementioned approaches regards the design and development of FMs in a distributed computing architecture, especially focusing in generating accurate models. Although the classifiers obtained are very accurate, the complexity of the models, in terms of number of rules or number of parameters of the fuzzy trees, is very high. The greater the complexity, the lower the *interpretability* [18]. However, the interpretability is a very important feature that characterize FMs, and it assume a special importance also in the contest of Big Data [15,30]. In order to generate FRBCs characterized by different trade-off between accuracy and complexity, a novel distributed *Multi-objective Evolutionary Learning* scheme has been proposed in [17]. The algorithm, denoted as *DPAES-RCS* is a distributed implementation, under the Apache Spark environment, of PAES-RCS [4]. PAES-RCS learns the RB through a rule and condition selection strategy. Moreover, also the parameters of the fuzzy sets are learnt concurrently with the RB. The accuracy and the complexity of the classifiers are concurrently optimized: the evaluation of the accuracy is calculated in a distributed fashion, in order to deal with big datasets. Very compact and accurate classification models can be obtained adopting DPAES-RCS. It is worth noticing that DPAES-RCS, represents, to the best of our knowledge, the first contribution in the field of *Multi-objective Evolutionary Fuzzy Systems* [10,13] for Big Data.

Even though a set of algorithms and tools are available for extracting useful knowledge from big data by means of FMs, we envision that the future directions in this context will regards: (i) *enhancing the interpretability* of the rules and of the fuzzy partitions, both at semantic and complexity levels [3], (ii) *handling data streams* [7] moving towards the more general granular computing framework [1]; and (iii) *reducing the computation efforts* for generating compact and accurate solutions. The three aforementioned challenges must be conducted in parallel as much as possible. Indeed, interpretable models, able to extract knowledge in almost real-time from huge amount of streaming and heterogeneous data, will be the actual added values for future research activities on FMs for big data mining.

References

1. Ahmad, S.S.S., Pedrycz, W.: The development of granular rule-based systems: a study in structural model compression. Granular Comput. **2**(1), 1–12 (2017)
2. Al-Ali, A., Zualkernan, I.A., Rashid, M., Gupta, R., Alikarar, M.: A smart home energy management system using iot and big data analytics approach. IEEE Trans. Consum. Electron. **63**(4), 426–434 (2017)
3. Alonso, J.M., Castiello, C., Mencar, C.: Interpretability of fuzzy systems: current research trends and prospects. In: Kacprzyk, J., Pedrycz, W. (eds.) Springer Handbook of Computational Intelligence, pp. 219–237. Springer, Heidelberg (2015). https://doi.org/10.1007/978-3-662-43505-2_14
4. Antonelli, M., Ducange, P., Marcelloni, F.: A fast and efficient multi-objective evolutionary learning scheme for fuzzy rule-based classifiers. Inf. Sci. **283**, 36–54 (2014)
5. Anuradha, J., et al.: A brief introduction on big data 5Vs characteristics and hadoop technology. Procedia Comput. Sci. **48**, 319–324 (2015)
6. Bechini, A., Marcelloni, F., Segatori, A.: A MapReduce solution for associative classification of big data. Inf. Sci. **332**, 33–55 (2016)
7. Casalino, G., Castellano, G., Mencar, C.: Incremental adaptive semi-supervised fuzzy clustering for data stream classification. In: 2018 IEEE Conference on Evolving and Adaptive Intelligent Systems (EAIS) (2018)
8. Chi, Z., Yan, H., Pham, T.: Fuzzy Algorithms: with Applications to Image Processing and Pattern Recognition. Advances in Fuzzy Systems - Applications and Theory, vol. 10. World Scientific, Singapore (1996)
9. Dean, J., Ghemawat, S.: MapReduce: simplified data processing on large clusters. Commun. ACM **51**(1), 107–113 (2008)
10. Ducange, P., Marcelloni, F.: Multi-objective evolutionary fuzzy systems. In: Fanelli, A.M., Pedrycz, W., Petrosino, A. (eds.) WILF 2011. LNCS (LNAI), vol. 6857, pp. 83–90. Springer, Heidelberg (2011). https://doi.org/10.1007/978-3-642-23713-3_11
11. Ducange, P., Pecori, R., Mezzina, P.: A glimpse on big data analytics in the framework of marketing strategies. Soft Comput. **22**(1), 325–342 (2018)
12. Elkano, M., Galar, M., Sanz, J., Bustince, H.: CHI-BD: a fuzzy rule-based classification system for big data classification problems. Fuzzy Sets Syst. **348**, 75–101 (2018)
13. Fazzolari, M., Alcalá, R., Nojima, Y., Ishibuchi, H., Herrera, F.: A review of the application of multi-objective evolutionary fuzzy systems: current status and further directions. IEEE Trans. Fuzzy Syst. **21**(1), 45–65 (2013)
14. Fernandez, A., Almansa, E., Herrera, F.: Chi-Spark-RS: an spark-built evolutionary fuzzy rule selection algorithm in imbalanced classification for big data problems. In: 2017 IEEE International Conference on Fuzzy Systems (FUZZ-IEEE), pp. 1–6. IEEE (2017)
15. Fernández, A., Carmona, C.J., del Jesus, M.J., Herrera, F.: A view on fuzzy systems for big data: progress and opportunities. Int. J. Comput. Intell. Syst. **9**(sup1), 69–80 (2016)
16. Fernández, A., del Río, S., Bawakid, A., Herrera, F.: Fuzzy rule based classification systems for big data with MapReduce: granularity analysis. Adv. Data Anal. Classif. **11**(4), 711–730 (2017)
17. Ferranti, A., Marcelloni, F., Segatori, A., Antonelli, M., Ducange, P.: A distributed approach to multi-objective evolutionary generation of fuzzy rule-based classifiers from big data. Inf. Sci. **415**, 319–340 (2017)

18. Gacto, M.J., Alcalá, R., Herrera, F.: Interpretability of linguistic fuzzy rule-based systems: an overview of interpretability measures. Inf. Sci. **181**(20), 4340–4360 (2011)
19. Han, J., Kamber, M., Pei, J.: Data Mining: Concepts and Techniques. Data Management Systems, 3rd edn. Morgan Kaufmann, Waltham (2012)
20. John Walker, S.: Big Data: A Revolution that Will Transform How We Live, Work, and Think. Houghton Mifflin Harcourt, Boston (2014)
21. Kim, Y., Shim, K., Kim, M.S., Lee, J.S.: DBCURE-MR: an efficient density-based clustering algorithm for large data using mapreduce. Inf. Syst. **42**, 15–35 (2014)
22. López, V., del Río, S., Benítez, J.M., Herrera, F.: Cost-sensitive linguistic fuzzy rule based classification systems under the MapReduce framework for imbalanced big data. Fuzzy Sets Syst. **258**, 5–38 (2015)
23. Ludwig, S.A.: Mapreduce-based fuzzy c-means clustering algorithm: implementation and scalability. Int. J. Mach. Learn. Cybern. **6**(6), 923–934 (2015)
24. Maillo, J., Ramírez, S., Triguero, I., Herrera, F.: kNN-IS: an iterative spark-based design of the k-nearest neighbors classifier for big data. Knowl.-Based Syst. **117**, 3–15 (2017)
25. Márquez, A., Márquez, F., Peregrín, A.: A scalable evolutionary linguistic fuzzy system with adaptive defuzzification in big data. In: 2017 IEEE International Conference on Fuzzy Systems (FUZZ-IEEE), pp. 1–6. IEEE (2017)
26. del Río, S., López, V., Benítez, J.M., Herrera, F.: A MapReduce approach to address big data classification problems based on the fusion of linguistic fuzzy rules. Int. J. Comput. Intell. Syst. **8**(3), 422–437 (2015)
27. Segatori, A., Bechini, A., Ducange, P., Marcelloni, F.: A distributed fuzzy associative classifier for big data. IEEE Trans. Cybern. **48**(9), 2656–2669 (2018)
28. Segatori, A., Marcelloni, F., Pedrycz, W.: On distributed fuzzy decision trees for big data. IEEE Trans. Fuzzy Syst. **26**(1), 174–192 (2018)
29. Wan, J., et al.: A manufacturing big data solution for active preventive maintenance. IEEE Trans. Ind. Inform. **13**(4), 2039–2047 (2017)
30. Wang, H., Xu, Z., Pedrycz, W.: An overview on the roles of fuzzy set techniques in big data processing: trends, challenges and opportunities. Knowl.-Based Syst. **118**, 15–30 (2017)
31. Zhou, L., Pan, S., Wang, J., Vasilakos, A.V.: Machine learning on big data: opportunities and challenges. Neurocomputing **237**, 350–361 (2017)

The Challenges of Big Data and the Contribution of Fuzzy Logic

Francesco Masulli[1,2(✉)] [iD] and Stefano Rovetta[1] [iD]

[1] DIBRIS, University of Genova, Via Dodecaneso 35, 16146 Genoa, Italy
{francesco.masulli,stefano.rovetta}@unige.it
[2] Sbarro Institute for Cancer Research and Molecular Medicine, Temple University,
Philadelphia, PA, USA

Keywords: Big data · Fuzzy logic · Data quality ·
Data Stream Modeling · Explainable Learning Machines

Extended Abstract

In recent decades we have witnessed a growing investment by all economic sectors in the acquisition of volumes of data characterized not only by ever larger cardinality, but also by increasing number of characteristics for each observed instance [1], and this led to the coining of the term Big-Data. It is estimated that the amount of data that will be produced in 2018 globally will amount to about twenty zettabytes, where a zettabyte, also referred to as ZB, corresponds a number of bytes difficult to imagine consisting of equal to a 1 followed from 21 zeros. Most of the data that is produced is stored in clouds or in data servers and only a small fraction of the information contained in it can be exploited using conventional processing techniques, as classical statistical methods are simply not designed to cope with the explosive growth of dimensionality of the observation vector.

Furthermore, the implicit informational content of Big-Data could be an asset of enormous value for companies, but to extract the more valuable content from them it is necessary to employ the most advanced Artificial Intelligence data-driven techniques, and in particular the methods of Computational Intelligence and of Machine Learning. Analysts point to Artificial Intelligence as a major technological challenge that can open new scenarios for companies and which could double the growth rate of developed economies by 2035 and increase labor productivity in increments until at 40%.

Lofti Zadeh departed on 6 September 2017, leaving us an arsenal of methodologies partly proposed by him directly and partly proposed by the community of researchers in Fuzzy Logic on his inspiration[1]. These tools allow us to face the technological challenges of today's world of Big Data that mainly concern the extraction of information, knowledge and value from the ever-increasing masses of data that our society accumulates every day.

---

[1] For a selection of the most representative papers of Zadeh, see, e.g., [2–10].

© Springer Nature Switzerland AG 2019
R. Fullér et al. (Eds.): WILF 2018, LNAI 11291, pp. 261–264, 2019.
https://doi.org/10.1007/978-3-030-12544-8_25

Building on this heritage, additional Fuzzy Logic-based tools, specifically focused to the modern challenges described above need to be developed. These contributions will have to operate on two separate dimensions: *data* and *contents*.

The *data* dimension regards data as an asset by themselves, while the *contents* dimension is concerned with the data as a means to carry information and knowledge, and consequently with its meaning and its use.

Along the data dimension, we will point out three areas relevant to Big Data analytics where new Fuzzy Logic-based tools may be developed.

The first area is the assessment of *Data Quality* that is a vital precondition for inductive learning [11]. As stated by Robinson [12], there are five components that ensure data quality: completeness, consistency, accuracy, validity, and timeliness. New unsupervised fuzzy and possibilistic clustering algorithms and clustering comparison techniques can be key-elements for evaluating data quality [13].

Another field where we see the need of the development of new tools is *Data Stream Modeling*. Data streams are becoming a major paradigm in the realm of data science, as they arise from seamlessly observed phenomena in an increasing number of fields [15]. They always depend on time, although to different degrees. They may represent actual time series, with a strong dependency on time, or quasi-stationary phenomena whose variability can be appreciated only in the long term. Moreover, the size of any collection of Big Data makes single-pass methods a necessity, turning these data effectively into a special case of streaming data. Unsupervised analysis in the form of data clustering using fuzzy and possibilistic techniques provides a useful toolbox to mine data streams [17,18,20].

The last area we highlight, where new Fuzzy Logic-based tools are feasible and extremely useful, is the one of the *Explainable Learning Machines* [21–24]. Two paradigms that are found in the toolbox of fuzzy modelling are centroid-based soft clustering and fuzzy rule-based systems. These two paradigms have been proved to be related [26,27] and feature two crucial properties: controlled rule support, that can be used to obtain non-statistical generalization guarantees, and the ability to easily incorporate prior knowledge, that can be exploited in a dual way for rule extraction to implement explainability.

Coming to the content dimension, there are several new issues that are raised by the pervasive use of data in all fields of everyday life and by its intersection with ethical and legal aspects. It becomes therefore increasingly important to understand these effects at the social level and to take responsibility for them, especially when they deal with human-related data. It is indeed well recognized that "technology solutions must play an important role in enabling our society to reap ever-greater benefits from big data, while keeping it safe from the risks" [19].

The main research question is: Is it possible to define a data processing workflow that can guarantee by design, as far as possible, that specific properties of the data related to sensitive information are preserved along the whole data processing chain (responsible data processing)?

The development of technological solutions satisfying non-discriminating requirements is currently one of the main challenges in data processing [28] and some preliminary proposals have already been provided for specific data processing tasks, namely ranking and set selection.

More generally, there are properties that can be measured and should be constrained to achieve specific goals. Among others, some important ones are diversity, fairness, and serendipity [16,25]. With respect to these, a desirable behaviour of data analytic processes is *compositionality*, i.e., the properties that are present in the original data should not be reduced when applying successive steps of data processing.

One particular step that is commonly found in many data-analytic workflows, where Fuzzy Logic-based approaches are particularly promising, is data clustering. This process can be used to enable exploration of query results, to provide automatic taxonomies, or as a first summarization step for further processing and interpretation. This task can operate on stored data, or on data streams, typically with different goals, but a common tool that enables this kind of analysis is the fuzzy clustering approach termed possibilistic clustering. The Graded Possibilistic C-Means (GPCM) method [14] intrinsically possesses a tunable outlier rejection property that has already been exploited for outlier analysis [17,18].

A possible approach is studying the decomposition of clustering models into a main component and marginal components, by exploiting the dual modeling capability of possibilistic clustering to distinguish between a typical component and atypical ones. This will allow to measure, monitor, and control the degree of satisfaction of the required constraints.

For stream clustering, compositionality of the required properties can only be strictly enforced in stationary regimes, not in the transient. To track non-stationary conditions, the statistics of the marginal components of the model could be used to monitor and control the process compositionality.

References

1. Donoho, D.L.: High-dimensional data analysis: the curses and blessings of dimensionality, plenary lecture. In: Mathematical Challenges of the 21st Century, Los Angeles, 6–11 August, pp. 1–33. The American Mathematical Society (2000). http://statweb.stanford.edu/donoho/Lectures/AMS2000/AMS2000.html
2. Zadeh, L.A.: Fuzzy sets. Inf. Control **8**(3), 338–353 (1966)
3. Bellman, R.E., Zadeh, L.A.: Decision making in a fuzzy environment. Manag. Sci. **17**(4), B141–B164 (1970)
4. Zadeh, L.A.: The concept of a linguistic variable and its application to approximate reasoning - I. Inf. Sci. **8**(3), 199–249 (1975)
5. Zadeh, L.A.: The concept of a linguistic variable and its application to approximate reasoning - II. Inf. Sci. **8**(4), 301–357 (1975)
6. Zadeh, L.A.: The concept of a linguistic variable and its application to approximate reasoning - III. Inf. Sci. **9**(1), 43–80 (1975)
7. Zadeh, L.A.: A fuzzy-algorithmic approach to the definition of complex or imprecise concepts. Int. J. Man-Mach. Stud. **8**(3), 249–291 (1976)

8. Zadeh, L.A.: A computational approach to fuzzy quantifiers in natural languages. Comput. Math. Appl. Spec. Issue Comput. Linguist. **9**(1), 149–184 (1983)

9. Zadeh, L.A.: Fuzzy logic, neural networks and soft computing. Commun. ACM **37**(3), 77–84 (1994)

10. Zadeh, L.A.: Fuzzy logic = computing with words. IEEE Trans. Fuzzy Syst. **4**(2), 103–111 (1996)

11. Cai, L., Zhu, Y.: The challenges of data quality and data quality assessment in the big data era. Data Sci. J. **14**, 2 (2015). https://doi.org/10.5334/dsj-2015-002

12. Robinson, A.: The 5 Key Reasons Why Data Quality Is So Important. Cerasis White paper, 29 June 2017. https://cerasis.com/2017/06/29/data-quality/

13. Rovetta, S., Masulli, F.: Visual stability analysis for model selection in graded possibilistic clustering. Inf. Sci. **279**, 37–51 (2014)

14. Masulli, F., Rovetta, S.: Soft transition from probabilistic to possibilistic fuzzy clustering. IEEE Trans. Fuzzy Syst. **14**(4), 516–527 (2006)

15. Zliobaite, I., et al.: Next challenges for adaptive learning systems. SIGKDD Explor. **14**(1), 48–55 (2012)

16. Binns, R.: Fairness in machine learning: lessons from political philosophy. arXiv preprint arXiv:1712.03586 (2017)

17. Abdullatif, A., Masulli, F., Rovetta, S.: Tracking time evolving data streams for short-term traffic forecasting. Data Sci. Eng. **2**(3), 210–223 (2017)

18. Abdullatif, A., Masulli, F., Rovetta, S.: Clustering of nonstationary data streams: a survey of fuzzy partitional methods. Wiley Interdisc. Rew. Data Min. Knowl. Discov. **8**(4), e1258 (2018)

19. Abiteboul, S., et al.: Research directions for principles of data management. Dagstuhl Manifestos **7**(1), 1–29 (2017)

20. Casalino, G., Castellano, G., Mencar, C.: Incremental adaptive semi-supervised fuzzy clustering for data stream classification. In: EAIS, pp. 1–7 (2018)

21. Alonso, J.M., Castiello, C., Mencar, C.: Interpretability of fuzzy systems: current research trends and prospects. In: Kacprzyk, J., Pedrycz, W. (eds.) Springer Handbook of Computational Intelligence, pp. 219–237. Springer, Heidelberg (2015). https://doi.org/10.1007/978-3-662-43505-2_14

22. Alonso, J.M., Ramos-Soto, A., Castiello, C., Mencar, C.: Hybrid data-expert explainable beer style classifier. In: IJCAI/ECAI Workshop on Explainable Artificial Intelligence, p. XAI-18 (in press)

23. Biran, O., Cotton, C.: Explanation and justification in machine learning: a survey. In: IJCAI-17 Workshop on Explainable AI (XAI). p. 8 (2017)

24. Fernandez, A., del Jesus, M.J., Cordon, O., Marcelloni, F., Herrera, F.: Evolutionary fuzzy systems for explainable artificial intelligence: why, when, what for, and where to? IEEE Comput. Intell. Mag. (in press)

25. Kaminskas, M., Bridge, D.: Diversity, serendipity, novelty, and coverage: a survey and empirical analysis of beyond-accuracy objectives in recommender systems. ACM Trans. Interact. Intell. Syst. (TiiS) **7**(1), 2 (2017)

26. Ridella, S., Rovetta, S., Zunino, R.: Circular back-propagation networks for classification. IEEE Trans. Neural Netw. **1**(8), 84–97 (1997)

27. Rovetta, S., Zunino, R.: Circular backpropagation networks embed vector quantization. IEEE Trans. Neural Netw. **4**(10), 972–975 (1999)

28. Zliobaite, I.: A survey on measuring indirect discrimination in machine learning. arXiv preprint arXiv:1511.00148 (2015)

The Future of Fuzzy Sets in Finance:
New Challenges in Machine Learning
and Explainable AI

Silvia Muzzioli[1,2]([⊠]) [iD]

[1] Department of Economics, University of Modena and Reggio Emilia,
Viale Berengario 51, 41121 Modena, Italy
silvia.muzzioli@unimore.it
[2] CEFIN, University of Modena and Reggio Emilia,
Viale Berengario 51, 41121 Modena, Italy

Keywords: Big data · Fuzzy sets · Finance · Machine learning ·
Explainable AI

Extended Abstract

Traditional statistical analysis is oriented towards finding linear relationships between the variables under investigation, often accompanied by strict assumptions about the problem and data distributions. Moreover, traditional analysis endorses data reduction as much as possible before modeling, and, as a result, part of the original information is lost. On the other hand, machine learning does not impose rigid pre-assumptions about the problem and data distributions since the underlying ratio is to "learn from data", without the need for data reduction or a priori knowledge before the learning. For these reasons machine learning has experienced a rapid dissemination in a large number of sectors including healthcare, finance, transportation, retail and social media services industry. Machine learning is the core technology of the new age of AI applications. Machine learning methods offer tremendous benefits, but are limited by their opaqueness, non-intuitiveness and difficulty to understand.

In finance in particular, machine learning methods have played a crucial role in improving the forecasting ability of financial models and trading systems, due to their ability to process a large amount of data and the peculiarity of capturing also non-linear relationships between variables. In recent years, the availability of sample data at very high frequencies (intraday or tick by tick) resulted in a fertile domain for their application, especially in the coding of indicators and patterns of technical analysis. Deep learning systems are the most advanced form of machine learning. They can match humans in recognizing images or driving a car, but why they come up with the solutions remains difficult to tell exactly. Businesses would have used machine learning more widely if they could understand how machines come up with their recommendations on trading, fraud detection, insurance and banking.

A challenge for AI in finance is the need to analyze and aggregate a large amount of information obtained from different sources. In the financial literature, the use of artificial intelligence (AI) and machine learning techniques is often limited to the coding of technical analysis indicators (such as moving averages or the flag pattern) for trading strategy purposes. As pointed out in [8], most of the contributions investigating

R. Fullér et al. (Eds.): WILF 2018, LNAI 11291, pp. 265–268, 2019.
https://doi.org/10.1007/978-3-030-12544-8_26

machine learning methods in' financial markets propose trading strategies that rely mainly on technical analysis and focus on a single stock market or market index.

In recent years, the financial literature has started to focus on the application of soft computing techniques to exploit the availability of sampled data at very high frequencies (intraday or tick by tick) to codify indicators for technical analysis (see [1] and among others [2, 9, 11, 13, 16, 21]). The main objectives include the forecasting of stock prices, direction of the market, and buy or sell signals. The techniques range from hybrid neuro-fuzzy systems [3], ANFIS (Adaptive Network-based Fuzzy Inference System) controller, recurrent neural networks [4]. A TSK type fuzzy rules is adopted also in [12], where the parameters of the rules were tuned by an ANFIS and the number of these rules was identified by means of fuzzy c-mean clustering. Different soft computing techniques are combined in [15], where a genetic fuzzy systems is integrated with self-organizing map neural networks.

On the other hand, there exist a few contributions in literature that aim to aggregate different information in a unique index. In particular, [20] propose an index of financial development, following a standard three-step approach, based on reducing multidimensional data into one summary index (the procedure generally follows the OECD Handbook on Constructing Composite Indicators (OECD, 2008)). The first step involves the normalization of input variables. In the second step, the aggregation of normalized variable is performed by constructing a limited number of sub-indices that represent a particular dimension. Finally, in the last step, the sub-indices are aggregated in the final index. Examples of application of the proposed methodology are the IMF Financial Stress Index [6, 7] and other financial inclusion indexes [1, 5].

At present the European Union lacks instruments for monitoring the financial risk of each Member State and of the European financial market as a whole. Only a small number of States have adopted a volatility index and none of the countries has developed a more advanced tail-risk index. A further factor to consider is that an aggregate European volatility index has not yet been developed.

The main obstacle to the construction of risk indices is the limited availability of option-based data for European peripheral countries. The European markets are in need of new techniques accounting for uncertainty in data and data processing going beyond the narrow focus of the existing indices, which treat financial markets as compartmentalized [14] and overlook important risk assessment determinants. On the other hand, investors and regulators need comprehensive risk measures able to aggregate and synthesize different types of information in a single indicator.

Another challenge is represented by the increasing dominance of computerized trading, which may cause more volatility during market downturns. The rising frequency of 'flash crashes' across many major markets, the increasing incidents of volatility such as the VIX spike on Feb. 5, 2018, the 10-year Treasury bond on Oct. 15, 2014, and the British pound on Oct. 6, 2016, are an important early warning sign that machines have to be closely supervised and understood. New measures and tools to control the volatility of financial markets [17–19] should be developed.

The semantic properties of linguistic fuzzy sets, their good coverage even in the case of lack of data, their management of the uncertainty, especially in Big Data [10], make them a very interesting tool for nowadays applications, especially when the practitioner need to understand why a given decision has been made.

References

1. Amidžić, G., Massara, A., Mialou, A.: Assessing countries' financial inclusion standing – a new composite index. IMF Working Paper 14/36. International Monetary Fund, Washington, February 2014
2. Ballings, M., Poel, D.V.D., Hespeels, N., Gryp, R.: Evaluating multiple classifiers for stock price direction prediction. Expert Syst. Appl. **42**(20), 7046–7056 (2015)
3. Bekiros, S.D.: Fuzzy adaptive decision-making for boundedly rational traders in speculative stock markets. Eur. J. Oper. Res. **202**(1), 285–293 (2010)
4. Bekiros, S.D., Georgoutsos, D.A.: Evaluating direction-of-change forecasting: neurofuzzy models vs. neural networks. Math. Comput. Model. **46**(1), 38–46 (2007)
5. Camara, N., Tuesta, D.: Measuring financial inclusion: a multidimensional index. Working Papers 1426, BBVA Bank, Economic Research Department (2014)
6. Cardarelli, R., Elekdag, S., Lall, S.: Financial stress and economic downturns. World Economic Outlook, October 2008 Issue, pp. 129–158. International Monetary Fund (2008). Chap. 4
7. Cardarelli, R., Elekdag, S., Lall, S.: Financial stress, downturns, and recoveries, IMF Working Papers 09/100. International Monetary Fund, Washington (2009)
8. Cavalcante, R.C., Brasileiro, R.C., Souza, V.L., Nobrega, J.P., Oliveira, A.L.: Computational intelligence and financial markets: a survey and future directions. Expert Syst. Appl. **55**, 194–211 (2016)
9. Cervello-Royo, R., Guijarro, F., Michniuk, K.: Stock market trading rule based on pattern recognition and technical analysis: forecasting the DJIA index with intraday data. Expert Syst. Appl. **42**(14), 5963–5975 (2015)
10. Choi, T.-M., Chan, H.K., Yue, X.: Recent development in big data analytics for business operations and risk management. IEEE Trans. Cybern. **47**(1), 81–92 (2017)
11. De Oliveira, F.A., Nobre, C.N.N., Zarate, L.E.: Applying artificial neural networks to prediction of stock price and improvement of the directional prediction index - case study of PETR4, Petrobras, Brazil. Expert Syst. Appl. **40**(18), 7596–7606 (2013)
12. Esfahanipour, A., Aghamiri, W.: Adapted neuro-fuzzy inference system on indirect approach TSK fuzzy rule base for stock market analysis. Expert Syst. Appl. **37**(7), 4742–4748 (2010)
13. Fadlalla, A., Amani, F.: Predicting next trading day closing price of Qatar exchange index using technical indicators and artificial neural networks. Intell. Syst. Account. Finance Manag. **21**(4), 209–223 (2014)
14. Frijns, B., Verschoor, W.F.C., Zwinkels, R.C.: Excess stock return comovements and the role of investor sentiment. J. Int. Financ. Markets Inst. Money **49**, 74–87 (2017)
15. Hadavandi, E., Shavandi, H., Ghanbari, A.: Integration of genetic fuzzy systems and artificial neural networks for stock price forecasting. Knowl.-Based Syst. **23**(8), 800–808 (2010)
16. Mabu, S., Hirasawa, K., Obayashi, M., Kuremoto, T.: Enhanced decision making mechanism of rule-based genetic network programming for creating stock trading signals. Expert Syst. Appl. **40**(16), 6311–6320 (2013)
17. Muzzioli, S., De Baets, B.: Fuzzy approaches to option price modelling. IEEE Trans. Fuzzy Syst. **25**(2), 392–401 (2017)
18. Muzzioli, S., Gambarelli, L., De Baets, B.: Towards a fuzzy volatility index for the Italian market. In: Proceedings of the IEEE International Conference on Fuzzy Systems, FUZZ-IEEE, Naples (2017)

19. Muzzioli, S., Gambarelli, L., De Baets, B.: Indices for financial market volatility obtained through fuzzy regression. Int. J. Inf. Technol. Decis. Making (forthcoming). https://doi.org/10.1142/S0219622018500335
20. Svirydzenka, K.: Introducing a new broad-based index of financial development. IMF Working Papers 16/5. International Monetary Fund (2016)
21. Wu, J.L., Yu, L.C., Chang, P.C.: An intelligent stock trading system using comprehensive features. Appl. Soft Comput. **23**, 39–50 (2014)

New Trends of Fuzzy Systems: Fintech Applications

Antonia Azzini[✉], Stefania Marrara, and Amir Topalović

Consortium for the Technology Transfer (C2T),
via Nuova Valassina, Carate Brianza, MB, Italy
{antonia.azzini,stefania.marrara,amir.topalovic}@consorzioc2t.it

Keywords: FinTech · Fuzzy systems · New trends

Extended Abstract

In the last years, the term *Financial Technology (FinTech)* has been adopted by literature to describe a wide range of services, aided by several financial technologies, for enterprises or organizations, which mainly address the improvement of the service quality by using Information Technology (IT) applications.

A continuous growth of the investment has seen the development of FinTech technologies in multiple areas, such as mobile networks [12], big data [18], trust management [1], mobile embedded systems [10], cloud computing [11], image processing [8], and data analytic techniques [13]. FinTech has become important due to several important factors, which include technical development, business innovation expectations (market), cost-saving requirements, and customer demands. A handful of financial technology (fintech) trends are expected to strengthen significantly in 2018 as envisioned in [16].

In particular, the growth of Artificial Intelligence (AI) looks to be instrumental for three specific reasons: hugely increased opportunities for improved customer centricity, ability to ease the regularity reporting burden through AI enabled 'RegTech', and massively improved cyber-security and data protection.

When it comes to AI, an important consultancy agency - DataArt, defines it as an "industry game changer", but one that at the same time "will not come without problems as the current industry wide skills gap turns into a *war for talent.*" Already we are seeing problems in one of the biggest users of AI, cyber-security, which is looking to the blockchains in search of a solution.

In this context several finance domains require an effective modeling of vague and imprecise information. Stock markets, for instance, which has been investigated by various researchers, are a rather complicated environment. Most researchers only concerned the technical indexes (quantitative factors), instead of qualitative factors, e.g., the political effects. However, the latter play a critical role in the stock market environment. Thus, it is important to study innovative models which can measure the qualitative effect on them. Other important studies are aimed at building and evaluating human skill based fuzzy expert systems

© Springer Nature Switzerland AG 2019
R. Fullér et al. (Eds.): WILF 2018, LNAI 11291, pp. 269–272, 2019.
https://doi.org/10.1007/978-3-030-12544-8_27

for decision making support in a stock trading process. The focus here, in which fuzzy systems may play a major role, is concentrated on envisioning a computer software capable to reproduce the knowledge from the skilled stock trader. Other important fuzzy systems applications would provide effective solutions in security issues as user recognition by mean of biometric techniques, like for example those presented in [6], in online banking, or customer satisfaction analysis in finance services supplying.

In this contribution our intention is to raise the attention of the scientific community on the needs that the FinTech domain proposes. Today, the prevailing theme in the financial world revolves around blockchains, but even in the contractual domain there is a dimension of uncertainty and flexibility that requires the analysis of appropriate solutions. The subject of this contribution will also be discussed at the 5th IEEE Int. Conf. on Data Science and Advanced Analytics (DSAA) Special Session on *Opportunities and Risks for Data Science in Organizations: Banking, Finance, and Policy* that will be held in Turin on October 1–4th, 2018.

1   Fuzzy Systems in FinTech

Over the years since the birth of the Fuzzy Sets Theory, Fuzzy systems have found wide use in expert systems, machinery, home appliances and robotics. Financial applications have also recently been discovered, taking advantage of fuzzy systems' ability to model vague and imprecise models. Fuzzy systems have been used with various technical indicators in previous studies. Zhou and Dong [19] model the cognitive uncertainty incorporated in technical analysis by using a fuzzy-logic approach. Their algorithm has been able to offer superior precision in detecting and interpreting technical patterns over visual pattern analysis done by experts. Lin et al. [14] make use of a fuzzy system with KD technical index to predict stock indices. KD index is a stochastic oscillator, which consists of two lines namely K and D, where D is smoothed version of the K line. Their research shows that the returns generated with the fuzzy systems are significantly larger than linear regression models, neural networks and other investment strategies. The results combining technical analysis and fuzzy logic were very promising.

Since different artificial intelligence methods have different strengths and limitations, hybrid systems have also been studied to obtain synergetic combinations of methods. Azzini et al. present in their work [5] a comparison between Nature-Inspired and Machine Learning approaches for detecting trend reversals in financial time series, while Abraham and Nath [2] provide an overview of different hybrid models and architectures. In particular, combinations of fuzzy systems with neural networks and/or genetic algorithms appear to be popular in real-world implementations. A neuro-fuzzy system to predict financial time series is described in [17]. The prediction of stock and option prices of the S&P and Dow Jones indices have been examined, which resulted in profitable trading strategies. The results from the paper show the potential of neuro-fuzzy modeling for finance and management. Neuro-fuzzy models have also been used to predict other time series such as the Greek manufacturing and Korea stock price indexes. They have also been applied for portfolio evaluation, while in a more recent work [4] Azzini et al. predict turning points in financial markets with Fuzzy-Evolutionary and Neuro-Evolutionary modeling. Hybrid combinations of fuzzy and probabilistic systems have also been proposed. For example, van den Berg et al. analyse a financial market by using a probabilistic fuzzy model [7], in which linguistic uncertainty is combined with probabilistic uncertainty.

The problem of finding desirable fuzzy rules is a very important process in the development of fuzzy systems. In practice, acquiring the rules from experts only is quite a difficult task. Alcala et al. [3] give an overview of different approaches of learning and tuning of a fuzzy system. Mohammadian and Kingham [15] develop a hierarchical fuzzy logic system by using genetic algorithm to predict the interest rates in Australia. Using a genetic algorithm as a training method for learning the fuzzy rules, the number of rules could be significantly reduced, resulting in more efficient systems. The results show that the system is able to give accurate prediction of the interest rates.

Finally, the approach presented in [9] describes a fuzzy system to predict market price movements for investing in portfolio of European, American and Japanese bonds and currency. The system has been developed with participation of traders and experts from a financial institution, whose knowledge forms a constraint on the design of the structure of the fuzzy system. The system takes a number of technical analysis indexes as input, which have been specified by traders. The system generates a buy or sell signal, but it can also be combined with portfolio allocation mechanisms for automated trading.

2 Conclusion

In this contribution we start a discussion on the potentiality of Fuzzy Set Theory, Fuzzy Logic and related techniques to boost IT applications in Finance, exploiting their ability to model the vague and imprecise information that may affect many finance domains. This topic will be also discussed at the DSAA 2018 Special Session on *Opportunities and Risks for Data Science in Organizations: Banking, Finance, and Policy* that will be held in Turin on October 1–4th, 2018.

References

1. Abawajy, J., Wang, G., Yang, L.T., Javadi, B.: Trust, security and privacy in emerging distributed systems. Future Gen. Comput. Syst. **55**, 224–226 (2016)
2. Abraham, A., Nath, B.: Hybrid intelligent systems design. Technical report, School of Computing and Information Technology, Monash University, Australia (2000)
3. Alcala, R., Casillas, J., Cordon, O., Herrera, F., Zwir, J.S.: Techniques for learning and tuning fuzzy rule based systems for linguistic modeling and their application. Knowl.-Based Syst. **3**(29), 890–942 (2000)
4. Azzini, A., da Costa Pereira, C., Tettamanzi, A.: Predicting turning points in financial markets with fuzzy-evolutionary and neuro-evolutionary modeling. In: Proceedings of the Applications of Evolutionary Computing, EvoWorkshops 2009, Tübingen, Germany, 15–17 April 2009, pp. 213–222 (2009)

5. Azzini, A., Felice, M.D., Tettamanzi, A.G.B.: A comparison between nature-inspired and machine learning approaches to detecting trend reversals in financial time series. Nat. Comput. Comput. Financ. **4**, 39–59 (2012)
6. Azzini, A., Marrara, S., Sassi, R., Scotti, F.: A fuzzy approach to multimodal biometric continuous authentication. FO DM **7**(3), 243–256 (2008)
7. van den Berg, J., Kaymak, U.: Financial markets analysis by using a probabilistic fuzzy modelling approach. Int. J. Approx. Reason. **35**, 291–305 (2004)
8. Castiglione, A., Santis, A.D., Soriente, C.: Taking advantages of a disadvantage: digital forensics and steganography using document metadata. J. Syst. Softw. **80**(5), 750–764 (2007)
9. Cheung, W.M., Kaymak, U.: A fuzzy logic based trading system. In: Proceedings of the Third European Symposium on Nature-inspired Smart Information Systems, vol. 59 (2007)
10. Gai, K., Qiu, L., Chen, M., Zhao, H., Qiu, M.: SA-EAST: security-aware efficient data transmission for its in mobile heterogeneous cloud computing. ACM Trans. Embed. Comput. Syst. **16**(2), 60:1–60:22 (2017)
11. Gai, K., Qiu, M., Zhao, H.: Energy-aware task assignment for mobile cyber-enabled applications in heterogeneous cloud computing. J. Parallel Distrib. Comput. **111**, 126–135 (2018)
12. Keke, G., Meikang, Q., Lixin, T., Yongxin, Z.: Intrusion detection techniques for mobile cloud computing in heterogeneous 5G. Secur. Commun. Netw. **9**(16), 3049–3058 (2016)
13. Lee, T., Kim, H.W.: An exploratory study on fintech industry in Korea: crowdfunding case. In: 2nd International Conference on Innovative Engineering Technologies (ICIET 2015), Bangkok (2015)
14. Lin, C.-S., Khan, H.A., Huang, C.C.: Can the neuro fuzzy model predict stock indexes better than its rivals? CIRJE F-Series CIRJE-F-165 (2002)
15. Mohammadian, M., Kingham, M.: An adaptive hierarchical fuzzy logic system for modelling of financial systems. Int. J. Intell. Syst. Acc. Financ. Manage. **12**(1), 61–82 (2004)
16. Moyce, C.: Ai, big data, digitisation, blockchain: the fintech that will dominate 2018. Bobs Guide (2018)
17. Pantazopoulos, K.N., Tsoukalas, L.H., Bourbakis, N.G., Brun, M.J., Houstis, E.N.: Financial prediction and trading strategies using neurofuzzy approaches. IEEE Trans. Syst. Man Cybern. Part B **28**(4), 520–531 (1998)
18. Yin, S., Zhu, X., Kaynak, O.: Improved pls focused on key-performance-indicator-related fault diagnosis. IEEE Trans. Ind. Electron. **62**(3), 1651–1658 (2015)
19. Zhou, X., Dong, M.: Can fuzzy logic make technical analysis. Bobs Guide (2004)

Author Index